THE UNHACKABLE INTERNET

How Rebuilding Cyberspace Can Create Real Security and Prevent Financial Collapse

THOMAS P. VARTANIAN

Prometheus Books

Essex, Connecticut

 Prometheus Books

An imprint of Globe Pequot, the trade division of
The Rowman & Littlefield Publishing Group, Inc.
4501 Forbes Blvd., Ste. 200
Lanham, MD 20706
www.rowman.com

Distributed by NATIONAL BOOK NETWORK

Copyright © 2023 by Thomas P. Vartanian

British Library Cataloguing in Publication Information Available

Library of Congress Cataloging-in-Publication Data Available

ISBN 9781633888838 (cloth : alk. paper) | ISBN 9781633888845 (epub)

∞™ The paper used in this publication meets the minimum requirements of
American National Standard for Information Sciences—Permanence of Paper
for Printed Library Materials, ANSI/NISO Z39.48-1992.

*This book is for the
president of the United States—today's and tomorrow's—
all of whom will eventually wish that they and
their predecessors had done more to secure cyberspace*

CONTENTS

ACKNOWLEDGMENTS

This book could never have come together without the valuable insights and contributions that I received from many friends and colleagues who are steeped in the knowledge that I have attempted to analyze and explain in this book. They unselfishly dedicated hours of their time to challenge and contribute to the quality of the thought and analysis herein. For that, I am eternally grateful. You are the beneficiaries of their insights and generous work.

Most important were the contributions that I received over several months from my colleague of almost forty years, Robert H. Ledig; my good friend and economist/lawyer Robert M. Fisher; and a who's who list of futurists, lawyers, regulators, academics, technologists, cryptographic mathematicians, software developers, and internet users and pioneers. They included Richard Borden, Brian P. Brooks, Jeremiah Buckley, Silvia Elaluf-Calderwood, John Carlson, James Dever, Adam Golodner, Peter Gutzmer, Richard Klein, Roslyn Layton, David Lesser, Josh Magri, Timothy McTaggart, Randall K. Quarles, Craig Schwartz, Richard Vartain, Karen Beth Vartanian, Scott Volmar, Peter Wayner, and Prof. Christopher Yoo. Special thanks to my agent, Andrew Stuart, and attorney, Lloyd Jassin.

It is impossible to write anything worth the time of any reader without the contributions, insights, criticisms, and inspiration of other experts. I was lucky to have people who were generous enough to assist. I hope they understand how valuable their comments were to the creation of this book.

ACRONYMS, ABBREVIATIONS, AND GLOSSARY OF TERMS

ABA	American Bankers Association
ACH	Automated Clearing House
AGE	authentication, governance, and enforcement
AGN	alternative global network
APT	advanced persistent threat
ARC	Analysis and Resilience Center for Systemic Risk
ARPANET	Advanced Research Projects Agency Net
ASN	autonomous system number
ATM	automated teller machine
AWS	Amazon Web Services
BIS	Bank for International Settlements
BPI	Bank Policy Institute
CA	certificate authority
CBDC	central bank digital currency
CCIP	Commission on Critical Infrastructure Protection
CFAA	Computer Fraud and Abuse Act
CFPB	Consumer Financial Protection Bureau
CFTC	Commodity Futures Trading Commission
CIRCIA	Cyber Incident Reporting for Critical Infrastructure Act of 2022
CISA	Cyberspace and Infrastructure Security Agency
CISA 2015	Cybersecurity Information Sharing Act of 2015
CNCI	Comprehensive National Cybersecurity Initiative
CPMI-IOSCO	Committee on Payments and Market Infrastructures–International Organization of Security Commissions
CRI	Cyber Risk Institute
DAO	decentralized autonomous organization
DARPA	Defense Advanced Research Projects Agency
DBOM	device bill of materials

DDoS	distributed denial of service
DHS	Department of Homeland Security
DISA	Defense Information Systems Agency
DLT	distributed ledger technology
DNS	Domain Name System
DOD	Department of Defense
DOE	Department of Energy
DOJ	Department of Justice
DONA	data-oriented network architecture
DOS	Department of State
ECB	European Central Bank
ETF	exchange traded fund
FASB	Financial Accounting Standards Board
FATF	Financial Action Task Force
FBIIC	Financial and Banking Information Infrastructure Committee
FCA	False Claims Act
FCC	Federal Communications Commission
FDIC	Federal Deposit Insurance Corporation
Federal Reserve *or* Fed	Board of Governors of the Federal Reserve System
FFIEC	Federal Financial Institutions Examination Council
FHLBB	Federal Home Loan Bank Board
FIED	financial improvised explosive device
FinCEN	Financial Crimes Enforcement Network
FISMA	Financial Information Security Management Act
FMI	financial market infrastructure
FSARC	Financial Systemic Analysis and Resilience Center
FSB	Financial Stability Board
FSCIG	Financial Sector Cyber Intelligence Group
FS-ISAC	Financial Services–Information Sharing and Analysis Center
FSLIC	Federal Savings and Loan Insurance Corporation
FSOC	Financial Stability Oversight Council
FSSCC	Financial Services Sector Coordinating Council
GAO	Government Accountability Office
GIFTC	Global Internet Forum to Combat Terrorism
GPS	Global Positioning System
GSE	government-sponsored enterprise (Fannie Mae, Freddie Mac, Federal Home Loan Banks)
HTML	Hypertext Markup Language
HTTP	Hypertext Transfer Protocol

ICANN	Internet Corporation for Assigned Names and Numbers
ICO	initial coin offering
IEC	International Electrotechnical Commission
IED	improvised explosive device
IEEE	Institute of Electrical and Electronics Engineers
IETF	Internet Engineering Task Force
IIF	Institute of International Finance
IMF	International Monetary Fund
IOS	International Organization for Standardization
IoT	internet of things
IP	Internet Protocol
IPTF	Infrastructure Protection Task Force
ISAC	information sharing and analysis center
ISAO	information sharing and analysis organization
ISP	internet service provider
ITERA	Identity Theft Enforcement and Restitution Act
MAD	mutual assured destruction
M2M	machine to machine
MFA	multifactor authentication
MICR	magnetic ink character recognition
MUT	malicious user of technology
NACHA	National Automated Clearing House Association
NCCIC	National Cybersecurity and Communications Integration Center
NCSRS	National Cyberspace Security Response System
NDN	named data networking
NETINF	network of information
NFT	nonfungible token
NIPC	National Infrastructure Protection Center
NIST	National Institute of Standards and Technology
NLP	natural language processing
NSA	National Security Agency
NSF	National Science Foundation
NSFNET	National Science Foundation Net
NSPD	national security presidential directive
NTIA	National Telecommunications and Information Administration
OCC	Office of the Comptroller of the Currency
OFDM	orthogonal frequency division multiplexing
OFR	Office of Financial Research
OMB	Office of Management and Budget
OMIG	Open Metaverse Interoperability Group

PDD   presidential decision directive
PIN   personal identification number
PIV   personal identity verification
PPD   presidential policy directive
PURSUIT   Publish/Subscribe Internet Routing Paradigm
QKD   quantum key distribution
RAA   registrar accreditation agreement
RFID   radio frequency identification device
SBOM   software bill of materials
SEC   Securities and Exchange Commission
SIFMA   Securities Industry and Financial Markets Association
SIRPENT   Source Internetwork Routing Protocol
SPN   secure private network
SWIFT   Society for Worldwide Interbank Financial
                    Telecommunication
TCP/IP   Transmission Control Protocol/Internet Protocol
TIBER   Threat Intelligence–Based Ethical Red Teaming
TLD   top-level domain
TOR   The Onion Routing Project
ToS   terms of service
TSA   Transportation Security Administration
URL   uniform resource locator
VPN   virtual private network
VPS   virtual private server
ZTA   zero-trust architecture

INTRODUCTION

Candlesticks Are Nice

In about 5 billion years, our Milky Way galaxy will collide with the Andromeda galaxy and send out massive gravitational shockwaves.[1] Five billion years later, our sun is expected to die, having burned up its fuel. Long before that happens, probably 1 billion years from now, humanity will have had to find a new planet due to the ever-increasing brightness of the sun.[2] But don't pack your bags yet; there are more immediate problems confronting us.

The invasion of Ukraine has launched a new world *disorder* that officially makes cyberspace a tool for malicious users of technology (MUTs) such as authoritarian nation-states and their affiliates, dictators, criminal cartels, fanatics, terrorists, and lunatics to dispense evil. War has a way of shining a brighter light on alternatives and priorities. People are beginning to realize that new technologies in the hands of people who are bent on control or destruction, and the growth of an internet that is increasingly insecure are enormous threats to this country's financial institutions, economic infrastructure, and ultimately democracy itself. And unlike the analog world, there is no cyber coast guard, no early-warning cyber defense system, and no cyber police or fire departments. As technology gets cheaper and finds its way into the hands of MUTs who do not play by the rules of civil society, the fragility of our financial and democratic systems will increasingly be on display as civilization finds itself on perpetual defense.

You would not have to be convinced of this if you had been in Estonia in 2007. Life stopped when the Russians pulled the cyber plug because of the placement of a statue in the capital. Havoc reigned as the Estonian parliament, banks, ministries, newspapers, and broadcasters were disrupted. Russia suffered no visible punishment for the attack—in retrospect, a massive mistake by world leaders. Similar cyberattacks attributed to a Russian advanced persistent threat group known as Sandworm crippled Ukraine's power grid in December 2015. Again, Russia paid little price for that attack.

In 2016, Brazil's banking system experienced one of the worst cyberattacks when all of Banco Banrisul's thirty-six domain name system registrations were taken over by hackers, controlling its online banking portal, mobile banking, automated teller machines (ATMs), point-of-sale terminals, and internal corporate domains and email system. Customers were routinely directed to websites that installed malware on their devices and stole their personal and account information.[3] In the United States, Capital One's networks were compromised in 2019, exposing data from credit card applications of approximately 100 million individuals dating as far back as 2005. Capital One was the thirteenth-largest bank in the United States at the time, but its systems were humbled by the malicious actions of one former engineer of its cloud provider.

In early 2022, the threat of a cyber war became even more real in the weeks before Russia invaded Ukraine. Hackers launched attacks on dozens of government websites, inserting a WhisperGate malware, something very close to the NotPetya Russian virus that had done $10 billion in damage around the world in 2017. The virus pretended to be ransomware but actually made machinery inoperable.[4] Online services of Ukraine's armed forces and its two largest lenders, PrivatBank and Oschadbank, were disrupted days later by cyberattacks as the world watched the country suffer the largest denial-of-service attack in its history.[5] PrivatBank's ATMs were down for an hour, and it experienced "instability" in its mobile communications.[6] The U.S. Cybersecurity and Infrastructure Security Agency (CISA) issued a "Shields Up" alert urging companies to stiffen security and aggressively monitor for unusual activity. It warned critical infrastructure operators in the United Sates to take "urgent, near-term steps" against cyber threats as U.S. banks ran through contingency plans to prepare for Russia-linked incidents.[7]

On February 23, the invasion began, and a wiper malware—HermeticaWiper—was unleashed on Ukraine, Latvia, and Lithuania computers. Experts believed that it had been embedded in those systems months before. Websites at government agencies and banks were disrupted for several days.[8] This was followed by an attack on the American satellite company Viasat hours before Russian troops invaded Ukraine. That resulted in an immediate loss of Ukrainian communications as "wiper" malware known as AcidRain attacked modems and routers, erasing data on the systems. As the machines rebooted, they were permanently disabled.[9] Several days later, Anonymous, the ad hoc international hacktivist movement, announced a cyber war against Russia for its invasion of Ukraine, and websites of the

Kremlin and the country's Ministry of Defense were hacked and went offline. Several Russian state TV stations began playing Ukrainian music and displaying images of Ukraine's flag and other nationalistic symbols.[10] Unfortunately, prankster-level shenanigans often serve to warn a target where its vulnerabilities are so that it can eliminate them.

In late March 2022, chilling revelations emerged when 200,000 messages between 450 members of the Trickbot group detailed its intention to attack and disable more than 400 U.S. hospitals. Trickbot is a Russian-affiliated cybercriminal gang. Its Conti ransomware was the most used in 2021 to attack and disable private and government computer systems in the United States. The messages became public only when Trickbot itself was penetrated by a Ukrainian hacker.[11] Several days later it was announced that hackers had stolen $625 million in cryptocurrency from the video game company Axie Infinity when the supposedly secure underlying blockchain that powered the game was infiltrated.[12]

Coincidentally, two Dutch researchers took home $90,000 and a hacking trophy for breaching software that ran critical infrastructures around the world, saying it was "their easiest challenge yet" because "security is lagging behind badly."[13] This was followed by the theft of $182 million worth of digital assets when hackers attacked an algorithmic stablecoin called Beanstalk, wiping out all of the Ether it held and collapsing the value of its "Beans." Beanstalk is a DAO, or decentralized autonomous organization. The hackers reportedly borrowed about $1 billion worth of different stablecoins through a flash loan to gain an overwhelming percentage of Beanstalk's voting power. They then proposed that money be donated to Ukraine, but that proposal included malicious code that diverted funds at Beanstalk to a wallet controlled by the hackers. They simply paid off the loan and pocketed the difference.[14]

These are just a few examples of an increasing and frightening misuse of the internet, its inadequacy when it comes to protecting critical infrastructures, and the helplessness we all experience when confronted with these attacks. No system can be 100 percent secure. That is not a legitimate goal to expect. But the insecurity of the many data sharing and transference networks that form the World Wide Web which operates on the digital foundation known as the internet, and the growing threats that they create to the integrity of our social, financial, and physical well-being, have become unacceptable. The holes in this brilliant architecture of the internet have become a perfect environment for bad actors, ranging from criminals to repressive governments, to abuse the money and data that we have entrusted to a digital world that few of us understand.

Anyone paying attention to events over the last several years readily appreciates how a terrorist, nation-state, criminal cartel, thief, or political operative can use technology to manipulate geopolitical dynamics, impact national elections, disrupt critical infrastructures, steal billions, or use fabricated news to foment discord on hot-button social issues. Once initiated, divisive social media, frenetic cable TV outlets, exploitive politicians, and profiteering activists jump into action to spin every issue into dollars and public sentiment to support them and their causes. Why wouldn't China, Russia, Iran, and North Korea nurture active hacker societies to manipulate reality and help America destroy itself? This is a treacherous path that the internet has put us on. Once confidence in governments, information, and the bona fides of our opponents evaporates, it is not clear where society turns for security and stability.

At the same time, the rising costs of cybersecurity services, computer maintenance, operational downtime, and related losses from cyberattacks are on a trajectory that is not sustainable. Prudence dictates that we reverse gears and adopt a more secure approach before we get so far down this path that there is no going back. That means we need real leaders who can do something besides run for election, wave, and make flowery speeches.

Every one of us is seriously exposed to the misuse of the internet. While writing this book, I experienced the frustration and helplessness that can result from our complete reliance on and trust in systems that are beyond our control. I could not access my online banking account for several days in September 2021. I did everything I knew how to do. I changed my password three times, deleted cookies, cleared my browser's cache, and reset my privacy and browser settings. I tried different browsers to see if that made a difference. After restarting my computer several times, I was still locked out of the account. I telephoned my bank's technical support desk and after twenty minutes with someone who seemed unsympathetic at best was told that the bank's systems were malfunctioning, perhaps undergoing periodic maintenance. I needed to be patient for another forty-eight hours. Six months later, I lost access to certain functions in the same online banking app that prevented me from transferring funds from some internal accounts to others. That lasted more than three weeks while the bank's tech people said they were working on it. Many of you are reading this and nodding your heads, having had similar experiences.

Each day, we are constructing a larger and more expansive insecure virtual world that we cannot control. In it, people and machines can assume fictional personas, act anonymously, and be as obnoxious or malicious as they wish without being easily detected. Things that used to be represented by paper and physical contact are being converted into microscopic universes of digital data that are completely portable. Every piece of information about us is being captured and sliced into digital packets that fuel an infinite number of commercial as well as nefarious purposes, stripping any last vestige of privacy from everyone as their data passes effortlessly throughout the digital world. For all its wonderful advancements, digital technology is also driving life-altering changes that are creating a form of societal dyslexia where cultural and financial dysfunction are becoming the norm. And for the most part, many freely participate in this effort because they expect their lives, financial or otherwise, to be easier, more profitable, and more connected. They probably think it is safe and that the "government" would not let it exist if it were not. They are wrong. It is time to say "STOP!" before we lose control of our futures.

The greatest risks are yet to come, and we are still unprepared. In addition to the risks posed by Russia, China's cradle-to-grave electronic surveillance society that uses apps on cell phones, facial recognition monitors, and an array of other behavioral and social scoring tools to identify citizens for "reeducation" is about as frightening a future as one can imagine. Such uses of technology are so pervasive that they may eliminate any ability for a population—current or future—to revolt against such repression. Thanks to its tracking technologies, the Chinese government knows when three people are meeting. But it is not just governments that can misuse this power. Hacker gangs armed with computers have formed around the world. In December 2021, Facebook reported that seven hacker-for-hire firms had targeted around 50,000 users of its various platforms.[15]

We may wake one morning and find that ATMs are not working, credit cards are not recognized, checking accounts have disappeared, and investment portfolios have evaporated, all due to the introduction of some malicious code hidden in a silly game that was downloaded. It is admittedly not easy to do given the fragmented and disjointed structure of cyberspace, but it is doable by an advanced persistent attacker. After death, there is no more frightening thought than the loss of everything that a person has worked their entire life to achieve. One person may eventually be reimbursed by their bank or broker if their accounts are emptied out. But if it happens to everyone, there may be no one left standing to reimburse anyone. What would you do if that happened? Do you know who the cyber police are? Do you know their phone numbers? Consider what your first instinct would be in such a situation.

This problem is not going away and seems to be getting more complex all the time. Technology will always push us forward, and there will always be evil and malicious people trying to use it to lie, steal, cheat, and kill more effectively. So, yes, we have made serious mistakes. We created an internet presuming that it would always be a virtual amusement park that would never house anything serious or of value. We have looked weak when countries like Russia, China, Iran, and North Korea have perpetrated the malicious use of technology around the world, treating such events as if our children had scuffed their knees at the playground after being chased by another child.

Not surprisingly, the invasion of Ukraine and the heightened threats of cyberattacks have forced U.S. officials to conclude that our cybersecurity efforts have failed. But they default to the need for stronger government regulations, which are rarely the most effective or efficient means to solve problems, particularly if they do not foster uniformity and delegate responsibility to dozens of agencies that are focused on their own narrow slices of cyberspace.[16] In 2017, the Bank Policy Institute (BPI) issued a report laying much of the blame for dozens of confusing bank cybersecurity rules that were "doing more harm than good" at the feet of federal and state banking regulators.[17] Similarly, the Transportation Safety Administration (TSA) has been roundly criticized for unwieldy and "baffling" cyber regulations that have jeopardized pipeline safety and fuel supplies.[18] We need to do better and think outside the same old internet box that has proven to have been ill constructed. The choice is entirely ours to make.

There is no reason to compound the mistakes that have already been made, particularly since we now know that every piece of data and penny of value is eventually going to be transported into this virtual world where it may potentially be misused. We don't have to continue to make it so easy for bad actors to take advantage of us. There are solutions that can neutralize the growing threats and create safer virtual spaces. None of them are easy or perfect. Many will be inconvenient, putting society in the position of having to choose the best worst option before it is too late. They will all require real leadership. And since whatever control of the internet does exist today is in the hands of many organizations and nation-states that have diverse long-term interests, it will take massive consensus building among democratic nations to effect change.

There are analogous efforts currently under way, for example, to improve the capacity of and better secure the Global Positioning System (GPS).[19] In 2017 and 2018, Congress mandated the development of a joint plan to deploy alternative GPS backup technologies.[20] That has proven to be a challenging task; a new internet will be even more challenging. But by a new internet, I am not suggesting that we need new pipes, cables, routers, providers, or digital technologies. That is unlikely to happen in the near term.

We can however enhance the use and security of private and offline network infrastructures that already exist by creating more secure gateway protocols that connect with the open architecture of the internet. We can regulate ownership and control of internet pipes, clouds, and infrastructures and ensure that they and their owners meet the highest security standards. We can better engineer the nuts and bolts of the internet and the software and hardware it relies on. We can create a parallel system of highly secure networks that rely on enhanced authentication, governance, and enforcement (AGE) standards. We can better segregate, decentralize, and secure data. Much of this has been under consideration for several decades by governments, the military, academics, and financial providers, but each effort has fallen short, confronted by a cyber status quo that prints money and creates jobs for millions of people. Frankly, the internet has worked well enough to discourage costly improvements and expenditures that would inevitably impact user experience and bottom lines.

The evolution of a new generation of the internet—Web3—may create greater data decentralization to dislodge the power that Big Tech has amassed. That would be a good time to consider an increased use of secure private networks (SPNs) that run parallel to the internet and whose security is monitored to ensure their reliability. Such private clouds and networks have worked in the financial services business, but to the extent that they are accessible through many points connected to the open architecture of the internet and are owned and controlled by third parties, they are vulnerable.

A more advanced system that requires virtual traffic to be authenticated and licensed to be able to move into and around SPNs is necessary. Such digital licenses and passports should require adherence to a code of internet conduct and the inclusion of "kill switches" that would disappear code, software, or any virtual traveler that violates the rules. Entrance to secure networks could be controlled through secure filtering applications that block out the bad actors—human and digital. Semi-decentralized versions of blockchain applications could provide more enhanced authentication and verification. What has been missing is the will or perhaps proof of necessity to build and enforce these standards on a universal basis. Again, there has been a massive lack of leadership.

A new world disorder may as a matter of market necessity foster the creation of a new internet or internets. The groundwork has already been laid with the actions taken by and against Russia. Russia blocked access to Meta (formerly Facebook) in early March 2022, but Apple, Microsoft, TikTok, Netflix, and a host of other prominent U.S. companies voluntarily withdrew from Russia. Some worry that this may be the beginning of a "splinternet" that leads to the end of the universal internet that we have come to know in favor of separate networks using different protocols and governing authorities.[21] The separatist actions of China, Iran, and North Korea have also been entirely consistent with the creation of distinct internets. Perhaps these events provide the opportunity to remake the internet and correct the mistakes we have made.

Society is at a crossroads. While perfect security will never be achievable, it should not take a digital Pearl Harbor—a difficult but not impossible feat to pull off—to convince democracies around the world to create a more *unhackable* internet to protect their economies and security. The alarms have sounded, and it is time for governments and businesses to take the lead. The potential advancement of quantum computing, artificial intelligence, and nanotechnologies that could dismantle current forms of online security have convinced the U.S. government to take steps toward creating a new secure quantum internet. Not surprisingly, however, China is already ahead in that race.

Building a new concept of the internet with offshoots and tributaries that have different rules, standards, and enforcement mechanisms will be difficult. Being the first to do so won't be popular, easy, or cheap. There will be significant hurdles in obtaining a global consensus on a new internet, and some countries will have to be left out. Users have gotten spoiled by the speed, freedom, and efficiency of the open architecture of the internet, so it won't be popular or politically expedient to change the rules to achieve more security that is difficult to see or measure. In addition, society will have to confront the question of who, if anyone, should oversee this new style of internet and enforce its rules. This challenge is fraught with complexity no matter how we turn. But the one thing that does seem clear is that the path we are on will eventually lead to virtual anarchy, erode democratic societies, and allow authoritarian governments to control their populations with a virtual pervasiveness that may make their grip on power uncontestable.

The Unhackable Internet is a story about the mistakes we have made, the possibilities for a more secure internet, and the tradeoffs it will require as seen through the prism of something near and dear to us—our money. This book reflects my forty-five years of experience in the financial services and technology businesses and is admittedly focused on the United States. I have not attempted the gargantuan task of evaluating the internet from any other perspective. But my experiences have led me to be concerned about our financial futures and offer some thoughts about how to navigate to a brighter digital world.

To test my theories for a new internet, I interviewed or had my draft manuscript read by experts, including software developers, cryptographic mathematicians, lawyers, academics, economists, government regulators, domain name providers, information sharing and analysis representatives, cloud services experts, and laypeople. Their feedback and contributions have made me even more convinced that the internet is broken and is in critical need of rehabilitation.

As I tried over the last several years to disassemble and then bring some organization to the chaotic virtual world that is the internet, I kept thinking about the scene from the 1988 movie *Bull Durham* when Crash Davis (Kevin Costner) ran out to the pitcher's mound during a game to give a pep talk to his pitcher, Calvin "Nuke" LaLoosh (Tim Robbins), as the infielders crowded around. Seemingly unaware of the threat confronting the team from the runners on base, the discussion bounced around between Nuke's father in the stands, his eyelids being jammed, a wedding gift for Jimmy and Millie, and the need for a live rooster that could take the curse off José's glove. When the coach finally trotted out to the mound to see what was going on, Crash laid out for him the list of issues they were trying to resolve, summing it up as "we're dealing with a lot of shit!" Between chews on a wad of tobacco and a thoughtful pause, the coach matter-of-factly offered: "Candlesticks always make a nice gift."

Fifty years into cyberspace, it's time to get serious, forget about candlesticks, and fix the internet. Ask yourself if you feel secure putting your personal, business, and financial data online? Do you know where technology will take us in ten years, and which country will be technologically superior? Most importantly, if the U.S. financial infrastructure becomes the subject of a cyberattack, do you know what to do? We all know the answers to these questions, and they are the best argument for reconstructing the internet now.

Part I

LOST IN CYBERSPACE

1

THE NEXT FINANCIAL BLACK HOLE

There is a financial panic coming—there always is. We just don't know when the ride ends and the collapse begins.

Every financial disaster, like a building, needs a sound foundation constructed over time. Over the last two centuries, except for the post-Depression-to-1980 period, the United States has averaged a financial panic every twenty years, the second-highest incidence of economic disaster of any country on the planet.[1] Each crisis has been influenced by the surrounding economic, political, and social circumstances, which ultimately collided in ways that had not been anticipated. That is something that I saw in the many banking crises that I lived through or studied for my book *200 Years of American Financial Panics*. Since my predictions there of the factors that could detonate the next financial crisis, things have only gotten worse. Inflation reached 9.1 percent and interest rates had doubled by July 2022. The national debt surpassed $31.5 trillion—119 percent of the country's gross domestic product—and is expected to double over the next three decades. By 2026, Medicare Part A is likely to run out of money, followed by the Highway Trust Fund the next year, and Social Security in 2033.[2] It is not hard to imagine dark financial days ahead.

Beyond the private sector excesses and governmental policy blunders, there are always unanticipated events that fan the flames of panic. Westward expansion, railroad construction, industrialization, securitization, and cyber technologies have all created backdrops that led to massive capital reallocation and investment excesses. Each time, our financial house of cards collapsed when a sudden loss of public confidence converted the energy of economic euphoria into a frantic race from risk. The United States is well along the path of building the next financial crisis through its traditional profligate monetary policies—creating too much money, too little market discipline, and too many misplaced expectations.[3] But unlike the past, the next financial panic and disappearance of confidence may come to us courtesy of forces beyond market or government mistakes. The use or abuse of technology, with some ingredients provided by a long list of foreign adversaries, is now fully capable of creating such an event.

THE ROLE OF CHINA

While it is by no means alone, there is no more formidable and frightening financial and security threat to the United States today than China. It could soon be the global leader in artificial intelligence, semiconductors, robotics, 5G wireless, quantum information science, biotechnology, facial recognition, voice recognition, fintech, and green energy. A loss of financial or technological superiority would have a dire impact on the U.S. economy. Trade balances, interest rates, liquidity, capital formation, and currency exchanges would all be impacted by a degradation in the economic stature of the United States. A dollar that was no longer the global reserve currency would hasten a devaluation of the U.S. economy, which would create a ripple effect that would translate into higher borrowing costs, stagnant incomes, and higher inflation for U.S. consumers.

Writing for the *Wall Street Journal*, Harvard professor Graham Allison and former Google CEO Eric Schmidt noted that in 2020, China produced 50 percent of the world's computers and mobile phones compared to the 6 percent produced by the United States, and seventy solar panels for each one produced in the United States.[4] China sells more electric vehicles, produces more computers and smartphones, has more 5G base stations and more networks, and has become a serious competitor or leader in artificial intelligence, quantum information science, semiconductors, biotechnology, and green energy.[5] Kai-Fu Lee, the entrepreneur and technologist author of *AI Superpowers*, describes a catch-up ethos in China that makes the lifestyle of Silicon Valley look lethargic.[6] China also makes the internet a far more dangerous environment for its adversaries. For example, beyond the repressive ways that technology is deployed in China to surveil and monitor the social behavior of its citizens, it is expanding its ability to use such modes of societal control in Africa and Latin America. Huawei has been selling its 5G technology to many of the countries there, taking advantage of what can at best be described as U.S. complacency and, at worst, geopolitical stupidity. Each country that adopts the Huawei Cloud and becomes a debtor nation to

Chinese interests has stepped onto a slippery slope that may be irreversible.[7] Is China a country that democracies should want to share an internet with?

The respected journalist and author David Sanger puts it in rather stark terms. He argues that a global conflict spurred on by America's waning influence is forcing allies to make a bet on which of the world's two largest economies will be more critical to their future. Nations understand the risks of betting on China, but they also understand the threats that already exist in betting against it. Countries like Germany sell 5 million cars to China every year. The 5G race only intensifies these challenges and the battle for global superiority.[8]

China may also soon catch the United States in semiconductor fabrication and chip design. It already controls much of the world's seventeen rare earth minerals that are indispensable to the manufacturing of chips, smartphones, electric vehicles, military weapons systems, and countless other advanced technologies. While it already has a significant share of the semiconductor chip manufacturing market, even more significantly, the world's largest and most advanced semiconductor manufacturers are approximately 100 miles off the coast of communist China on the island of Taiwan, which increasingly seems to be coveted by China. A November 2021 report by the U.S.-China Economic and Security Review Commission summed it up by concluding that China is a "menacing adversary determined to end the economic and political freedoms that have served as the foundation for security and prosperity for billions of people," putting the safety and security of the United States and its partners, friends, and allies in peril. Without immediate action, it warns that we will continue to see "the slow but certain erosion of the security, sovereignty, and identity of democratic nations."[9]

In late February 2022, Daxin, a malware described as the "most advanced" that hackers affiliated with China have ever used, was revealed to have been deployed against governments around the world for at least a decade. China also maintains a thriving "bug bounty" business that generates revenue from companies, many of which are American, in return for identifying and reporting security vulnerabilities. What a great business model—launch malware and charge the targets money for identifying it! And to make it worse, these vulnerabilities are also being reported to and cataloged by Chinese authorities. President Xi Jinping's intention is to make China a cyber superpower, a process begun with the reorganization of China's military and intelligence agency, the prioritization of cyber warfare, and the blending of military and civilian organizations focused on cyber capabilities. Today, experts describe China's offensive cyber capabilities as rivaling or exceeding those of the United States.[10]

A recent article in the *Atlantic* put a finer point on Chinese strategy, arguing that the growing confrontation between the United States and China is over who will make the rules on trade, technology, climate change, and public health that guide how countries, companies, and individuals interact. That conflict may determine whether the world adheres to democratic principles driven by the rule of law or follows the Chinese model that uses law to create rules to support the government.[11] Toward that end, China's ability to use cheap labor to monopolize production and resources around the globe is creating a long list of nations on several continents economically beholden to it. A reasoned analysis of the dangers of the internet could stop right here and not even have to proceed to discuss the endless incidences of hacking, spying, and disruption that China and its surrogates engage in.

THE ECONOMIC THREAT OF CYBER FLARES

The growing cybersecurity risk to the country's economic infrastructure is the latest and most significant financial risk that has been added in the last twenty years. China is a big part of that threat but is by no means alone in planting financial traps for the U.S. economy. Thanks to MUTs, the financial landscape is littered with not only traditional financial improvised explosive devices (FIEDs), but also an infinite collection of technological threats that can intentionally or unintentionally cripple any system and stop the movement of money and commerce.

Consider a well-funded, persistent cyber attacker such as a hostile nation-state or terrorist organization that is determined to disrupt the financial sector or destroy an economy. It might take a year or more to plan and deploy such a sophisticated attack that included network penetrations, the use of logic bombs, various forms of malicious software scattered throughout financial networks, bribery, the conscription of people to infiltrate financial companies and government agencies, ransomware attacks, and technological reconnaissance efforts. But it is entirely feasible today.

The goal of such an attack would likely mimic the nature of the attacker, whether it be a hostile nation-state, criminal cartel, terrorist, fanatic, or prankster. It could range from disrupting the economy to skimming one cent from every financial transaction processed by payments systems for months or years to finance illicit operations. It might simply involve replication or alteration of data, spying, or locking up data in return for cryptocurrency ransoms. Such cyberattacks on the financial services sector could be initiated directly or through the computer networks of law firms, through consultants, or even by cleaning crews that provide services to banks, the Federal Reserve, the Treasury, securities exchanges, the Clearing House, Mastercard and Visa, ATM networks, and other Automated Clearing House (ACH) and payments systems. The only thing that would be required is access to any intermediate network that eventually communicated with the target's network.

Surveillance apps, spying software, and logic bombs could be set to ignite at a particular moment and take control of some or all of a system. Recruits working in financial institutions may have provided network passcodes to allow for silent access to or the downloading of malicious code, applications, and other software in the target's systems. Such a panoply of malicious analog and digital tools and strategies could wreak havoc on the financial services system or simply watch it and wait. Prevention is by no means perfect, so the best we can hope for is early detection, systemic resiliency and redundancies, and secure backup facilities to repopulate hacked networks. That is a huge bet to make on the functionality of the entire economy.

The increasing use of all forms of artificial intelligence greatly complicates this already challenging landscape. A typical threat scenario enabled by artificial intelligence might begin with the deployment of hundreds of chatbots seeking to befriend targets through social media. The bots would study and learn about their targets' social media profiles and create believable online content, including pictures of nonexistent people. Having gained the trust of the targets, the intelligence gathered by the bots could be used to craft convincing phishing attacks to trick them into downloading malicious apps or documents containing dangerous links to exploitive servers. Once in the system, those forward reconnaissance apps would troll for vulnerabilities and seek to blend in with regular network operations to launch attacks to harvest account credentials and data. Intelligent neural network features could enhance the efficiency of the attack by ensuring that only relevant material was exfiltrated. As an AI-augmented attack, hundreds of similar parallel assaults could be run simultaneously.[12]

An attack could start slowly and invisibly but then grow as thousands of ransomware notes were sent providing three hours for funds, for example, in the U.S. Treasury general account, to be transferred to accounts around the world through a maze of routers, IP addresses, and interconnected networks. Whether that transfer by the Treasury did or did not occur, every piece of malicious software could be triggered, disrupting major banks, payments systems, the Federal Reserve, and the Treasury. If anyone were lucky enough to get access to their accounts, they might see a zero balance. Large banks, brokers, and exchanges would call the Treasury, their regulators, and a gaggle of government and private sector organizations, all of whom would deploy plans and backup systems that had already been created. Whether they meshed, whether anyone would be in charge, whether private networks and fragmented systems could protect some portion of the infrastructure, and whether systems could be put back on their feet in a reasonable time is unknown.

The race would be on to find and bring back to life the various information sharing and system rebuilding efforts that the country had presumably spent the last twenty-five years constructing. In the meantime, commerce could cease, as no one would have access to any money beyond the cash in their pockets. In 2020, the Federal Reserve Bank of New York estimated that a cyberattack on a money-centered bank that suspended its ability to make payments could cause 6 percent of the country's banks to breach their end-of-day thresholds.[13] In 2022, the Federal Reserve System paper confirmed the disruptive impact that a multiday cyberattack event could have on the payments systems as well as the Federal Reserve System.[14] More than a single day of dislocations could be catastrophic to confidence in the system. Panic, disorder, and civil unrest would surely occur within a reasonably short amount of time if systems were not regenerated.

This may seem like a script from *Mr. Robot*, a TV series about the ultimate cyberattack on the corporate world. Years ago, I would have bet that such a tale was pure fiction. Today, I don't believe it is fiction. It would be difficult to pull off, but how close we are to this being reality—virtual or otherwise—is not quantifiable and not important. The critical fact is that today or perhaps tomorrow, all of this is technologically possible. And the scale of the potential damage is larger than it has ever been. We are not talking about a robbery of a local bank branch.

Knowing this, and guessing that there is a 5 percent chance that something like this could happen, shouldn't we try to reduce the chances of it occurring given what is at stake? Unfortunately, the country, the Federal Reserve, the Treasury, and every financial institution, securities broker, payments system, financial utility, trading market, and securities exchange—and you too—are taking that risk, intentionally or not, every single day. We have known about this risk for at least twenty-five years and have not been able to eliminate it. We all continue each day assuming or hoping the worst won't happen. Perhaps that will continue to be the case. But what would you do if it did happen? Who would you call?

There are many technical enhancements that could be implemented to alter the way that people, hardware, software, and networks act, react, operate, and are governed to make the current internet more secure. Everything from internet hygiene to the security of the most insignificant software programs and hardware would have to be upgraded, and the free-for-all open nature of the internet would have to change. The use of fragmented and private networks would have to be increased, likely changing the operation and efficiency of the internet.

THE PATH TO A SOLUTION

The internet seems broken. By that I mean it is not up to the challenge of the current dynamic and fast-moving uses it is being put to, and it certainly was never created to be super secure. It has been playing catch-up and has never been able to completely bridge the gap by adding to and modifying existing systems.[15] The result has been an increasing array of high-profile cyberattacks around the world. These attacks represent present and future dangers as malicious code is embedded throughout cyberspace. At the same time, the increasing first-mover financial benefits of technology have conditioned markets to reward innovation significantly more than insecurity is punished. That of course creates a self-fulfilling prophecy that encourages innovative products to come to market, leaving the quality and the security of the coding and software as an afterthought. It is a "sell first and patch later" philosophy. As a result, internet vulnerabilities are being created at a more rapid rate than the identification of their solutions.

This makes it imprudent to continue to convert every inch of data and ounce of value into digital form to be stored and transmitted on insecure networks susceptible to abuse by MUTs. That is a radical and controversial conclusion given the global dependency by individuals and businesses on the internet. But something must be done to change the trajectory the country and its economy are on before a critical infrastructure like the U.S. financial system is severely compromised. There are several paths that can be pursued separately or in tandem.

Improved authentication, governance, and enforcement (AGE) enhancements of the internet are a first line of defense. Many of these improvements would be viewed as antithetical to the philosophy and operation of the internet to the extent that they would reconstruct it and the way it functions to buy more security. A more feasible path is one that leaves the internet largely as it is and increases the use of parallel private, proprietary, and offline networks for communications and transactions that require a high level of security. Gateways that allow their traffic to cross over to them from the internet would be subject to strict security protocols. This path raises significant issues such as the feasibility of critical businesses like banking owning their own clouds, pipes, and internet infrastructures instead of relying on third parties to provide the highest levels of security. Alternatively, if clouds, pipes, and internet infrastructures continue to be owned and controlled by third-party tech and telco companies, some entity would have to have the authority to oversee the security they offer, a result that would be seen as equally controversial.

All of this is a tough sell—consider how you react when a web page loads too slowly. Moreover, a divided Congress would have to pass legislation that actually made a difference in the face of opposition from technology companies and users. Republicans and Democrats would have to agree or compromise on issues such as the regulation of Big Tech, the status of China, and the future of products manufactured outside the country's borders. Nations around the world would have to reach some consensus on the changes.

Having lived and thought through these scenarios and the options that are available for many years, I believe that we make a serious, life-alerting mistake in choosing to continue to build the current internet. We don't need to start over and build a completely new internet. But to reduce the possibility of the next financial crisis turning into a financial Armageddon, we need to develop an internet that is unhackable—or at least less hackable.

2

THE ILLUSION OF
ONLINE SECURITY

In his prescient 2005 book *The Singularity Is Near*, Ray Kurzweil predicted that technology will fundamentally change the future of life as it gallops toward intellectual and biological singularity with machines in 2045. Perhaps this assimilation of man and machine is what evolution is about. The question is, will human consciousness become more machinelike, or will machines become more animate?

Technology is a two-edged sword. Great technology in the hands of benevolent people can greatly advance the quality of life. That same technology in the hands of malicious people can lead to the destruction of commerce, morality, cultures, nations, and humanity. Where technology is leading us is one of the great mysteries of life. We tend to consider ourselves as masters of the universe, but human civilization has been but a mere speck in the known quantum of time, space, and dimensions of the universe. There are several hundred billion to a trillion galaxies in the observable universe, which is thought to be 14 billion years old and 93 billion light years in diameter and growing. Though it is a subject of anthropologic and archeologic debate, human civilization may have begun only 12,000 years ago, making our involvement with what we can perceive to be the universe a microscopic fraction of its known existence. That makes the time that humans have had to impact the universe roughly equivalent to a single grain of sand on a beach. This suggests that there is a much larger game afoot than humans care to recognize. Physicists are only beginning to scratch the surface of quantum mechanics and the vibrating strings thought to make up the intrinsic matter underlying the invisible atoms, electrons, quarks, neutrinos, photons, gluons, and gravitons that are the building blocks of everything. If they are correct about there being as many as eleven different dimensions, it starts to become apparent how little we know as we function in less than half of them, and perhaps how small a role we play.

We are convinced that we are the puppeteers in our world. The final vignette in the movie *Men in Black* suggests an alternative when it is revealed that the massive alien universe that was being sought had been neatly tucked into a small jewel hanging from a cat's collar. As technology and its mesmerizing inevitability continue to shape the future of humanity, the critical question is who—man, machine, or some other entity—will end up in charge as this cosmic hoedown unfolds? The best way to ensure that humans stay ahead of the curve is to choreograph the evolution of technology by planning and establishing rules of order and social governance as much as possible. That should be the way we approach the internet.

It is difficult to determine what role virtual reality ultimately plays in the process of human and machine development. We know that the internet was created under circumstances that did not anticipate what it has become. It should be no surprise, therefore, that virtual networks are not sufficiently secure given how they are used today. With the advent of 5G and the internet of things (IoT), the landscape will change yet again as everything will be connected—toasters, coffeemakers, light switches, medical devices, automobiles, dams, aircraft, satellites, and thermonuclear devices. Since everything that is connected is vulnerable, everything will be vulnerable.

Cyberspace is just another step in the human journey. It is an amalgamation of the most innovative and intellectually developed computer sciences foraging and interreacting in a virtual environment that seems to become more real every day. The internet and its World Wide Web, which connect computers and data, provide greater abilities to cure diseases and advance the human state through the collection, storage, analysis, and transmission of data in ways that were never anticipated. But just as in the analog world, the use of technology to achieve scientific and cultural achievements must be balanced against the potential for its misuse and abuse.

This stark contrast of good and evil suggests a challenging journey ahead. Emerging from this world of technological extremes are ever-increasing threats on the analog world. This has been vividly demonstrated by an increasing number of cyberattacks that suggest that the internet is not functioning the way it should. Each of our daily clicks through cyberspace creates a different set of risks and lays the foundation for future attacks. Our personal data and money seem to be easily accessible to a wide array of commercial and criminal enterprises.

Even when the process is done properly and terms of service (ToS) agreements are presented to internet users, they blindly agree to enjoy immediate online gratification, not appreciating or caring how every millimeter of their data will be sliced and diced by data brokers, social media companies, and others and used in ways that should make their toes curl. Even reasonable ToS agreements are often illusory given that some participants in the virtual assembly line of data will simply snatch whatever they want no matter what users think they may have sanctioned. Either way, unless we begin to reconfigure cyberspace to assure personal, national, and global safety and security, we are all at risk.

THE ELUSIVE GOAL OF SECURITY

We seem to have become numb to the catastrophic threats growing in cyberspace. But that complacency is beginning to dissipate as ransomware attacks take center stage. In 2021, after a string of cyberattacks that impacted Fortune 100 companies and government networks all the way up to the White House, the Biden administration issued rules to prevent this growing and most current reminder of internet insecurity. Those rules are likely too little, too late. Their imposition of policies, procedures, and record-keeping requirements on the law-abiding participants in a cyberspace that is insecure will not eliminate the vulnerabilities and threats that exist. While perfect security can never be guaranteed, until we decide that this internet has become obsolete and affirmatively incentivizes misbehavior, we will continue down a perilous path.

Sixty years ago, Italy was a hotbed of corporate kidnapping. In 1991, Italy passed a law allowing the government to freeze the assets of families of kidnapping victims so that they could not pay a ransom. It may have seemed harsh, but by removing the economic payoff in the kidnapping business model, kidnappings for profit in Italy dropped from twenty-nine to five a year.[1]

Data kidnapping began in 1989, when a Harvard-taught evolutionary biologist named Joseph Popp sent floppy disks to addresses all over the world. Once inserted, the floppy disks locked up computers and told the owner that by printing out the invoice provided and sending money, the computer could be unlocked. Popp was arrested and charged with multiple counts of blackmail, becoming known as the inventor of ransomware.[2] Once that new form of extortion had been invented, corporate CEOs could rest easy—they just weren't worth the aggravation anymore relative to the ease that the ransomware business offered.

Ransomware is today the most visible symptom of an insecure internet. The introduction of cryptocurrencies has spawned malicious nests of hackers around the world engaged in "for-profit" extortion businesses focusing on extracting ransom in return for kidnapped data. In 2020, 2,400 high-tech ransomware attempts were reported to the FBI using disarmingly simple tactics.[3] Malicious code is slipped into a network system through a variety of traditional (internal human error or conspiracy) or digital (embedded in emails, software, updates, or flash drives) methodologies to steal or lock up a network's data. The end user receives an email through an anonymized source advising that the data can be retrieved or unlocked if a ransom is paid. Since there isn't going to be a physical handoff between the parties in a nearby park, a briefcase of cash won't do. That is where cryptocurrencies come in as unintended (hopefully) and untraceable (presumably) abettors of the crime. In return for an electronic transfer of value represented by a sum paid in one of the thousands of cryptocurrencies in circulation, the data is freed, and the hospital, food processor, bank, municipality, dam, power company, or nation is back in business.

In August 2021, the FBI underscored its increasing high level of concern with the impact that ransomware can have, releasing a bulletin on the dangers of "Hive," a form of ransomware that uses phishing emails and fake robo-calls to distribute and trigger malicious attachments to blackmail companies.[4] Hive had gained access at the time to the networks of twenty-eight organizations that joined an ever-increasing club of businesses and government agencies that have had their virtual pockets picked in full view of the world by who knows who in China, Russia, North Korea, Iran, or some criminal cartel or terrorist group. Can anyone prevent these attacks from occurring? It seems not. General Paul M. Nakasone, the head of Cyber Command, disclosed in December 2021 that a new cross-functional military effort had been gathering intelligence to combat criminal groups targeting U.S. infrastructure and acting contrary to its prior more hands-off approach.[5] That is a startling admission after almost three decades of growing internet insecurity, which underscores the fact that innovative global solutions will be necessary to stop criminal enterprises that are often situated in countries with no interest in stopping them. As I will discuss later, extradition treaties for hackers and sanctions for countries that harbor them are the first steps that could be taken to get at the problem.

Cyberattacks and the responses are falling into predictable patterns. Pick any significant recent attack or data breach that has occurred—Solar-

Winds, Colonial Pipeline, JBS, T-Mobile, or Log4j. The hack—usually months or years before—is disclosed and is often explained as a "one-off" breach, the functional equivalent of "pilot error."[6] Various government agencies announce that they are investigating as they indirectly apologize for the corporate targets that were somehow surprised by "undetected" deficiencies in software, hardware, networks, clouds, or the behavior of insiders. Third-party providers are often blamed, but in any event the company announces that it and its providers are fixing the software and establishing an online portal with information for potential victims. Customers receive a year of free credit reports to ensure that they can watch their lives being stolen in the future.

The script is very often the same: apologize, rinse, and repeat. A sense of inevitability and "it can happen to anyone" seems to create a force field that protects companies and governments from the harsh criticisms and economic penalties that might otherwise compel them to want to make the internet secure. "It wasn't that bad—no one died, so let's just move on." It is an intriguing wag-the-dog scenario that incentivizes spending money on getting to market quickly rather than ensuring security. Every person and company has essentially been deputized to be responsible for their own cybersecurity. It is as if the secretary of defense advised 7-Eleven to purchase ballistic missiles to defend the land on which its stores sit from attacks by nation-states. Unfortunately, in cyberspace, 7-Eleven is not permitted to launch any of those missiles.

DEFINING THE PROBLEM

It is difficult to dispute that insecurity in cyberspace is a problem, but it is even harder to find a consensus on what security should look like and how serious a problem insecurity is. Insecurity impacts users, networks, network providers, communications, hardware, software, nodes, links, routers, servers, and databases in different ways, so it will mean different things to different constituents. Security is complicated by the fact that networks are only as secure as the least secure link in the system, and most software is "chained" to a past legacy system that can't be abandoned and must be integrated into new systems, vulnerabilities and all.[7] Moreover, cyberspace is "continuously contested territory," which can be controlled and secured only some of the time. As I wrote in *21st Century Money, Banking & Commerce*, secure systems must deflect as close to 100 percent of attacks as possible, detect the ones that get through, and be able to remediate the situation and return the network to full operability as quickly as possible. In that regard, the goal of cybersecurity is indefinable unless you know the user's and system operator's risk tolerance. This creates an enormous challenge for policy makers who must decide how and when to intervene to balance efficiency and effectiveness with security.[8]

A fundamental building block of any system of defenses in the digital world is encryption. Encryption is constructed from cryptographic algorithms, which are basically instructions for solving a mathematical problem. Data can be encrypted by applying a key to camouflage it. So, for example, in its simplest form, if A = 1, B = 2, and C = 3, etc., the word "cat" would be seen as "3 1 21." Making the keys significantly longer and more complex makes the process of breaking them more difficult. If there are two different keys—asymmetric keys which include a public key for anyone to use and a private key for only the recipients of the data to use—the transmission will be more secure. But when all is said and done, it is all just about mathematics. Passwords, multifactor authentication, biometrics, and other tools can be used to enhance security, but the security of the internet is largely riding on the hope that a thief won't be able to solve a mathematical problem. Ironically, that is exactly what Bitcoin requires miners to do to mine a coin—solve a mathematical problem. Every encryption code is theoretically breakable with enough computing power and time. The most difficult way to break a code is by running a brute force attack—every permutation possible is attempted by a computer. The easiest is to obtain a password or key from an insider who is sloppy or has been bribed or recruited.

Networks include a sender, transmitter, and receiver, each of whom impacts the overall integrity of a message, website, or transaction. If the sender's computer is easily infected with viruses and malware, it increases the likelihood that the receiver's systems can be penetrated and corrupted. Deficiencies in transmission networks and encryption interruptions or compatibility issues interject a level of vulnerability between the sender and the receiver. While all these parties must use standard communication protocols, they are not required to employ standard security measures. So the transmission of data and information on the internet resembles something akin to a car driving from a paved road onto a dirt road and then onto a cobblestone driveway before reaching its ultimate destination, with each having a different impact on the condition of the vehicle.

The challenges grow each day given that the increasing size and complexities of hardware and software have only multiplied the places where hostile actors can hide. Richard J. Danzig underscores the point, noting that some graphics processing systems now use more than a billion transistors. The Linux operating system grew from 176,000 to 15 million lines of code in twenty years, and there were reportedly 8.6 million lines of code in the Pentagon's joint strike force fighter. There are increasingly more entry points complicated by the fact that secure interfaces reside on open-source programs like Linux kernal.[9] Danzig cites the director of the CIA, who characterized the problem of computational vulnerabilities this way: "Think about it for a moment—we share the same network with our adversaries."[10]

Most views about online security seem to fall into four broad categories:

Security Is Good Enough: Those in this group, the "good-enoughers," believe that security has become stronger, networks more difficult to breach, and even when breached, the resulting damage is usually manageable and not worth building a new internet to attempt to reach higher levels of protection. Their approach would say that when the confidentiality of data is breached, for example, and someone takes sensitive information about customers, that is a situation that can eventually be remediated. If the operability of a computer network is halted for some period of time, stopgap methods, backup facilities, and institutions will usually prevent that network from being offline for very long. In short, there are reasonable ways to manage our way through cyberattacks, and the cost and inconvenience of building new systems, training new people, and obtaining new equipment and advisors is not justified by the extent to which such risks would be reduced.

Disaster Is Waiting to Happen: These "disaster disciples" generally understand cyberattacks from an up close and personal vantage point. They believe that current defenses are illusory and that in the case of a potent first strike, large segments of critical infrastructures, including banking and finance, would be disabled for unacceptable periods of time. They see cyber war as a real possibility, and one in which the United States and democracy may not be able to prevail. They would implement a long list of changes as described later in this book to repair the internet.

Whatever Is Broken Can Be Fixed: With the cybersecurity business growing exponentially, cybersecurity experts, real or imagined, make up an increasing part of the cyber market. These "fixers" have a solution (for a fee) for every security problem and can patch them one at a time as they arise, much as municipalities repair potholes in roads. They are so busy fixing things that they have little time to consider how the internet might have been built differently, or how it should be rebuilt in the future.

There Is a Better Model: This is a category of futurists and new web architects—the "clean slaters"—who are concerned about the aggregation of corporate and political power that accompanies the centralization of data, the general insecurity of the internet, the undemocratic way it works, or all of the above. They believe that knowing what we all know now, the internet would never have been constructed the way it was and should be rebuilt or, more precisely, supplemented with other parallel networks, assuming that someone else will pay for it.

I seem to fall into a fifth category. I think the internet is now ill fitted for the role it has been given. While it will be difficult to remake it into something much more highly secure, it could be supplemented for transactions and communications requiring high degrees of security and confidentiality with connecting sidecar networks—secure private networks (SPNs)—that operate according to an enforced set of rules and require digital IDs. In effect, we would simply transfer the rules and principles that bring order to our analog lives to our virtual lives.

Internet pioneers freely admit that the internet that they built was constructed of layers of components that accepted failure, thus requiring each layer to be designed to anticipate the failure of the layer below.[11] As a result, current applications driving internet functions have always been known to be risky, making the entire system "insecure by design."[12] Some argue that such systemic insecurity was not the result of a failure to consider the issue, but an inability to understand at the time how security should be considered.[13] Compounding this problem is the economic reality that markets financially reward technology for its speed and efficiency, building a natural but often irrational exuberance about how it can improve our lives. At the same time, many government entities are slow to oversee and create safety and security buffers, lacking the technological resources and not wanting to be skunks at the technological garden party. Add to this the increasing threats that arise from the insecurities caused by badly coded software, improperly tested hardware, network fault lines, sloppy internet service providers, and other virtual sinkholes, and we have a complex set of variables corrupting the race toward technological progress simultaneously as it is being run.

It is unfair to assume that progress has not been made with regard to cybersecurity; security is a moving target, and progress will be difficult to measure over time. In the last two decades, VPNs (virtual private networks) have gained popularity as a means of encrypting internet traffic in real time. When a computer executes a VPN link, it creates a secure encryption tunnel. VPNs disguise online identities by directing data through specially configured servers that hide IP addresses from ISPs and others that track and collect data to market to third parties. When a secure VPN connection is interrupted, VPNs reduce the likelihood that data will be compromised by closing off certain features.

Cryptographic developments allow modern-day encryption to use extraordinarily long keys to protect data. If a brute force attack (essentially trying every possible permutation) on encrypted data could take hundreds of millions of years to succeed, the economics of directly attacking such encrypted data may simply not be worth the time. Blockchain applications boast a new generation of impenetrable security, though cryptocurrencies that rely on them seem to be constant victims of security breaches and thefts. As new generations of technology such as quantum computing are introduced, they will call into question the efficacy of prior generations of security. That is why the Chinese are reportedly stealing and storing as much sensitive encrypted data today as they can, hoping to have the holy grail to decode it when quantum computing arrives.[14]

WHAT FINANCIAL DATA ARE WE PROTECTING?

Financial institutions generally have three buckets of data to protect: (1) customer information, (2) company information, and (3) customer and company financial value. As visualized in figure 2.1, financial institutions and those buckets of data are susceptible to three general types of risks:

Disclosure involves the theft and/or unauthorized release of sensitive data. The Capital One attack by a former cloud engineer is a good example. Customer information was taken and customers were compromised, but there were remedial steps that could be taken to reduce the impact. The bank did not stop operating, and systems continued to function. The seriousness of

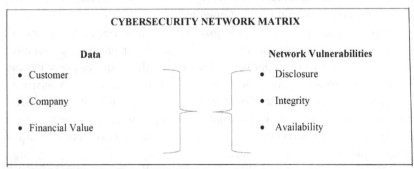

Figure 2.1.   Risks faced by financial institutions. *Thomas P. Vartanian, 2022 (with thanks to Richard Borden for his input)*

the problem from a company perspective increases when the *integrity* of data is attacked. Perhaps a system has been penetrated and every account balance has been increased by several cents, which are then withdrawn over time to avoid immediate detection. Alternatively, data may be destroyed or changed in ways that cause significant financial disruptions. But the company will still be operating and can begin to fix the issues.

The most serious event for a company arises when the online *availability* of a network is compromised. Imagine a major money-center bank not able to move money, acknowledge obligations, or provide customers with access to their accounts. A few days of that would likely create financial panic as the contagion from the removal of a large bank from payments systems cascades through the financial services sector and impacts customers. A report by the Federal Reserve Bank of New York speaks to the seriousness of that risk.[15]

ACTION VS. PROGRESS

The Clinton administration largely identified the range of security issues and critical infrastructures at risk twenty-five years ago. We knew then that we needed sophisticated partnerships between the public and private sectors and a more organized system of governance to protect the security of users and networks. Since then, government and nongovernmental organizations around the world have produced dozens of reports targeting the threats to banking and finance as well as potential solutions. Executive orders by every president have recycled the same warnings and constructive ideas. But beyond the rhetoric, there has been limited comprehensive action taken to reduce the escalating threats. Moreover, beyond unidentifiable hackers, we can't even decide who the prime enemies are in this chaotic digital mess. Is it Big Tech, software developers, government rules, hostile nation-states, criminal cartels, terrorists, or uninformed users? Most prophylactic recommendations can't seem to get beyond a strategy that requires every person and company to be largely responsible for their own defense in cyberspace. Given the rapid advancement of technology as more and more commerce has moved to the internet, vulnerabilities are exploding at exponential rates, like cyber canaries screeching in virtual coal mines. How many canaries must die before we close the mine?

We find ourselves in a difficult position when trying to balance the advantages and risks of virtual life. Big Tech wants to control every piece of data and every user for profit. We may think that these behemoth companies are American, with loyalties to the country, but that is naive. They are global companies whose financial interests are impacted by markets and countries beyond the borders of the United States. National adversaries and criminals have darker geopolitical and financial goals, often using online incursions as reconnaissance missions to embed digital bugs in every computer, network, and critical infrastructure in the country in preparation for more serious geopolitical events to come. The United States is likely doing the same, which dampens its willingness to blow any whistles on the practice and establish rules to prohibit it. So, as we naively click through virtual highways littered with digital spies and FIEDs as we store every scrap of data about us and our businesses, we are increasingly living on the razor's edge between prosperity and devastation.

Emerging "billionaire-making" technologies, an irrational exuberance for the virtual enhancement of our lives, and the never-ending desire to amass power have combined to construct an all-encompassing digital financial bubble that almost no one wants to burst. For three decades, companies around the world have migrated businesses into cyberspace, holding their noses about the threats being created because of the efficiencies and profits they could realize. Countries that could not otherwise have ever dominated geopolitical policies and events from an economic or military perspective are seeking to achieve virtual superiority that gives them a disproportionally larger influence on world events.

The scale of the threats and the potential ramifications have become enormous as we run toward innovation at a breakneck speed without due consideration for the insecurity of new products, systems, and networks. The threat of losing our money, freedom, and selves will become even more pronounced as the next stages of technological evolution deliver even greater advancements built on or by artificial intelligence, new 5G infrastructure capabilities, quantum computing, robotics, facial recognition, biometrics, synthetic biology, new communication networks, nanosciences, nonbiological neurons and neural networks, self-replicating nonbiological cells, and the inevitable march toward Ray Kurzweil's point of singularity where human biology and silicon-based machines merge.[16] Throughout this book I will refer to this collective group of futuristic technologies as the "super sciences." This cavalcade of near-term leaps will bring about transformative changes that require more intellectual imagination to visualize than most of us possess at the moment. They demand that we reimagine the internet if we want to create a safer virtual space to live in.

3

IS VIRTUAL ORDER POSSIBLE?

Virtual reality is a driver of a type of cultural dystopia. Mesmerized by the pot of gold at the end of the internet, too many people seem to have tunnel vision when it comes to recognizing the dangers it poses to society, global stability, and financial security. At the same time, a variety of exposures, including distributed denial of service (DDoS) attacks, brute force attacks, impersonation, digital manipulation, replication of data or value, malware, viruses, worms, logic bombs, ransomware, and internal threats create a myriad of business and personal battlefronts that are increasingly difficult to defend.

This increase in and severity of cyber evil can perhaps be measured by the deployment of pernicious zero-day exploits. Zero-day exploits are forms of malicious code that expose vulnerabilities in software or hardware and create havoc or destruction before anyone realizes something is wrong. They leave no opportunity for detection. In 2021, about seventy zero-day exploits were found, twice the number in 2020. According to journalist Patrick Howell O'Neill's reporting in *MIT Technology Review*, it is difficult to know what this really means.[1] Whether more zero-day software is being deployed (a very bad thing) or more are being caught and neutralized (a very good thing) is difficult to know. At the top of the zero-day food chain are government-sponsored hackers like China, which is suspected of being responsible for nine of the seventy zero-days in 2021 according to Jared Semrau, a director of vulnerability and exploitation at the American cybersecurity firm FireEye Mandiant. At least 2020 and 2021 can serve as baselines to determine the frequency of these exploits in the future.

The financial sector must work overtime just to roll with the punches. You won't be surprised if I tease you with the fact that the U.S. government's report card averages an A on recognition of risks and a D+ on effectiveness or prevention. But don't just believe me. A senior cybersecurity official at the Pentagon quit his job on September 2, 2021, frustrated that the Pentagon was not making cybersecurity a priority. He described the capabilities and cyber defenses of some U.S. government departments as "kindergarten level" with no "fighting chance against China in fifteen to twenty years."[2] Five months later, the chief innovation officer at the Federal Deposit Insurance Corporation (FDIC) resigned. He had concluded that the lack of technology expertise in the agency was contributing to a national security risk.[3] Similarly, a March 2021 report from the National Security Commission on Artificial Intelligence warned that the United States is simply "not prepared to defend the United States" in the era of artificial intelligence.[4] That is an incredibly shocking statement. What if it had said the United States cannot defend itself in the case of nuclear war? Things keep moving toward dangerous and irresponsible terminal points as if we have all been infected with a virus that obscures reality. This trend of public officials jumping ship continued in April 2022, when Preston Dunlap, the chief architect for the Space Force in the Pentagon, resigned from his job, citing his concern that the United States had lost its technological edge to adversaries such as China.[5] Like the *Titanic*, the country is gleefully steaming toward a digital iceberg, and no one seems to be at the helm.

Consider the moneymaking aspect of cyberattacks. According to a November 2021 study, the overlap between the world's arms trade and the surveillance industry is creating the potential for even greater abuse. The Atlantic Council studied more than 200 surveillance companies in the largely subterranean cross-continental surveillance industry and concluded that there are numerous companies that are "irresponsible proliferators" of cyberattacks who market their products internationally to both North Atlantic Treaty Organization (NATO) members and NATO adversaries with little to no regulation.[6] The European Union and the United States are slowly moving toward stricter rules on surveillance technology exports and new licensing rules for cyber-intrusive tools in the face of what some experts are calling the "growing use of mercenaries in cyberspace."[7] Profit is a main driver of cyber insecurity.

To make the challenges a little more complicated, the world is looking at the possibility that the United States may one day no longer be economically, militarily, and technologically superior. That is the geopolitical D-day that is in the script that China has written. It intends to become the world's largest economy with a central bank digital yuan that will replace the dollar as the world's reserve currency. The most stunning development in that regard is not the recent issuance of a digital yuan by the Chinese central bank, but the announcement that China is accelerating discussion with Saudi Arabia to pay for some of the oil it purchases each day in yuan rather than dollars.[8] It also intends to dominate the fields of artificial intelligence and quantum computing and construct a new quantum internet. Perhaps those are all pipe dreams that will never be realized. But China is outspending the United States by many multiples on the way to these goals and has already created the foundation for a new quantum internet that dwarfs what the United States has been able to build.

The internet has brought about a seismic economic, cultural, and social shift in the United States similar to what accompanied the country's transformation from an agrarian to an industrialized and then a computerized society. As we transition to a networked society, even the lines between humans and nonbiological intelligence are beginning to blur. Sure, the opportunities created by digital technologies are substantial, and the financial benefits can make life more prosperous. But those benefits should not obscure the unprecedented threats being created by converting all data and money into virtual form without assurances that they will be safe. In fact, the opposite is true. They are not safe. This is precisely the plan one might draw up to facilitate the next great American financial collapse.

Yet leaders continue to drag their feet when it comes to the protection of America's critical financial infrastructures. More than a quarter of a century into this dance for superiority and perhaps survival, we have an enormous amount of rhetoric, but with the gap between vulnerabilities and solutions ever widening.

DEFENDING CYBERSPACE IN A FLAT WORLD

Current strategic approaches to cyber defense are built around traditional analog assumptions. The "good-enoughers" tend to rely on the fact that progress evolves over time, providing adequate warnings and lead time for those relying on fading technologies to rearm themselves and protect the integrity of their data and systems in a new environment. That may be antiquated thinking that lulls us into a false sense of security given the increasing speed at which developing technologies are reaching realization, and the possibility that they may be perfected or purchased first by adversaries.

"Disaster disciples" find some solace in a second assumption that derives from the nuclear era and relies on the reciprocal fear of mutual assured destruction (MAD). While some nation-states and their agents could today collapse large segments of the economy of a country like the United States using malicious technologies, they refrain from doing so to avoid sowing the seeds of their own destruction. There is no doubt that even a temporary collapse of the U.S. economy could seriously damage China and other world economies, providing a significant disincentive to similarly situated nation-states to pull that trigger. China would not likely launch a cyber-attack to collapse the U.S. economy given the financial interdependence between the two countries unless it thought that it could pick up the pieces and emerge the winner.

Given the enormous stakes and the power of technology, would you bet your life and your money on the likelihood that MAD will protect you? Perhaps, but logic may simply not prevail in a world where the relatively modest cost of the technology needed to disproportionally inflict wide-scale destruction becomes too much of a temptation to terrorists and fanatics. MAD may fall short as a shield when technology provides the opportunity to snatch and control the largest economy in the world. Perhaps the elixir is really the concept of mutual unassured destruction (MUD), as Richard J. Danzig proposes in his seminal article titled "Surviving on a Diet of Poisoned Fruit." He suggests that from a geopolitical perspective, because of "uncertainties for nuclear command, control, and attack warning," we are actually becoming less confident about our offensive and defensive abilities, which should provide an incentive for the United States, China, and Russia to reach agreement on the use of cyber weapons.[9]

ORDER FROM CHAOS

For the most part, everyone participates in this new, imperfect, virtual world of their own free will. There are few internet hostages. Everyone wants their lives, financial or otherwise, to be easier, more profitable, and more socially connected. But how does it make any sense not to create the laws, governance structures, security, and enforcement mechanisms that are required to ensure that order prevails? If the police were ever actually defunded in a city in this country and disbanded, do we believe that anyone would voluntarily live there? So why do we continue to live in a virtual world without standardized rules, a police force, or the assurance of law and order?

Global partnerships between the public and private sectors in democracies around the world must begin to orchestrate the future of technology and the security of virtuality to ensure that order prevails in cyberspace, the metaverse, or whatever technological nirvana follows them. That should include stronger AGE standards—enhanced authentication, increased governance, restricted access, consumer transparency, network security standards, safety rating systems, secure software, resilient hardware, service provider obligations and liabilities, and the establishment of strong incentives and enforcement mechanisms. Governments can't do it alone, and the private sector won't do it unless pushed. The United States must exert its economic and political leadership to lead democratic nations toward the creation of a new, safer internet.

It is far too easy to be seduced by the bright shiny newness and exuberance created by new technologies and rush headlong to be the first to market or use them. We don't have to look very far to see the results of that seduction over the last quarter century. Software, hardware, servers, and networks that are the portals through which all online commerce and our virtual lives must pass do not generally need to satisfy security standards. We think we can trust them, but they have not been approved or declared safe by anyone in authority. It is frightening that products that may impact, control, or end your physical and economic lives are being deployed simply because two young innovators think they're cool.

We might want our virtual world to be the result of agreement among the nations and the private sector regarding minimum levels of internet hygiene tethered to rules of conduct, standards of governance, and meaningful enforcement mechanisms. It is not. When violations of conduct occur, shouldn't there be efficient and effective ways to identify the offending parties and punish them, the networks used, and any country where the network and its servers are located? There are not. Are resources being appropriated to significantly increase online offensive arsenals and defensive capabilities to incentivize law and order in cyberspace? Not nearly enough. Shouldn't cyberspace governing organizations establish clear lines of cyber defense responsibility built around a single point of contact or single source of response? They have not. For the last twenty-five years, everyone has agreed that governments and the private sector must collaborate in the most digitally intimate of ways at every level to defend against attacks and ensure that backup systems are capable of immediately making financial systems operable after an attack occurs. We are likely to find out soon if that has happened.

The current internet seems unfixable at this point. But it is not too late to create a better one. Christopher Yoo, the internet legal scholar, points out that "an increasing amount of traffic has begun to deviate from the traditional pattern," transmitting data over proprietary networks that bypass the public backbone and exchange data through alternative technologies that support greater functionality.[10] Such shifts to dedicated networks reduce central efficiencies that still seem to be paramount to concerns about insecurity. In a safer internet, traveling online could be considered a privilege attached to a license that is collateralized by meaningful value that would be surrendered when rules are violated. It could be monitored by some type of a cyber police force, and every form of digital conveyance could include a kill switch to take it offline or destroy it if its user failed to adhere to the rules.

Virtual reality must also have a clear set of definitions and rules of engagement if there is to be greater order. What is "cyber war," how is it officiated, and who can retaliate in what ways? The *Tallinn Manual 3.0*, the closest thing the world has to a cyberspace rule book, is simply an unofficial collection of rules written by two groups of international experts in their "personal capacities."[11] Democratic nations could create similar standards so that everyone would know whether crashing the stock market or using an electronic pulse that burns out electronic components in a financial network would be considered a crime, an act of war, or a war crime.

In my 2021 book *200 Years of American Financial Panics*, I argue that the United States consistently hurtles headlong toward even bigger and more ferocious financial crises because of the way that increasingly larger and more complex financial systems, markets, and politically driven government policies collide over time. The internet increases the stakes underpinning this formulation and can plant the seeds of even more devastating financial scenarios. And the pace at which technology is changing our financial and social lives has been exponentially accelerated by the global COVID-19 pandemic. Future financial crises could lead to economic annihilation if defensive and remedial capabilities do not keep pace, or if we fail to harness technology in a way that ensures that benevolent humans who are part of civilized societies continue to be the stewards of change.

4

FOR OUR NEXT ACT!

Y ou can't but wonder where this is all going and who is in charge. We
have seemingly been launched onto a technological flight path that is
accelerating toward a terminal point that we don't see or understand. Every
day we encounter new forms of artificial intelligence making decisions and
predicting futures in almost every aspect of life. Behind it, new 5G capabilities plan to take the way things and people connect to the next level and
enable the IoT and alternative virtual realities such as the metaverse.

THE 5G REVOLUTION

The implementation of 5G technologies will have an enormous impact on
the advancements that technology can bring to the human condition. It
would be shortsighted to view 5G simply as the next generation of mobile
phone technology. 5G will create ubiquitous connectivity in human-to-
human, machine-to-machine, and human-to-machine interfaces. It will
also increase the number and scope of threats and vulnerabilities that accompany the use of new virtual and mobile networks.

There have been five generations of mobile network technology. In
1972, electrical engineer Amos E. Joel Jr. created the wireless "handoff"
that allowed callers to stay connected as they moved from one transmitter
cell to another. That led to the invention of the first cell phone a year
later when Motorola developer Marty Cooper produced the DynaTAC
brick phone in 1983.[1] That first generation, 1G, delivered analog cellular
in the 1980s. The second, 2G, introduced the first generation of fully
digital cellular technologies in the 1990s enabling texting. 3G introduced
mobile data connections to the internet through an exponential increase

in speeds and greater bandwidth in 2001, and 4G introduced mobile high-speed broadband in 2009.[2]

The 5G networks of 2021 rely on territorial cell sites that are connected to a network backbone, whether through a wired or wireless backhaul connection. However, 5G wireless technology is based on orthogonal frequency-division multiplexing (OFDM), a method of modulating a digital signal across several different channels to reduce interference and deliver a higher degree of flexibility and scalability to connect more people and things. 5G mobile networks can transfer data at twice the speed of 4G networks—more than a gigabit per second—with half the latency. It does require transitioning from a cell tower system to a denser grid of "small cells" that can be mounted onto streetlights or buildings and which eventually may be installed in homes to create "private" networks that will operate much faster and more efficiently than current Wi-Fi systems. Through a feature called "sidelinking," 5G devices will be able to communicate directly with each other. 5G has already been deployed in more than sixty countries.[3]

5G technology will have an impact on the global balance of power. Its rollout has already been intertwined with geopolitical intrigue given the focus on Chinese companies like Huawei that manufacture and deliver 5G networks to countries around the world that the Chinese government may use for surveillance purposes. In the last decade, Chinese handset vendors like Huawei, ZTE, Xiaomi, Vivo, and Oppo have rapidly grown in global market share. Today, four of the top ten internet companies by revenue are Chinese, an enormous shift from where things were just fifteen years ago when every one of them were U.S. companies. 5G has the potential to skew future networks and global power even further in the direction of China.[4]

The capabilities of 5G will not only heighten the stakes in the race for technological, economic, and political superiority. They will also be a game changer that transforms military strategies and capabilities.[5] The Department of Defense (DOD) sees 5G as transforming diverse sensors and weapons into connected battlefield networks that can enable unmanned and autonomous weapons systems.[6] It also understands the risks, anticipating that competitors will access networks and exploit data through vulnerabilities that may be difficult to detect and prevent.[7] Those challenges are formidable, with DOD stating that networks providing 5G connectivity will be considered to be untrusted given that they may create "significant security vulnerabilities that can be exploited by adversaries at a global scale."[8] DOD believes that the scale, complexity, and decentralized design of 5G architectures make it "infeasible to depend upon perimeter security, which assumes that only trusted devices have been allowed inside the network."[9] So it is developing a zero-trust model for 5G.

In addition, to the extent that DOD relies on mobile and other networks around the world—even among allies—that are built with Chinese components in the supply chain, it sees that reliance as not only increasing security risks but economic ones as well. Given China's first-mover advantage in 5G, its handset and internet applications and services could become globally dominant as they leverage off the size of the Chinese market, even if they are excluded from U.S. markets. Any diminished market power that the United States experiences in the development, manufacture, and deployment of wireless technologies would allow China's 5G suppliers to increase market share globally and become financially stronger, and enable them to control markets and create a downward competitive and economic spiral for U.S. and DOD suppliers. That could leave them increasingly unable to invest enough R&D dollars to provide the kind of quality, pricing, and availability of products that would allow the DOD to maintain its position of superiority.[10]

5G may be the gateway to a massive shift in global political, economic, and military power relationships. In that regard, a 2019 report on these concerns by the Defense Innovation Board produced four top-level recommendations composed of more than two dozen sub-recommendations.[11] The board argued that the United States cannot depend on "stumbling its way to a lead merely by assuming that the wizards of Silicon Valley will stay ahead. The country that owns 5G will own many of these [critical] innovations and set the standards for the rest of the world." Many question whether that country will be the United States.[12]

There are those who envision 5G technology as splitting the internet into two different virtual networks. The Western version would continue to be used for commerce, with the inevitable abuses that come with an unregulated cyberspace. The other would be a Chinese-controlled "authoritarianet"—an internet that would pass through Chinese undersea cables where content would be controlled by the government, facial recognition would ensure that control, and artificial intelligence would search and destroy dissent.[13] In this new bifurcated cyberspace, the data would still have to be exchanged between free and authoritarian, and secure and less secure, internets. The United States may have to figure out how to control its digital environment in a virtual world where it may not be the dominant player.[14]

THE INTERNET OF THINGS

5G technology will also expand the mobile ecosystem to new areas such as seamless IoT capabilities. The IoT is a descriptive term for an enhanced state of virtual reality that will be created by embedding sensors in objects connected to a network so that they can exchange data with other devices and systems within the network. The IoT has been enabled by a combination of advances in low-cost, low-power sensor technology, 5G technology, cloud connectivity and computing platforms, Big Data, and advanced machine learning and analytics. Connected "devices" ranging from people and pets to ordinary household appliances and industrial tools will be able through a range of chip technologies to be connected in seamless, low-cost communication and computing networks to operate, record, monitor, adjust, and interact between themselves with minimal human intervention. Once again, the microchips, RFIDs (radio frequency identification devices) and the related antennas, transceivers, and a transponder required to sew all things into the fabric of the internet are not all made in the United States, making compliance with applicable U.S. laws and security standards difficult.

Advances in neural networks and natural language processing (NLP) have already made digital personal assistants such as Alexa, Cortana, and Siri quite popular in homes. But they are only the tip of the IoT iceberg. As different and higher forms of artificial intelligence continue to evolve, machines and algorithms will increasingly be in control. By connecting products such as cars, owners will be able to operate and service their vehicles remotely, allowing continuous relationships between car manufactures, dealers, and owners. IoT wearables, whether they be chips or sensors built right into clothing will enable people to better monitor their health and safety.[15] But beware the communication of so much data; if it can be communicated, it can be collected, stored, used, and abused.

The application of IoT technology in industrial settings (industrial IoT, or IIoT) through machine-to-machine (M2M) communication will help create preventive predictive maintenance programs, smart power grids, smart cities, and smart digital supply chains. As everything made and shipped is identified and tracked at all times, and the route and method of transportation down to the last step of the delivery person is mapped out by artificial intelligence, businesses will be able to operate at significantly lower expense ratios. For these reasons, the IIoT is sometimes referred to as the fourth wave of the industrial revolution.[16]

The applications are endless. But so are the risks. The IoT enhances the quality of life when health data is seamlessly shared with physicians, employers, or health and life insurance companies and they all work to remotely monitor and improve people's health. But the collection and storage of this data also raises enormous privacy, moral, and ethical issues as we envision how a cavalcade of entrepreneurial or unfriendly entities steal or abuse that information. The increased threats to privacy, freedom, and democracy from the theft of IoT data or unauthorized control of connected devices are stunning. Kai Strittmatter's amazing recitation of offensive uses of technology by China in his book *We Have Been Harmonized: Life in China's Surveillance State* is mandatory reading. Everything that is connected will be vulnerable to abuse, and everything will be connected in the IoT. There will always be governments ready, willing, and able to use that data for authoritarian purposes.

THE QUANTUM LEAP

Quantum computers use the quantum state of an object to produce what's known as a qubit. These represent the undefined properties of an object before they've been detected, such as the spin of an electron or the polarization of a photon. Instead of relying just on 1s or 0s, they can process exponentially more data compared to classical computers because quantum data can be superimposed in the state of qubits where it can leverage the behavior of particles in quantum states as both zeros and ones simultaneously. Quantum computing is comparable to the analytical complexity of a coin spinning through the air where the superpositions are entangled with those of other spinning coins and plugged into advanced algorithms to solve problems that would take a classical computer forever, if it could do it at all.[17]

In 2019, Google announced that its 53-qubit machine had achieved "quantum supremacy." Its claims about achieving contrived quantum speedups were only for specific benchmarks.[18] That same year, IBM launched its 53-qubit quantum computer followed by IonQ's 32-qubit system in 2020. In November 2021, IBM launched a 127-qubit quantum processor followed promptly by QuEra Computing demonstrating a 256-qubit quantum simulator with enhanced programmability.[19] More qubits means more information can be stored and processed, and positioning the qubits improves programmability and real-time adaptability. In the next few years, QuEra expects to reach 1,000 qubits, nearly one-fifth of the 4,099 qubits estimated to be capable of breaking RSA-2048 encryption in just 10 seconds rather than 300 trillion years. QuEra is also working on quantum algorithms that can solve extraordinarily complex computational optimization problems.[20]

Some argue that quantum computing may never materialize. Others suggest that quantum computers may someday solve specific cryptographic and physics problems in a fraction of the time classical computers would take but may only be capable of modestly assisting with other important practical day-to-day problems. There are still practical difficulties in building quantum computers because of unwanted interaction between a quantum computer and its environment that results in premature "measurement" of the qubits.[21] That requires quantum error correction, a field in which significant advances were made in late 2021.[22] With the United States, Russia, and China racing to build quantum computers and internets, there is much talent and money chasing that future. In addition, the capital and credibility of several large companies and countries seem to be riding the quantum wave, telling us something about its prospects.

The most notable advancement of quantum computing will be the speed at which it will process data, train neural networks, and accelerate machine learning. Quantum cryptography will likely accompany its emergence using the same principles as standard cryptography to communicate in encrypted forms over dedicated communications links. Unlike standard cryptography, it will rely on photons to transmit a key in an unbreakable cryptosystem that is not measurable unless it is disturbed.[23] This would theoretically allow for the instantaneous detection of any party into the process that isn't supposed to be there. As we will discuss later, the verdict is out on the ultimate practicality and feasibility of quantum computing, but progress is being made to remove the remaining hurdles such as the recent discovery of quantum error-correcting codes.[24]

The development of quantum key distribution (QKD) should theoretically be able to generate encryption algorithms that provide confidentiality assuming that the original QKD transmission can be authenticated. According to the National Security Agency (NSA), this will require users to have dedicated fiber connections or to "physically manage free-space transmitters." The reasons are that QKD is hardware based, it lacks flexibility for upgrades or security patches, and it cannot be easily integrated into existing network equipment. The hardware used to perform QKD can introduce other vulnerabilities, however, such as increased risk of denial of service.[25]

The financial products and networks being created through distributed ledger technology (DLT) such as the blockchain and the many cryptocurrencies that use them could theoretically become cannon fodder for quantum computing or whatever generation of computing follows it. Central bank digital currencies (CBDCs) could either become more secure or destroyed, stolen, or manipulated by the next generation of computers. In short, the "immutable" nature of DLTs in the current technological era may be susceptible to infinitely faster and more efficient computers that will make them "mutable."

If that is a possibility, we should be cautious about creating financial products such as cryptocurrencies and CBDCs that could be undone by technology and put the financial systems of countries at risk. Simply stated, the extent to which quantum computing or advances like it are ultimately a threat or an enormous enhancement of the human condition turns in some measure on who gets there first and what they do with it. Google prematurely claimed to have achieved "quantum supremacy" when its Sycamore solved a mathematical calculation in 200 seconds that was estimated to require 10,000 years for a current supercomputer.[26] China is the country that plans to get there first, and it is spending like it means it.

In May 2022, the Biden administration took an important step forward in the quantum race, albeit much too late in the game. Recognizing the potential threats to current cryptographic security measures posed by the speed and capabilities of quantum computing and the progress being made by China to dominate the field, it issued two executive orders.[27] Together, they established a new National Quantum Initiative Advisory Committee consisting of the director of the Office of Science and Technology Policy and up to twenty-six members appointed by the president, and an open working group including critical infrastructure owners and operators chosen by the director of the National Institute of Standards and Technology. They required the cataloging of all federal government cryptographic-reliant systems, an analysis of the risks posed by quantum computing, and the deployment of quantum-resistant cryptography throughout government information systems by the end of 2023.

WELCOME TO THE METAVERSE

When it comes to evaluating the security of the internet, it seems hard in hindsight not to conclude that Mark Zuckerberg's oft-quoted mantra, "move fast and break things," has turned out to be quite a dangerous strategy. It takes advantage of the propensity for people to leap off virtual cliffs in return for the transitory high delivered by the next hit of technology. Now his company, Meta Platforms, Inc. (formerly Facebook), wants to escort us into the metaverse—the next generation of the internet that will create a new virtual chamber of pleasures and horrors where we will have no apparent sense of order or user control.[28] In reality, like most new things, it is not so new.

The concept of a metaverse has been around for twenty years. In 2003, an online multimedia platform that allowed people to create avatars and have a second life in an online virtual world was launched by the San Francisco–based firm Linden Lab. By 2013, it had approximately 1 million regular users. Second Life users, also called residents, could interact with places, objects, and other avatars, explore the world (known as the grid), meet other residents, socialize, participate in both individual and group activities, build, create, shop, and trade virtual property.[29]

Perhaps the most solid proof of when something has blossomed into a real thing is the existence of a trade association and a way for value to be created and transmitted. The metaverse already has at least one trade association. The Open Metaverse Interoperability Group (OMIG) hopes to establish technological standards to connect virtual worlds in the metaverse.[30] But wait, there's more. Metaverse currencies also exist and are being used for high-value transactions. One of them, MANA, is used in a virtual world called Decentraland where its value reached as high as $1.60 after trading for years around $.10, bringing the aggregate value of MANA tokens to a stunning $2.4 billion. Decentraland even has casinos where MANA can be gambled. Sotheby's ran an art show in Decentraland.[31] For all its hype, the metaverse is not an answer to internet insecurity and will only likely exacerbate the problem by exposing more data and people to the insecurities of yet another virtual space.

The genesis point of the metaverse seems to be the ever-growing gaming industry which is laying the foundation for virtual reality to morph into a quilt of interconnected online avatars engaging in a virtual version of our analog lives played out through cartoonish representations in video games, social media relationships, online shopping, 3-D renderings of products, business meetings, and sporting events where cryptocurrencies will be the coin of the realm. Companies like Axie Infinity are promoting a play-to-earn model where gamers win prizes translatable into marketable characters and other nonfungible tokens (NFTs) that can then be traded or sold using various forms of cryptocurrencies. Axie claims to have facilitated $4 billion in NFT trades.[32]

But like any other world, there will be crime, unanticipated consequences, and other darker aspects of life that will inevitably creep in. The security of NFTs is in question, as is the interoperability of these platforms, which necessarily impacts the long-term existence and value of the characters and prizes there.[33] For better or worse, metaverse rules will likely be made by innovators driven by "progress" and the billions of dollars that can be made from selling technological snake oil and collecting even more data about users. But there is little doubt that the metaverse will increase the stakes and the degree of difficulty in creating a more secure online environment. This suggests that from the very inception of being embraced by mainstream users, it should be built and adopted to be secure. As of now, that won't happen.

People and businesses work mightily to ensure the safety and stability of their analog worlds. They establish rules, police, borders, and armies. They do not invite strangers into their homes or offices and willingly share personal, professional, medical, sexual, financial, trade secret, and other sensitive information with them. But vast numbers of us now often willingly participate in virtual spaces that effectively do just that. Cyberspace has become a virtual Wild West of insecurity, unwanted surveillance, and anonymous anarchy. The dark underbelly of technology creates parallel universes where crime is often undetectable, and supremacy belongs to whoever possesses the most advanced computers and skills.

It does not make sense to dive headlong into new virtual worlds and cede even more personal space to progress when we have not figured out how to manage our security and safety in the virtual world we already have. Cyberspace changes everything we have known about geopolitical superiority and negotiation. MUTs are working overtime to achieve the virtual or technological equivalence of geopolitical superiority that they could never have attained in the analog world militarily. We all want to believe that the government is all over this problem and that we are somehow protected. It is not, and we are not.

The government is incapable of fixing the insecurity of the internet by itself, and businesses won't do it until there is a crisis. And there are no silver bullets. Only the consensus that can emerge from a better public appreciation of how technological benefits and threats must be balanced and how governments and private businesses must work together to do it will eventually produce the leadership required to forge a safer online environment.

5

HOW THE CYBER GOT INTO SPACE

Most of us have the same relationship with the internet that we have with our automobiles. We turn a key or touch a button, and if all goes well, it runs. Occasionally the car breaks down and we throw open the hood to look inside as if we know what we're doing. We may even dare to tinker with a few things. But ultimately we are left waiting for a tow truck.

When you boot up a computer, do you ever wonder how within seconds it can connect to every person, machine, and scintilla of data that has ever existed? It seems simple but very mysterious. Whatever is housed within the confines of that laptop can instantly connect us with Aunt Suzie in Des Moines and the New York Yankees' website. All we need do is click on their name or icon that they seemingly have placed there to provide us that access, and there they are. That is what makes it all so disarming and seemingly trustworthy. But as we know from periodic crashes and infections from unwanted malware, there is very little on this planet that is more complex, technologically advanced, or dangerous.

The internet is a connection of computers, routers, nodes, internet service providers (ISPs), and other equipment and transmission providers. Though it may seem like a monolith, the internet is more accurately a network of many layered, interconnected networks exchanging information through a uniform standard known as the Internet Protocol (IP) as data travels through backbone providers that exchange it and hand it off.[1] To understand the insecurities of the internet and consider how to eliminate them, we must first appreciate its complex architecture, how its various access points work, who makes it function, and the extent to which the components of its infrastructure connect with and impact each other. Many different parties are involved in every click through cyberspace. But though we are steadily constructing an alternative life online, there is no president of the internet or prime minister of cyberspace. There aren't even dedicated cops on the beat. Whatever syllabus of rules and protocols does exist applies to the back-end commercial entities and internet associations that oversee name assignment and create the pipelines that make up its plumbing. These attributes are both its beauty and its weakness.

A BRIEF HISTORY OF THE INTERNET

One of the most memorable years in my life was 1969. I met my wife while I was still in the seminary studying for the priesthood. That is another story for another book. Also tucked away in 1969 was an event that most people never took note of and perhaps don't today. On October 29, the Advanced Research Projects Agency launched ARPANET, the first iteration of the internet that connected four computers at universities so that they could access files and transmit information from one organization to the other. When more universities were added, the ARPANET was renamed the "Internet." Email was introduced in 1972 as libraries across the country were linked and data was exchanged through Transport Control Protocol (TCP) and Internet Protocol (TCP/IP) architecture.

In 1986, the National Science Foundation funded NSFNET, a network of supercomputers throughout the country that connected academic and personal computers. The University of Minnesota developed the first user-friendly internet interface in 1991. The National Science Foundation discontinued its sponsorship of NSFNET in May 1995. This lifted all commercial use limitations on the internet and ultimately led to the launch of companies like AOL, CompuServe, Prodigy, and Delphi to offer rudimentary internet service. By 2030, experts project there to be 7.5 billion internet users and 500 billion connected devices.[2]

THE INTERNET'S SKELETAL STRUCTURE

There are endless access and choke points to the internet. They include internet service providers, financial institutions, web hosting companies, search engines, browsers, domain name registrars, and internet connecting, routing, and network providers. The policies and the protocols that they establish among themselves are the backbone of whatever governance exists as this digital Wild West was cobbled together by computer engineering pioneers, government agencies, military divisions, and universities. Layered on top of these are terms of service (ToS) agreements that are presented to users who usually have little understanding of what they are agreeing to and virtually no leverage in negotiating their terms. Most people mindlessly click "I Agree" to launch their virtual lives.

To communicate over the internet, every device needs a unique identifier linked to an internet address domain name and an autonomous system number (ASN). Governmental and private standards setting organizations such as the Internet Corporation for Assigned Names and Numbers (ICANN), a nonprofit organization, oversee the assignment of IP addresses and domain names and the operation of the root server system. The Domain Name System (DNS) is a decentralized naming system that provides a worldwide directory under the supervision of ICANN. It is also responsible for the addition of new top-level domain names and for the maintenance of universal connectivity standards.[3] The assignment of domain names is delegated to those who register with an ICANN-accredited registrar and agree to be bound by the terms of the registrar accreditation agreement (RAA).

Domain name registrants list their domain names on hosted name servers so that they are reachable on the internet. Registries are responsible for maintaining, among other things, the registry for each top-level domain.[4] ASNs are unique binary numbers assigned to network operators. Top-level domain names are segmented into subdomains represented by a uniform resource locator (URL). A system of root zone servers around the world under the supervision of the National Telecommunications and Information Administration (NTIA) in the Department of Commerce converts top-level domain names into IP addresses.[5] The system is more complex in operation, and for those who want to understand them, there are abundant resources available.[6]

Routers around the world connect autonomous system network operators such as ISPs, which connect computers and the internet so that packets of data can be disassembled, transmitted, and reassembled once they reach their destination.[7] Currently thirteen district root server implementation operators oversee a network of hundreds of servers in countries around the world. They include Verisign, the Universities of Southern California and Maryland, Cogent Communications, NASA, the Internet Systems Consortium, Inc., the Department of Defense, the U.S. Army, Netnod, RIPE NCC, ICANN, and Wide Project.[8]

CONNECTING TO CYBERSPACE

Once a computer has a unique IP address assigned by an ISP, it can connect to the internet and communicate with other computers through a TCP/IP stack built into the computer's operating system. When a domain name is typed into the address bar of a browser, the letters of the URL—*www.thomasvartanian.com*—are converted into a binary set of numbers that are recognizable to all computers, routers, and servers. An IP address is based on binary digits or "bits" of "0s" and "1s" based on standards established by the American Standard Code for Information Interchange. Thus, each letter in *www.thomasvartanian.com* is assigned a string of 0s and 1s, beginning with the *w*, which equals "01010111."

Internet Protocol version 6 (IPv6) is the most recent version of the communications protocol developed by the Internet Engineering Task Force (IETF). Using a 128-bit address theoretically allowing 2^{128}, or approximately 3.4×10^{38} addresses, IPv6 addresses are represented as eight groups of four hexadecimal digits. Each letter of the address *www.thomasvartanian.com* is separated into groups of eight bits so that *w* translates into a binary code that sums to 71, the first two digits in the IP address.[9] Figure 5.1 demonstrates how the calculation of 71 is derived.

**Translating www.thomasvartanian.com
Into an IP Address**

"w" Binary Equivalent:

01000111

Multiplier:

128+64+32+16+8+2+1

Binary Code for W:

0 + 64 + 0 + 0 + 0 + 4 + 2 + 1
(128x0) (64x1) (32x0) (16x0) (8x0) (4x1) (2x1) (1x1)

Sum of Binary Code:

71

Figure 5.1.   The calculation of 71. *Thomas P. Vartanian, 2022*

Primary and secondary DNS servers determine how a web browser locates an address by connecting to the primary DNS server, which routes the message and requests the web page being sought.[10] Hypertext Markup Language (HTML) is the language used to write web pages. Hypertext Transfer Protocol (HTTP) makes the web work. Once a user enters an address (URL) into a web browser, the browser connects to a domain name server and retrieves the corresponding IP address and connects to the web server, sending an HTTP request for the desired web page. The web server searches for the desired page and, if it is located, forwards it on.

The internet does not have a centralized home office or coordinated governing structure. It is a collection of interconnected tiered networks, network providers, content companies, and content delivery companies contractually linked to each other to operate its transmission facilities, cables, satellite links, and other hardware. Some arrangements produce revenue through fees, while others rely on mutual agreements where each network exchanges their respective customer traffic without payment.[11] These networks generally connect at shared exchange points—the connection points that exchange packets of data and route them toward their destination.

The bottom line is that this is extraordinarily dense and complicated stuff that only a small percentage of the humans on this planet can describe in detail from beginning to end. Perhaps even fewer know how to rearrange and reconstruct packets, routers, IP addresses, network architecture, and protocols to improve the internet from a technological point of view. Technological know-how is not, however, the stumbling block that is preventing the world from having a more secure internet.[12]

20,000 LEAGUES UNDER THE SEA

In 1858, Abraham Lincoln delivered his "House Divided" speech, and Charles Darwin went public with his views on evolution. The two ends of a 2,500-mile-long cable a little more than a half inch wide were also connected that year to link Europe and North America by telegraph. The inaugural cable sent by Queen Victoria took over seventeen hours to reach President James Buchanan. The MAREA cable, completed in 2017 and partly owned by Microsoft, Meta/Facebook, and Telxius, a subsidiary of the Spanish telecom Telefónica, stretches 4,100 miles to connect Bilbao, Spain, with Virginia. It boasts transmission speeds 16 million times faster than the average home internet connection to transport most international communications, according to TeleGeography.[13]

There are now around 380 underwater cables equaling almost 800,000 miles in length transporting data around the world to support the internet. These cables can be damaged, creating system downtime. Such disruptions occur about 200 times a year and are largely the result of human error.[14] Earthquakes, storms, and intentional and unintentional man-made events can disrupt global commerce over these cables, making the internet highly vulnerable. On January 15, 2021, an underwater volcano destroyed the single undersea cable that connected Tonga to the rest of the world through a line that runs 514 miles between it and Fiji. It was estimated that it could take weeks to repair. Two years before, the country had experienced a similar event when its undersea cable was cut by a Turkish-flagged ship dropping its anchor on it.[15]

These types of disruptions underscore the vulnerability of networks that have so many single points of failure that can compromise communications and commerce. This is encouraging significant redundancy to be built into this invisible highway for global internet communications. Though it may be difficult to plan and execute a malicious attack on these undersea cables, it is feasible for an attacker that has the resources to do so. This includes a limited group of nation-states, their affiliates, and criminal cartels. And the point of attack determines the degree of difficulty caused. For example, divers from U.S. submarines attached devices to Soviet cables in the Sea of Okhotsk to intercept a decade of communications until news of the surveillance was sold to the Soviets by a former NSA employee.[16] James Griffiths, a journalist for CNN, wrote about the vulnerability of the point where undersea cables connect to land and the disruption that can be perpetrated by their disruption.[17] In 2013, documents provided by NSA whistleblower Edward Snowden indicated that a British spy agency had secretly gained access to this network of cables and was monitoring telephonic traffic as NSA's operation Upstream is alleged also to have done.

Recently, however, this picture has become more complicated and troubling. In the last ten years, four large tech companies, Microsoft, Alphabet/Google, Meta/Facebook, and Amazon, have become the dominant users of undersea cable capacity, with their share being about 66 percent today.[18] It is estimated that in the next three years, notwithstanding the enormous costs of constructing and maintaining undersea cable systems, Big Tech will become the primary financiers and owners of undersea internet cables. They are being driven by an insatiable demand for ever more terabytes of bandwidth. Industry analysts have raised concerns about whether the world's most powerful providers of internet services and virtual marketplaces should also be permitted to own the infrastructure on which they are all delivered.

By expanding vertically in the internet business, Microsoft, Alphabet/Google, Meta/Facebook, and Amazon are saving money over time as they create vertically integrated tech empires. At some point, these expansions may raise antitrust concerns or result in these companies being regulated as telecommunications companies. The stranglehold on undersea cables and the junction points that these companies might control could facilitate the ability to establish choke points to increase and enforce online governance and security. It could also give dominant tech companies overwhelming market power, which would likely be attractive to repressive governments seeking geopolitical control.

Adding to the complex geopolitical factors playing out underwater, China is expanding its construction and maintenance of undersea cables in the East China and South China Seas. In addition, Chinese companies are building a new undersea Peace Cable that will eventually travel overland from China to Pakistan and then snake underwater for about 7,500 miles via the Horn of Africa before emerging in the French port of Marseille. Google and Meta/Facebook claim that they won't be using the Peace Cable because they have enough capacity already. If they needed to, it would be difficult because of the U.S.-led national security boycott of many Chinese telecommunications equipment makers, including Huawei.[19] China's increasing efforts to control data-carrying pipes and cables is thought by many to create enormous additional security and surveillance threats.

Now that we understand a little about who and what is behind the curtain, we can begin to appreciate the complexities and points of vulnerability that exist in the internet's financial infrastructure.

Part II

THE VIRTUALIZATION OF MONEY

6

THE NEXT GREAT TRAIN ROBBERY

In the year 1855, England and France were at war with Russia in the Crimea. . . . Once a month, twenty-five thousand pounds in gold was loaded into strongboxes inside the London Bank of Huddleston and Bradford and taken by trusted armed guards to the railway station. The convoy followed no fixed route or timetable . . . to the coast and from there to the Crimea. . . . Each safe weighed five hundred and fifty pounds . . . and was fitted with two locks requiring two keys, or four keys altogether. For security, each key was individually protected. . . . The presence of so much gold in one place naturally aroused the interest of the English criminal elements. But in 1855 there had never been a robbery from a moving railway train.[1]

This opening narrative from *The Great Train Robbery*, a 1978 movie starring Sean Connery and Donald Sutherland, romanticizes their theft of gold from a moving train as it traveled from London toward the British coast on its way to the Crimea. Such a heist was thought to be unimaginable, just as many of the virtual thefts and ransom attacks that we see today once were. This is a trend line that I lose sleep over because I believe that my control over my money is continuing to erode as the internet evolves.

Stan Stuart understands why I have been losing sleep. On his way to work one sunny summer day, he stopped at Sam's Coffee Stop in downtown Atlanta, part of a global chain. After ordering a cappuccino, he used the "Sammy" app on his phone to pay for it. Simultaneously, malware sitting on his phone from a popular game app he had downloaded two days before was activated and transferred to Sam's system where it quietly ran in the background, sending several cents from every purchase made around the world to an account in the Cayman Islands. From there the money was immediately moved through several hundred accounts in twenty-two countries.

The malware was also designed to shut down Sam's payments systems and infect its cloud services if it was detected or the company took remedial actions. Sympathetic cyber pirates operating in different parts of the world had been recruited to disable redundant backup systems, servers, and clouds, for the moment completely crushing the company's ability to operate.

Stan left the coffee shop thinking there had been a momentary technical glitch in Sam's systems. It happens, he thought. He went to a nearby ATM to get cash, only to find a screen message saying it was out of cash. He tried another ATM around the corner, not realizing that every ATM in the southeast had been simultaneously programmed through logic bombs embedded months before to stop operating and display the same message. Meanwhile, the cameras on the ATMs and in every branch of several large New York banks began to record and livestream on social media sites through a server in eastern Europe the unfolding panic of customers who increasingly could not access their money. This greatly fed into the growing panic as millions of people began to experience what Stan was experiencing. Stan called a 1-800 number on the back of his ATM card to report these issues, but that number and his phone were now disabled, as was his new car, which was loaded with three dozen microchips connected to various diagnostic, navigation, and self-driving networks. The security system in his home refused to unlock the doors after his ten-minute walk there. His refrigerator had turned itself off.

As the things that Stan relies on began to shut down, he was not aware of the damage being done behind the scenes. Eight leading financial clearinghouses that make up the plumbing of the financial system and move money, securities, and investments every minute of the day also went offline and seized up. Employees at the U.S. Treasury realized that they could not issue new notes, transmit funds to investors, or make electronic benefit transfers (EBTs) to U.S. citizens. When the Federal Reserve was unable to transfer funds between and among banks, businesses, and consumers, the economy effectively collapsed.

A half-dozen federal regulators and various cybercrime-related agencies that had prepared for this moment leaped into action. But the systems they used were overwhelmed by the massiveness of the attack and the new breed of malicious technology and supercomputers that were encrypting, destroying, replicating, disabling, and stealing data at will, as well as disappearing and diverting trillions of dollars. For the moment, almost no one could get their hands on their money. The economy stopped as each company and governmental entity launched its own separate, unconnected, and often contradictory financial detection and remediation plans.

Federal, state, and local government agencies and a host of private organizations that had planned for just this occasion sought to analyze and help companies rebuild their systems through backup facilities and inter-company consortiums that had been created. They had difficulty communicating with each other, locating their resources, and coordinating responses. The chaos was similar to 9/11, when police, fire, state, municipal, and federal authorities realized that they had never created a seamless communication network that would work in the kind of attack that had occurred. Notwithstanding years of preparation, all of these organizations seemed unprepared, with only limited and equally vulnerable backup facilities to put in place to allow money to keep flowing and commerce to continue.

CALL CYBER 911

Stan's is a hypothetical story. That kind of a coordinated attack would admittedly be difficult to pull off given the resources that would be required. But many of the pieces of this story have already occurred at different times and places. This string of recent cyber events tells us that there are significant holes in our defenses against a well-heeled advanced persistent threat (APT) by a nation-state or state-sponsored group that gains unauthorized access to a computer network and remains undetected for an extended period. There are no cyber police or virtual firefighters to call. On a macro level, *no one* is establishing the rules of engagement as we zip down this waterslide of technology. *No one* is building a cyber financial coast guard or early-warning system. *No one* is responsible for the protection and resurrection of the country's economic infrastructure. There are no financial bumper guards to ensure that economic anarchy does not prevail.

Let me be clear. When I say *no one*, I do not suggest that there are not thousands of dedicated governments, people, agencies, companies, consortiums, commissions, and legislators actively thinking about, urging, and in many cases taking action to deal with these kinds of scenarios. Thankfully, there are, and we will consider many of them and the efforts they have made over the last twenty-five years. When I say *no one*, however, I mean precisely that. Compared to military defenses, no one *single* person, agency, company, consortium, or commission is responsible for deploying defenses, arming us offensively, and preparing to rebuild the economic infrastructure of the country in the event of a cataclysmic digital event. No *one* person or agency is in charge. There is no chairman of the Economic Joint Chiefs and no financial commander in chief for cyber war. There is still a lack of the kind of centralized, coordinated defense and responsiveness that exist to respond to kinetic military attacks when it comes to the economic infrastructure of the United States in cyberspace. That should concern us as much as if the country defunded its military and delegated national defense to every corporation in the country.

Technology makes money more transportable, commerce easier, and life more manageable. It also makes money easier to steal. Much like the cultural ripples that flowed from the marriage of Madison Avenue and computers sixty years ago and revolutionized the marketing of merchandise and politics, we are witnessing an unprecedented digital consolidation of finance, technology, and behavioral science that will change American finance forever and create threats that require new financial defense systems. Technology has pounced on its users. The natural and more gentle evolution of technology that usually occurs and allows necessary defensive and offensive systems to be developed was sidestepped by the COVID-19 pandemic. As humans struggled to survive in an infected analog world, everything moved to a virtual world that seemed less toxic. Tech pioneers and companies enthusiastically assisted in that effort for both altruistic and financial reasons. But this accelerated pace makes even those entrepreneurs apprehensive about whether their products can stay ahead of the next new wave of technology that could deconstruct them. It is time for people to pounce back and seize control.

While this new world was unfolding over the last two decades, financial policy makers, regulators, academics, politicians, and financial industry experts have often been distracted by yesterday's issues, the crises of the day, and a sense of technological euphoria that they do not want to subvert. Enormous amounts of resources continue to be expended debating analog-world issues that date back nearly a century, such as the economic impact of companies that are "too big to fail" and whether a new Glass-Steagall Act should be put into place to once again separate banking and commerce. Too many "experts" are trapped by an "anchoring" problem, hopelessly intoxicated, and limited by their own familiar stale policy positions, trying to force-feed them into a world that has passed them by.

WHO IS MINDING THE STORE?

We tend these days to believe that the digital sky "probably" won't fall. And either too many or not enough people are minding the internet store. There are endless distractions. Legislators make the laws and direct what regulators focus on, but when it comes to technology and particularly financial technology, they are lagging well behind. And even though some policy makers grasp the severity of the issue, it is nearly impossible to find agreement or consensus given the brand of polarized politics practiced these days that seems to elevate the interests of political parties over those of people. Watching many congressional hearings on technology and financial services gives me the same feeling I had when I attended my children's school plays. They were entertaining and cute, but not exactly prime-time Broadway material. Moreover, Congress is easily distracted by big money that representatives need to finance elections, and only a few are willing to take on difficult or scary issues like cybersecurity until they have already exploded and can attract campaign funds.

The combination of inconsistent leadership, politics, increasing cybersecurity costs, and the collision of financial services and technology are creating economic threats that we have never encountered before. The Clinton, Bush, Obama, Trump, and Biden administrations all saw the growing risks and, for a variety of political, financial, and resource reasons, failed to advance solutions over the last twenty-five years that could assure that the United States would remain in control of its own financial future. The private sector saw the growing problem, and as far back as the early 2000s, some experts decided that we needed a new and more secure internet given how it had pivoted to handling communications and transactions that had not been anticipated when it first burst onto the scene.

In that regard, China's announced aspirations include supremacy in quantum computing and artificial intelligence by 2030, the creation of the largest economy on the globe by 2050, widespread use of a digital yuan as a global currency, control of the microprocessor semiconductor chips necessary to support technology, and a monopoly on the world's rare earth minerals that are needed to make batteries, microprocessor chips, and components used in machinery and computers. Domination in those areas equates to world domination, and it is spending multiples of what the United States is spending in many of these areas, advantaged by an economy that is fueled by cheap labor. In this way, China is financially co-opting the economies of countries around the globe by making them financially dependent on cheap goods. At the same time, it is accused of unleashing ferocious barrages of corporate espionage and cyberattacks that are seeding cyberspace with malicious applications that can be activated at a moment's notice.

So here we sit today knowing that if a country like China attains its aspirations, America and other democracies around the world could begin to become subservient to its technological, economic, and military superiority.

A fundamental lack of confidence by other nations in a government like China's that is constantly accused of using authoritarian tactics, theft, espionage, and aggression to advance its goals may undercut its ability to earn enough global respect to reach the top of the geopolitical ladder. But it doesn't take respect to be an economic or technological tyrant, and those are bets that we should be hedging in the United States.

Make no mistake, the country's economic infrastructure is protected against cyberattacks and hostile control, but not in the comprehensive way that its shores, skies, and seas are defended by the DOD. Each computer and individual effectively protects its own territory but is unable to fire back if attacked. Admittedly, it is difficult to translate military defense into a world where most of what is being protected are digital networks and data under the control of private companies and individual users. But a responsible government should lead the way to comprehensive public-private offensive and defensive strategies to deal with this unique landscape.

Let's assume that you have been hacked and suffer a devastating breach that results in the loss of sensitive personal data and the money in your banking and brokerage accounts. Who do you call when such a financial cyber robbery occurs? The local police, the FBI, the Secret Service, the Cyberspace and Infrastructure Security Agency (CISA), U.S. Cyber Command, the NSA, the Federal Reserve Board, the Department of Homeland Security, the Treasury, the FDIC, the Financial Crimes Enforcement Network, the air force, the navy, or all of the above? There are protocols and interagency agreements in the financial services industry that determine how a cyber event should be handled. Once detected, a bank would call its regulator which would contact Treasury. The Treasury would then contact the National Security Agency, which would communicate with U.S. Cyber Command and the Department of Homeland Security.

Perhaps such a telephone tree approach will work, but between December 2020 and March 2021, the U.S. Post Office could not manage to deliver more than a few dozen pieces of mail to my home, offering COVID-19 and "complex challenges at this time" as excuses. It did, however, somehow find a way to scan and send me an email picture each day throughout that period of every piece of mail that was not being delivered. While we may have talented people staffing federal and state agencies, having an alphabet soup of government agencies competing to coordinate a response to an attack on the integrity of a financial company or the U.S. government should not create a sense of confidence.

The chances today that a great cyber robbery can be stopped are not high enough. Consider astronomer Cliff Stoll's recitation of the hoops he described having to jump through in his book *The Cuckoo's Egg* to find any-one in the government willing to react to a hack at his Lawrence Berkeley Lab computer network in the 1980s.[2] Various federal and state agencies and law enforcement authorities ignored Cliff since he could not identify a loss of at least $1 million or prove a disclosure of classified data. His problem all started with a $.75 accounting discrepancy related to the use of computer time, something so small that no government agency was willing to pay any attention to it. Only due to his dogged persistence at pulling the thread to pursue the hacker did he eventually uncover with the help of authorities in Germany an international spy ring linked to the KGB that had penetrated the Berkeley computers and from there assessed military bases and other logistic targets across the United States. By the time any prosecution was brought years later, the damage had already been done and many of the protagonists were long gone.

Government systems and reactions have been substantially improved since the 1980s, but there are still too many different agencies to negotiate in an emergency. I am told that the $1 million damage and classified infor-mation thresholds are still likely to be thrown up as a hurdle in most cases. As in Cliff Stoll's situation, the initial cost of the hack may just be the tip of an enormous iceberg lying beneath the water.

VULNERABILITIES ARE GROWING FASTER THAN SECURITY

Hackers have become the new guerilla mercenaries in the cyberspace jungle, available to the highest bidder—terrorist or nation-state. That should suggest that along with the instructions for launching a nuclear attack, the president's black bag should also include codes for the release of U.S. cyber infobombs and defenses.[3] The military learned in the 1990s that an "infowar" arms race could be one that the United States would lose because its technological dominance also created its vulnerabilities. An article in *Time* magazine in 1995 blew the whistle on any further military complacency. It concluded that an attack against the United States could be prepared and exercised under the guise of "unstructured hacker activities" and cripple U.S. operational readiness and military ef-fectiveness, while the country might not even know it was under attack. If there is "no nationally coordinated capability to counter or even detect a structured threat,"[4] it is understandable that it could go unnoticed long enough to create devastating havoc. These concerns laid out in the late

1990s led to the modernization of U.S. military defenses, which are increasingly focusing on the use of algorithmic intelligence rather than tanks to defend against and attack military targets.

This same influential *Time* article in 1995 also noted the vulnerability of the financial services systems, revealing the NSA's concern that banking systems and stock exchanges could "easily be crippled by determined hackers."[5] But an equivalent amount of urgency has never been shared by policy makers about the vulnerabilities of the country's national economic security. Over the last quarter century, agreements, structures, information sharing vehicles, protocols, and procedures have been put in place by the financial services industry to maximize the ability of individual companies and systems to store data, replicate it, and stand it back up after an attack. But they are diffused, not organized around a clear chain of command, and have never been meaningfully tested.

The threat that technology poses to the creation, transmission, and investment of money may be the quintessential existential threat. But it doesn't feel like that in Washington, DC, where the world is portrayed as something other than it really is. In the business world, where decisions are tethered to profit and loss statements, companies stop producing products when no one wants to buy them. In Washington, DC, where politics make and break businesses every day, it may take decades for bad ideas to shrivel up and die and the real ones to be moved to the fore. That usually only happens when politicians can no longer use the issue to dragoon donors into giving them money.

Understanding all this, why wouldn't economic terrorists and financial pirates try to disrupt societies by manipulating or disrupting national financial infrastructures? Why wouldn't rogue nations use technology to leapfrog over U.S. military superiority, particularly if they thought they could dominate or take advantage of whatever was left after the digital bubble exploded? This set of challenges only increases to the extent that intelligent machines armed with artificial intelligence are dispatched to flourish unrestrained by globally accepted rules and standards. Won't intelligent machines eventually try to reap the profits of their own brilliance, particularly as they progress toward artificial general intelligence? After all, that is what humans would do, so why wouldn't intelligent machines and synthetic nonbiological creatures created in the image of humans not do the same thing? The super sciences will only create bigger risks and blur more lines between machine and biological intelligence. Assuming that the worst will not occur because we have governments to protect us is, unfortunately, wishful thinking.

Financial markets are increasingly being impacted as a new breed of technology titans seek to squeeze every penny out of this digital revolution. Confident that it must be all right if the government allows it, consumers climb on the digital bandwagon not wishing to miss the next best thing. How else could a single Bitcoin that has no intrinsic value and which the CEO of JPMorgan describes as "worthless"[6] have ever sold in the $68,000 range? In one respect, Jamie Dimon is wrong about crypto's worthlessness. People love to steal it and are doing so at an alarming rate!

The total of cryptocurrency thefts in 2021 was a whopping $14 billion despite the "immutability" of the blockchain networks they operated on.[7] That is likely more than has been stolen from all banks in the world since the beginning of recorded time. In April 2022, it was announced that hackers had stolen $625 million in cryptocurrency from the video game company Axie Infinity when the supposedly secure underlying blockchain that powered the game was infiltrated.[8] This was followed by the theft of $182 million worth of digital assets when hackers attacked an algorithmic stablecoin called Beanstalk, wiping out all of the Ether it held and collapsing the value of Beans.[9] Perhaps the most troubling crypto theft occurred in October 2021, demonstrating the "mutability" of blockchain when $16 million of an indexed crypto fund started by two twenty-somethings went missing when an eighteen-year-old mathematician outsmarted the blockchain-enabled fund with a series of often illogical trades that essentially tricked the entire trading system with what he claimed were entirely legal trades.[10]

The stablecoin landscape was further muddied when TerraUSD broke the buck and subsequently crashed after a series of large withdrawals from a decentralized bank for crypto investors in May 2022, creating significant investor fear and dislocations in the fragile crypto market.[11] By the end of the month, the price had dropped to fractions of a cent as $42 billion of market value was erased in a very short time. Tether also lost its dollar peg almost immediately thereafter when its price dropped below $0.95 briefly around 3:30 a.m. on May 12. Like almost every financial crisis, Tether's drop was attributed to a drop in market confidence (i.e., lingering questions concerning the existence and quality of the assets backing it one-to-one with the U.S. dollar). If crypto is where you want to park and grow your money, good luck with that.

We need only look back at history, particularly the 1920s and 1980s to appreciate that there is always a reckoning for rampant speculation, unprincipled financial behavior, and regulatory mistakes. Each day that passes and every step taken on this path proves that moving fast and breaking things is stupendously dangerous. The combination of increasing digital insecurity and potential reliance on imports such as rare earth minerals and semiconductor chips under the control of China has created a true existential challenge of our day. An individual, company, or country will be capable of being dominated digitally without the movement of a troop or dropping of a bomb. And yet we continue to approach these threats with too many analog tools, thought processes, and solutions. It was one thing when a thief in the analog world stole a strongbox of money, a vault full of silver, a bundle of bearer bonds, or household electronic equipment. The extent of those incursions always had natural physical limits. Digitization puts the future of everything that can fit on a flash drive in jeopardy, altering the risk-reward ratio significantly. If we set rules and anticipate the threats, they can be managed. But history suggests that we may not be willing to put our technological houses in order and establish global rules, standards, and enforcement processes until a digital Pearl Harbor occurs.

The insecurities of the internet underscore that democratic societies are increasingly at risk of virtual domination and control as markets run toward innovation without due consideration for the insecurity of new products, systems, and networks and the absence of standards of conduct.[12] We have not done enough to protect money, data, infrastructures, and networks from malicious use, nor has there been enough oversight of the collection and theft of data about people, businesses, societies, and nations. We should not continue to drift aimlessly toward progress, almost matter-of-factly merging biological and nonbiological elements without rules, standards, or protocols that can protect people, privacy, data, secrets, or defenses from abuse. The result of continued complacency may be the next great cyber robbery.

THE CHALLENGES

The absence of a comprehensive early-warning economic defense system and the nonexistence of a cybersecurity police force create obvious risks. The technology exists today to separate you from your money and accounts, disrupt the Federal Reserve System, and hinder the Treasury's distribution of Social Security, Medicare, welfare, food stamps, and other forms of assistance. Capital markets are susceptible to wild gyrations from the use and abuse of program trading algorithms or can be made to engage in millions of ghost trades, causing confidence in them to evaporate. Why wouldn't adversaries and fanatics try to rob the equivalent of a moving train in cyberspace if they could, particularly if they had nothing to lose? As in 1855, the fact that such a robbery has never occurred should be little comfort in a world that is moving so fast. There will always be someone who is bold or irrational enough to believe that there is more to gain than lose from committing an act that no one else has ever done. All they need is the means.

Consider the words of General Mark Milley, chairman of the Joint Chiefs, before the Senate Appropriations Committee in June 2021:

> Adversaries continue to use operations in the cyber domain as a means to compete with the U.S. and pursue a position of advantage in crisis and conflict. Malign cyberspace actors increasingly exploit supply chain vulnerabilities, such as commercial software, to gain network access and conduct cyber operations against U.S. citizens, organizations, and institutions. The low cost/barrier to entry and anonymity that cyberspace provides make this domain a priority focus for adversaries to asymmetrically compete without escalation.

7

WILL THE REAL SATOSHI
PLEASE STAND UP?

People have an emotional relationship with their money. That tends to create generational inertia in the evolution of financial routines and the products that are used. Consumers often stay with what they were familiar with when they entered the employment world. This is changing as developments in technology have accelerated and made behavioral inertia a ticket to financial isolation. But as the pace of financial change increases, the security of products and networks never seems to catch up, having started behind to begin with. In a business where trust is a critical selling factor, this is a big problem.

To complicate the matter, crypto mania (both positive and negative) is now in full swing, playing a starring role in the changing dynamics of banking and finance. While cryptocurrencies began as a darling of the social media set, the possibility of getting rich oh so quickly has made it interesting to every generation of investors. Crypto is a shape-shifting financial instrument—it can resemble whatever its miner, purchaser, trader, investor, or exchange wants it to. And even without clear definitions of what it is, there is already $2–3 trillion of cryptocurrency trading throughout the world that are the basis for a market of asset-backed and derivative securities estimated to equal almost another $3 trillion.[1] Even if only half of these financial instruments are purchased on margin or otherwise leveraged, we are talking about potential global financial positions approaching $10 trillion—positions that are ultimately supported by no underlying intrinsic value outside the stablecoin market. Sure, a van Gogh painting is no more than a few dollars of paint and a canvas and can still sell for millions. The dollar bill is no longer backed by gold, so it is really only worth the paper it is printed on. But the van Gogh and the dollar are backed by established forms of confidence and boundaries that transcend whatever aspirational hopes currently support cryptocurrencies.

Crypto markets are creating their own special brand of threats as a speculative, largely unregulated investment transmitted over insecure networks that can attract hackers from every part of the planet. Cryptocurrencies have no capital, reserve, liquidity, or other prudential protections that are built into traditional financial systems. That is their beauty to many. But as they grow and are increasingly the subject of well-funded, high-powered cyberattacks, the reactions and ripples throughout global markets will increase, creating panics not only in the virtual but also the analog world. A loss of confidence in a $10 trillion market would be devastating. There has never been a time in America when speculation in an asset that had no underlying value did not end badly. It is unlikely that this time will be different. If the market begins to collapse, there is no alternative but for holders to sell into it since there is no collateral to liquidate to create any residual value once cryptocurrencies trade to zero. The only way out is to find the greater fool to buy your coins.

Digital money did not begin with Bitcoin in 2009. It has been around for quite a while, but just not quite in crypto form. Federal Reserve banks, clearinghouses, and exchanges have traditionally "moved" money without physically transporting it using electronic book entries and the daily netting of transactions between businesses and financial intermediaries on proprietary networks. The Fed creates money simply by crediting accounts at large banks when it purchases securities. These forms of "digital money" grew out of necessity to some degree to avoid the physical relocation of bags of cash and checks across towns, states, and nations.

No significant effort to create retail digital money for consumers occurred until companies such as Mondex and DigiCash introduced smartcards and digital wallets to transmit electronic money in the late 1990s. I participated in and wrote about the Mondex electronic money and smartcard beta test in Swindon, England, in the mid-1990s and subsequently handled endless issues that these new products raised on behalf of clients before an alphabet soup of federal, state, and global regulators. There was a sense of digital euphoria driven by fabulous innovations that captured the imagination of young and old, blinding many to the hidden dangers of digital finance played out in an open-architecture network. There was a tendency to believe the statements made by the products' salesmen about their security, but in hindsight we understood very little about it then. It was not until about 2009 that retail digital money actually began to gain traction with consumers when Bitcoin was introduced, relaunching another period of digital euphoria. Again, the benefits of technological innovation and the capital and profits they generated overshadowed the threats of financial insecurity and instability that were being created.

To appreciate the new vulnerabilities that financial technologies are creating, consider the fundamental nature of money, the different forms it can take and the infrastructures that move it. Every user, systemic seam, junction, handoff, and intersection in each system creates vulnerabilities that can be taken advantage of. Cash, paper checks, electronic checks, credit cards, Fedwires, ACH transactions, and a wide range of payments systems, clearinghouses, financial utilities, and securities exchanges currently make up the traditional internal plumbing of the United States that moves value. Figure 7.1 lays out the different uses and vulnerabilities of money and money equivalents.

As a companion to this chart, appendix A explains the derivation and role of the products and players in traditional financial services markets, payments systems, and exchanges. It is worth noting the vast number of participants and infrastructures that comprise these systems and create potential vulnerabilities as they convert to a virtual, digitized world. But for the moment let's focus on the digital expansion of money driven by cryptocurrencies, crypto exchanges, and the risks they are creating.

U.S. MONEY & MONEY EQUIVALENTS – USES & VULNERABILITIES

	CASH (FIAT CURRENCY)	CHECKS	CREDIT CARDS (Debt Instrument)	CRYPTO-CURRENCY	STABLECOIN	CBDC-DIGITAL DOLLARS
PROS	• Ease of Use • Anonymous • Real Time • Finality of Transfer • Unhackable	• Ease of Use • Standard Rules • Legal Protections	• Ease of Use • Standard Rules • Legal Protections	• Ease of Use • Real Time • Anonymous (?)	• Ease of Use • Real Time • Value Stability • Fiat Backed • Finality	• Ease of Use • Real Time • Government Backed • Finality
CONS	• Can Be Lost/Stolen • Bulky • Destruction • Counterfeiting	• Not Real Time • Limited Finality – Stop Payment • Forgery • Destruction • Systemic Failure	• Can Be Lost • Fraud • Theft • Systemic Failure	• Acceptability • Value Fluctuation • Security • Theft • Manipulation • Destruction • Counterfeiting • Systemic Failure	• Acceptability • Security • Theft • Manipulation • Destruction • Counterfeiting • Systemic Failure	• Security • Theft • Manipulation/Destruction • Counterfeiting • Systemic Failure
TRUST FACTOR	• High	• High	• High	• Moderate	• Moderate	• High
BACKING	• Government	• Issuer • Banks • Clearinghouses • Federal Reserve	• Banks • Processors	• None	• Fiat Collateral • Marketable Securities	• Government
TRUSTED INTER-MEDIARY	• Banks • Federal Reserve	• Banks • Federal Reserve	• Banks • Processors	• Crypto Exchanges	• Crypto Exchanges	• Federal Reserve

Figure 7.1. Different uses and vulnerabilities of money and money equivalents. *Thomas P. Vartanian, 2022*

AND THEN ALONG COMES BITCOIN

The *Summit* by Ed Conway is an excellent historical rendition and analysis of world dynamics surrounding the events at Bretton Woods that effectively declared the dollar to be the world's reserve currency despite the objections of Britain's John Maynard Keynes, Russia, and others. Because the United States emerged from World War II as the only superpower and had the most gold in its vaults, it could essentially dictate the terms of the Bretton Woods Agreement. It was the political and economic elephant in the room, and try as they might, other countries could not muster the arguments or economic power to do anything to derail the dollar from becoming the preeminent form of money in the world. The International Monetary Fund and the World Bank also emerged from the Bretton Woods Agreement, both headquartered in Washington, DC.

In the early 1970s, President Nixon terminated the dollar's linkage to gold, leaving it backed solely by confidence in the U.S. economy. The delinkage of the dollar from gold was ironically a precursor of cryptocurrencies like Bitcoin that are not backed by the government or anything of tangible value. Putting aside the creditworthiness of the United States, which has continued to erode each year over the last thirty years, the dollar bill represents nothing that a user can cash in for anything else. It is the genesis point of its own value; it cannot be taken to the Federal Reserve in return for $1 worth of the Grand Canyon. But if money does not need to be anchored to any metal, commodity, or government backing, can't it exist only in cyberspace and have the value that markets give it?

In some of the darkest days of the financial panic of 2008, a person or group of persons using the pseudonym "Satoshi Nakamoto" posted a message—"Bitcoin P2P e-cash paper"—which embedded a link to a white paper called "Bitcoin: A Peer-to-Peer Electronic Cash System." It was the blueprint for a decentralized digital currency. Bitcoin's genesis block was mined on January 3, 2009. Within it was a message: "The Times 03/Jan/2009 Chancellor on brink of second bailout for banks," a headline from that day's London Times about the desperate shape of the traditional financial services sector. The first transaction occurred about one week later.[2] A little more than a year later, a Florida man bought two Papa John's pizzas valued at $25 for 10,000 Bitcoins, establishing the initial real-world value of four Bitcoins per penny. In 2011, miners and coders started to build other networks as the price of Bitcoin crossed the $1 threshold. By the end of November 2013, it was worth more than $1,000. Four years later it was $10,000.[3] At the time this book went to print, one Bitcoin was worth about $19,500, down from its high of $68,000.

Three general types of digital money have now emerged: floating rate cryptocurrencies, stablecoins, and central bank digital currencies (CBDCs). For definitional purposes, floating rate cryptocurrencies are digital coins generated by miners using computer programs running established protocols and rules. It seems ironic to me that while complex mathematical formulas in the form of cryptographic algorithms are used to securely encrypt everything that flows in cyberspace, crypto miners are required to solve similar complex mathematical formulas to successfully produce crypto coins. While those coins are branded as money, retail consumer acceptance has been slow in coming. Several things work against them in that regard. They are not backed by or linked to any fiat currency or collateral, their value floats in the market as investor interest ebbs and flows, and they are often used for illicit purposes ranging from ransomware to human trafficking. In addition, cryptocurrencies often require large amounts of energy to mine them, making them an environmental problem. Bitcoin is the most well known example of a cryptocurrency, but by 2017, dozens of other cryptocurrencies were emerging and doing public initial coin offerings (ICOs). With thousands of cryptocurrencies in circulation today, Bitcoin is still flying high as a speculative investment at best and a hope certificate at worse. And still, no one knows who Satoshi Nakamoto is.

There were hopes in 2021 that would change. After his death in 2021, the family of David Kleiman sued his former business partner, a fifty-one-year-old Australian programmer living in London named Craig Wright, claiming that the two men were Satoshi Nakamoto and that they jointly controlled the partnership's assets, which included Satoshi Nakamoto's 1 million Bitcoins.[4] Those coins were equivalent to approximately $64 billion at the time. Proof that the two men were Satoshi Nakamoto would require the surviving one to produce the private key that controlled the account where the 1 million Bitcoins were held, or simply move one coin. Over the years, many people claimed to be or denied that they were Satoshi Nakamoto. It's the kind of story that Indiana Jones movies are made of. The Kleiman case fizzled out when a Florida jury rejected claims that Wright and Kleiman had launched Bitcoin.[5] The world was disappointed not to have learned the truth, but happy to have the mystery continue.

Cryptocurrencies such as Bitcoin operate on top of blockchain algorithmic applications where there are no financial intermediaries, go-between fees, clearing systems, or regulation. Blockchain technology has itself evolved into a symbol of the cultural way of life the internet represents. The use of blockchains seems to speak to one's sense of privacy and the role of government. Blockchains are an advanced generation of digital signatures that encrypt packets of information to form connected encrypted blocks of data simultaneously shared among all users, making them difficult to be created, modified, or replicated since every user's program would have to agree to any change.

Cryptocurrencies such as Bitcoin are viewed as a cultural form of egalitarianism—they may be created and controlled by anyone as long as they have the computing power and electricity to do so. There are no financial intermediaries, a characteristic that creates a challenging and unfamiliar model for financial institutions and their regulators. The rationale and purpose behind cryptocurrencies are impacted by a variety of economic, political, social, and political instincts and aspirations. At one end, innovators seek to make money, while at the other, users want to live free, with as little government centralization and intervention in their lives as possible. This latter category of users tends to be those who see cryptocurrencies as the symbol for a way of life.

Cryptocurrencies can provide innovative advantages, including the ability to (1) transact business (legal or otherwise) anonymously; (2) escape government oversight or intrusion; (3) execute and pay for transactions in real time, even across borders; (4) reduce costs by extracting costly go-betweens; and (5) avoid systemic control by any one or a small group of entities. Some of these features, such as lower costs and anonymity, are illusory. In the case of Bitcoin, the mining of new coins takes significant computing power that requires large amounts of electricity, which translates into large carbon footprints and transaction fees for using, trading, or storing it that can reach $60.

The challenges posed by cryptocurrencies include known and unknown issues and threats to the extent that cryptocurrencies and supporting blockchains (1) do not enjoy the confidence provided by government involvement; (2) are too often used to facilitate illegal business transactions, including terrorism, ransomware, money laundering, the purchase of drugs, and human trafficking; (3) create fiscal and monetary control issues to the extent that they may be mined without an offsetting deposit or reserve requirement; and (4) may be subject to hacking, replication, or theft, particularly as technology evolves and cryptographic security systems become more vulnerable to attack.[6] In this regard, of the more than twenty instances in 2021 where a digital robber stole at least $10 million in digital currencies from a crypto exchange or project, in at least six cases, hackers stole more than $100 million. In December, a Bitmart company account lost almost $200 million, causing the company to freeze customer transactions for three days.[7]

So far, cryptocurrencies are more like aspirational money—sort of digital lottery tickets. Nevertheless, by the end of 2021, approximately 16 percent of Americans owned one of the thousands of different global cryptocurrencies,[8] which had a market value of over $2 trillion.[9] While these numbers and valuations can vary widely depending on who is providing them and what the market looks like on any given day, there is something important going on in crypto—it has become more than a fringe commodity. Perhaps crypto coins are a hedge against the weakness of a national currency or economy. In that regard, they may simply function as financial canaries in economic coal mines. Cryptocurrencies claim to be more secure than other forms of value because they are decentralized and transmitted by a peer-to-peer blockchain that is highly impervious to being hacked. So far, though, hackers have somehow stolen a lot of the stuff, and only time will tell if the security boasts attributed to blockchain systems have been overestimated, as some experts contend.[10]

If you wonder why anyone would see value in cryptocurrencies and why Congress and financial regulators would allow them to become so large without imposing some oversight, you are in good company. In some ways, cryptocurrencies represent the antithesis of a safe and secure financial system much as an anonymous internet does. But those arguments have been systemically drowned out so far under the weight of technological euphoria and the billions of dollars being made in this twenty-first-century version of the tulip bulb mania. But the tide appears to have turned with the collapse of some cryptocurrencies, forcing a reckoning with reality. In the first quarter of 2022, there was a 20 percent decrease in crypto stocks and a 60 percent decrease in crypto-focused companies even before the crypto reckoning that would follow.[11] Is crypto today's tulip bulb mania?

Tulips were introduced into Europe from Turkey in the Middle Ages. Demand for colored varieties of tulips eventually began to exceed the supply, causing prices for individual bulbs in 1610 to grow so large as to be acceptable as dowries for brides or considered valuable enough to purchase a brewery. Holland's tulip trade was originally limited to professional growers, but as with many rapidly growing markets, nobody wanted to be excluded. Rising prices coaxed families to speculate in the tulip market and mortgage everything to purchase them. In 1637, when doubts and rumors about the continuous escalation of prices circulated, confidence in them turned shaky and the price structure quickly collapsed, wiping out the fortunes of many Dutch families.[12]

Time will tell if cryptocurrencies follow this pattern or lead to the next phase of money and payments. El Salvador is perhaps providing some answers to this question. When China's central bank effectively banned Bitcoin in 2021, miners were welcomed to El Salvador, which boasted of its abundance of energy that its volcanoes could create as a mining advantage.[13] Its experience has been a bumpy one, however, leading to demonstrations in the street and an "Adopting Bitcoin" summit in November 2021. By February 2022, experts labeled the move a "disaster" as the El Salvadoran treasury lost $200 million due to falling Bitcoin prices and consumer adoption proved anemic.[14] Securities and Exchange Commission (SEC) chairman Gary Gensler doesn't see "much long-term viability for cryptocurrencies," likening cryptocurrencies in existence to the so-called wildcat banking era that took hold in the United States from 1837 until 1863 in the absence of federal bank regulation.[15]

STABLECOINS

Stablecoins are cryptocurrencies that are supposed to be linked to or backed by fiat currencies such as the U.S. dollar or Treasury notes to provide them a stable value and translation platform. That is supposed to make them safer. The predominant users of stablecoins are those engaged in crypto trading and decentralized finance who seek a stable-valued currency to trade cryptocurrencies that do not have a stable value. How they are backed, secured, or collateralized by marketable assets and fiat currencies varies according to each stablecoin.

Stablecoins usually rely on blockchain applications, but neither issuers nor exchanges are prudentially regulated or required to carry capital or liquidity to backstop any devaluation of assets to stave off a run and prevent a collapse of the system. As a result, stablecoins may not always live up to their lofty commitments to back up their tokens with reliable value instruments, making them even more risky.

Yet another variation of cryptocurrency stablecoins has recently emerged called "algorithmic stablecoins." They aren't necessarily backed by any assets at all but rely on financial engineering to maintain their link to the dollar. Supporters argue that algorithmic stablecoins are superior because they run autonomously on blockchain-based networks rather than a single centralized entity relying on traders to keep them tied to the dollar. This naturally makes it more difficult for anyone, including regulators, to control them.

TerraUSD, launched in 2020, was for a short while the most popular algorithmic stablecoin, with about $17.3 billion in circulation. It relied on market dynamics to keep the price stable. If its price dipped below $1, traders could remove it from circulation in exchange for $1 worth of new units of another cryptocurrency, theoretically raising the price of the remaining Terra coins. In short, traders seeking arbitrage profits are supposed to keep TerraUSD close to a $1 value. What could possibly go wrong? As previously mentioned, it did go wrong in May 2022. Lots can go wrong when there is little regulation of a financial product, no government backing, and little to no screening of the people that operate them.

An investigative report by *Bloomberg Businessweek* in October focused on the opaqueness of stablecoin disclosures about the existence and valuation of qualifying reserve assets. The report asserted that Tether, the leading stablecoin issuer, which would be one of the fifty largest banks in the United States if it were a bank, was not able to demonstrate how $69 billion of its coins in circulation were backed by fiat currencies, Treasury securities, or other marketable assets.[16] The reporters tried to fact-check Tether's claims by canvassing Wall Street traders to see if any had seen Tether buying anything. None had. Their research found that the chief executive officer listed on Tether's website was a Dutchman who lived in Hong Kong and had never given an interview or spoken at a conference. The chief financial officer was a former plastic surgeon from Italy who was described on Tether's website as the founder of a successful electronics business. The reporters' search of Italian newspapers showed that he was once fined for selling counterfeit Microsoft software. This description of Tether's operations by the report is worth noting:

> Just last month, traders bought $3 billion in new Tethers, presumably sending billions of perfectly good U.S. dollars to the Inspector Gadget co-creator's Bahamian bank in exchange for digital tokens conjured by the Mighty Ducks guy and run by executives who are targets of a U.S. criminal investigation.[17]

A Tether representative reportedly declined to say where the $69 billion was invested and what backed its tokens in circulation. That is not surprising since companies have no obligation to report to journalists. But on October 15, 2021, the Commodity Futures Trading Commission (CFTC) issued an order simultaneously filing and settling charges against Tether for making untrue or misleading statements and for omissions of material fact in connection with the issuance of its stablecoins. The order required Tether to pay a civil monetary penalty of $41 million,[18] asserting that Tether held sufficient fiat reserves in its accounts for only 27.6 percent of the days in a 26-month sample period from 2016 through 2018.

The *Bloomberg* report gives the impression that pinning down the specifics of how some stablecoins operate is akin to stapling Jell-O to a wall. Even when stablecoins are fully backed by fiat currencies and/or baskets of liquid financial assets such as government-issued notes, that should not be a "get out of jail" card. Given the complexities of matching assets and liabilities, hedging bets, managing liquidity, and monitoring collateral values, the business is extremely complex. Moreover, the value and acceptability of stablecoins will always be impacted by public confidence, which can be a fickle friend at best.

In light of the increasing significance of and concern over stablecoins, the Committee on Payments and Market Infrastructures of the Board of the International Organization of Securities Commission invited comment on a consultative report in October 2021, recommending that stablecoin systems have appropriate governance arrangements, regularly review material risks, provide clear and final settlement, and create little or no credit or liquidity risk.[19] The G20's Financial Stability Board also released a report that month noting that vulnerabilities in the crypto and stablecoin space continue to grow. It produced ten recommendations, including the need for comprehensive functional regulation; appropriate governance frameworks; effective risk management frameworks; robust systems for collecting, storing, and safeguarding data; appropriate recovery and resolution plans; and transparent information for stakeholders.[20] While strongly focused on financial safety and soundness, few words in these reports mentioned the increasing security risks to financial infrastructures being created by these new forms of "money." The President's Working Group on Financial Markets, with the FDIC and the Office of the Comptroller of the Currency[21] as well as the Financial Stability Oversight Council,[22] also jumped into the fray, announcing various studies of stablecoins and recommending that legislation be enacted by Congress to, among other things, address systemic prudential concerns and regulate stablecoin issuers as banks.

In a thoughtful piece by Greg Baer, CEO of the Bank Policy Institute, Baer identifies four central problems with stablecoins. First, stablecoin issuers have failed, and consumers have lost their money. In some cases, the underlying value of the coins declined. In others, the money was stolen through hacking or defalcation. Second, the "reserves" backing stablecoins are undefined and have varied greatly, as one might expect in an unregulated business. Third, even if stablecoins are regulated at the state level as money service businesses, there is no obligation for the issuer to disclose what is backing them. Fourth, there is significant financial stability risk in the case of the failure of a major stablecoin issuer that ignites runs on other unregulated stablecoins, causing a wide array of asset liquidations in the market.[23]

I would add a fifth factor—increased security concerns. I attribute those to the insecurity of the internet, increasing exposures that result from more private participants in the creation and movement of economic value that are not prudentially regulated, the lure of all that value to hackers, and the increasing likelihood that the "immutable" nature of blockchain technologies will be more vulnerable to being breached by future generations of computer geniuses and machine technologies. The drumbeat by regulated institutions to bring stablecoins within the regulatory tent is reaching a fevered pitch and getting traction with both regulators and Congress.

CENTRAL BANK DIGITAL CURRENCY

Central bank digital currencies are the third piece in the development of new forms of digital money that will create new sets of systemic and consumer vulnerabilities. Central banks around the globe have decided that they must compete with private cryptocurrencies and each other in the race to issue CBDCs for a variety of competitive, monetary control, and jurisdictional reasons. In the United States, CBDC is being studied by the Federal Reserve to determine whether it should represent a digital form of the nation's fiat currency. As the roles of central banks continue to expand beyond financial necessities because of the increasing incidents of financial crises thrusting them into greater and greater control of national economies around the globe, they have seemingly become almost as infatuated by technology as private sector markets. In some cases they appear more interested in keeping a competitive pace with financial companies than in regulating the threats that they and the financial sector are taking on in this new world.

U.S. banks are apprehensive about a digital dollar assuming a position in the markets that competes with them and further separates them from their traditional role as financial intermediaries, a role they earn significant money performing. No decision has been made on whether to issue a CBDC in the U.S. payments system.[24] But the deployment of a token or account-based digital dollar that directly links retail users and the Federal Reserve will create a kaleidoscope of complicated monetary, policy, and financial stability issues that require a careful balancing of the related benefits, costs, and risks.[25] High among them should be the risk created by making available a digital form of central bank money to individuals around the world that may double as a wormhole into the Federal Reserve System.

CBDCs have become a sort of financial Rorschach test. Everyone sees something different when they look at them. They represent modernization, efficiencies, and cost savings to some, while others focus on the advantages, real or imagined, of real-time global payments transfers. There are hopes that they will better serve low-income and unbanked citizens, while less benevolent nations view them as an efficient means of surveilling and controlling their own citizens and defanging U.S. dollar diplomacy and sanctioning ability. Bankers fear that CBDCs will provide the Federal Reserve with an even more direct role in setting retail interest rates, transforming the flow of consumer funds that banks rely on to make consumer and business loans, and altering the risk profile of U.S. banks and the economy itself.[26]

CBDCs will alter current banking and payments models by eliminating or recasting some if not all of the tasks that traditional financial intermediaries perform in various payments systems depending on whether the central bank offers CBDC accounts directly to consumers or requires them to have CBDC wallets issued by banks, savings institutions, or credit unions. Even bank-operated CBDC wallets will transform payments markets in fundamental ways. But the most serious gating issue is the new security risks CBDCs will create, a factor largely ignored in the Bank of International Settlements June 2022 "Annual Economic Report" extolling the virtues of CBDC.[27] Before governments become intoxicated with technological happy talk and deploy CBDCs, they must consider new systemic vulnerabilities.

This exposure will become even more significant if and when quantum computing becomes mainstream, making current forms of public-private key cryptography an insignificant challenge for them. The NSA is hoping to develop quantum-resistant algorithms by 2024, a project bolstered by two executive orders issued by President Biden on quantum computing in May 2022.[28] Consider the enormity of the challenge. A quantum computer with 4,099 perfectly stable qubits could reportedly break standard RSA-2048 encryption in just ten seconds, rather than the 300 trillion years required today.[29] While such quantum computers do not yet exist, as we will discuss later, they may be on the way, and with them, alterations to the security of every piece of data and value that has ever been digitized. CBDCs should not be launched just because they can be. They should be deployed because they are the most secure form of money ever developed.

On April 21, 2021, the *American Banker* newspaper published my op-ed on CBDCs, throwing some cold water over the rush toward innovation without an equal and more balanced concern for the insecurity of the internet and the number of online entry points that CBDCs might create.[30] The vice chair for supervision of the Federal Reserve agreed two months later, wondering whether CBDCs were a solution in search of a problem.[31] Vice Chair Quarles said that the potential benefits of a CBDC were unclear compared to the significant risks it could pose to the structure of the U.S. banking system by sucking up deposits that banks rely on to support the credit needs of households and businesses. He also noted that CBDCs could present an appealing target for cyberattacks and would need to be "extremely resistant" to such threats, requiring that the far greater number of "entry points" that it would create for an attack be secured.[32] Federal Reserve governor Christopher J. Waller echoed similar concerns in his own speech about a month after that, expressing skepticism as to whether there was a compelling need for the Federal Reserve to create its own digital currency.[33]

Several months later in January 2022, the Federal Reserve sought public comment on a discussion paper summarizing its progress. The paper was general in nature, listing the obvious benefits and hurdles. Its initial analysis was that a U.S. CBDC might best serve the needs of the United States by being privacy protected, intermediated, widely transferable, and identity verified.[34] But the paper finally highlighted its cybersecurity concerns, noting that a CBDC could become an attractive target, would create an almost limitless number of entry points that might be vulnerable to attack, and could benefit from offline capacity.[35]

In March, the White House released a long-anticipated executive order on crypto assets.[36] It added very little, adhering to the long-standing time- and resource-eating approach of directing some two dozen different agencies to coordinate action and create dozens of reports. The order missed two fundamental points. First, before considering the usefulness and regulation of crypto assets—including CBDCs—the impact of such massive reallocations of liquidity and capital on the economy and financial services infrastructures must be identified and evaluated. We seem to have forgotten the damage that was caused by the innovations created by money market funds in the 1980s as they sucked up consumer dollars that otherwise might have gone into bank deposits because the interest rates they could pay were not capped by law as bank and savings and loan deposits were. Second, policy makers must address the critical need for better security to support crypto assets, which are and obviously will become even bigger targets of MUTs.

Unfortunately, the country's more temperate and deliberative approach to crypto assets will have to hold firm in the face of more aggressive actions taken by China and other countries that are seeking to "get in the game" first and deploy their own CBDCs.[37] Let them race ahead. Building new forms of digital currencies on top of insecure networks will be a reckless step that those economies will eventually come to regret.

REGULATION IS ON ITS WAY

In 2021, given its prominent role in high-profile ransomware attacks, U.S. banking regulators began to close in on the oversight of cryptocurrencies with a renewed sense of purpose. Speeches given and actions taken in 2021 by the Federal Reserve, Treasury, federal bank regulators, CFTC, and the SEC all began to crystallize the government's growing concerns. In November 2021, the President's Working Group on Financial Markets, the FDIC, and the Office of the Comptroller of the Currency issued an extensive report on stablecoins concluding that the time had come for them to be regulated and for rules to be established allowing them to be issued only by insured depository institutions.[38]

China has gone several steps further. After having banned crypto exchanges several years before, its central bank declared all cryptocurrency-related transactions illegal to maintain national security and social stability. Perhaps this was clearing the way for the digital currency that China's central bank had already launched. The action was indeed a curious reaction by a country that had housed 50 percent of Bitcoin miners. The value of cryptocurrencies was impacted by China's actions as the price of Bitcoin fell more than 8 percent and Ether more than 11 percent.[39]

BALANCING THE GOOD AND BAD

Whether the ledger will balance over time and crypto assets will do more good than harm to financial infrastructures will play out over the next decade. Tulips took nearly three decades to implode. To the extent that crypto assets purport to be money or a medium of payment, I have my doubts that any instrument that lacks a scintilla of government backing or oversight will be able to succeed long term. In that regard, the small margin for error and significant chance of dire financial consequences from a loss of confidence as the crypto business grows beyond $10 trillion of value should be of concern. We should also be looking through the euphoric veneer of crypto products to the underlying science that is reconstructing the business of financial services using peer-to-peer validation instead of trusted inter-mediaries. Very smart leaders in the crypto asset business argue that such a shift democratizes financial services by allowing crypto miners and investors to "own" a portion of the network that they create through the blockchain, effectively turning current Big Tech and bank-centric financial services sys-tems into cooperatives. They argue that such systems should today be safer by dint of the way that blockchains eliminate the centralization of data and require every user's approval.[40]

Similarly, financial technology experts rightly raise the issue of global competition and tech forum shopping. Fidelity Investments of-fers a Bitcoin exchange traded fund (ETF) in Canada that it can't in the United States. Physically settled crypto ETFs are safe and legal in Ger-many, Brazil, and Singapore, but not in the United States. And unlike the United States, crypto exchanges, stablecoin issuers, and others are eligible to receive e-money licenses to access the payments system in the United Kingdom. Crypto talent is moving from Silicon Valley to Portu-gal, Dubai, Abu Dhabi, Singapore, and other jurisdictions "that are not at all unregulated but that have a more positive posture toward innovation and growth."[41] These factors impact the future development and safety of financial technology everywhere given the global impact of financial technology products and policies.

There is both a lot of truth and salesmanship to the arguments for and against cryptocurrencies and assets. I tend to look at it with the eye of a financial services pragmatist who has authorized or lived through the failure of about 3,500 financial companies and four or so financial panics. I equate cryptocurrencies to firecrackers. If abused, they can blow a person's fingers off. However, the gunpowder that firecrackers require to work is a commodity that changed the world. All the benefits, efficiencies, inclusive-ness, and cost savings that blockchains and crypto assets can produce must be vigorously pursued, but not at the cost of creating the largest financial conflagration the world has ever known. The tokenization of financial and other products can lead to significant advances in the movement and delivery of real-time financial services. Almost any instrument of value can be tokenized, communicated, and traded via blockchains, whether they be money, green stamps, loyalty points, or money market funds. But before we tap the benefits of such a potent new product, everyone—entrepreneurs, users, governments, and system providers—should understand and underwrite the risks.

Beyond the obvious systemic risks embedded in potential insecurities in cryptocurrencies, stablecoins, and CBDCs that may be taken advantage of by hackers of all stripes, consider the risk created by haphazardly reconstructing the global financial services industry and allowing economies to migrate from traditional money and lending sources to private or government-controlled entities without knowing what the potential risks are. If money did begin to migrate in that way, banks would have less funding with which to finance loans, and the cost of credit would increase as availability decreased. Demand for government securities would likely increase as those would be the only assets available to fund growing stablecoin issuance. Treasury yields would presumably be lower, creating a threat to economic security.

Suffice it to say that such systemic reconstructions and shifts of capital and liquidity require deep thought and analysis about who should be regulated and under what circumstances. Even more impactful perhaps would be the realignment of private and public sector touch points that would have to be secure to ensure the stability of financial systems in the face of what would undoubtedly be a constant onslaught of cyberattacks. Given that the crypto world is not prudentially regulated from a financial point of view or held to globally accepted security standards, it is hard not to imagine vastly greater security risks being injected into the financial system. The question is whether there would be counterbalancing financial benefits to the impact of such organic changes.

Given the digitization of everything and the corresponding change in the scale of things, the haul from one online theft can be enormous and the chances of getting caught much lower than robbing a bank wearing a ski mask and brandishing a gun. We should expect that various forms of cryptocurrency, and particularly CBDCs, will be prime targets for hackers. If we had a safer and more secure internet, some of these complex issues about the potential security of various cryptocurrencies and how they can change the world would be easier to deal with. But all of these products are being layered on an internet that is insecure by nature. That is a serious problem.

8

DATA LOLLAPALOOZAS

In his story about tracking Russian hackers in *The Cuckoo's Egg*, Cliff Stoll worried about protecting financial networks. His fears some thirty years later were prophetic: "I wonder what happens on financial networks, where millions of dollars slosh around every minute. I am ticked off that the Feds don't seem to be minding the mint. And I am upset that looters have proliferated."[1] Things have improved since 1989, but not nearly enough.

The depository financial services industry in the United States is a complex combination of large and small interconnected banks, savings institutions, and credit unions whose deposits are insured by either the FDIC or the National Credit Union Share Insurance Fund. They are the backbone of the multitrillion-dollar payments systems that transmit and clear physical and electronic checks, credit card data, ATM transactions, debit card charges, and ACH transfers, along with the Federal Reserve's wire services and various other clearinghouse activities.[2] In addition, insurance companies, trust companies, mortgage bankers, nonbank lenders, and online and peer-to-peer lenders complete an intricate financial system. Everything is centered around the U.S. Treasury and the Federal Reserve System, transferring enormous sums between individuals and institutional entities around the world. Investments in and transfers of equity and debt securities create yet another parallel universe of global markets, exchanges, investment banks, broker-dealers, clearinghouses, traders, institutional firms, hedge funds, private equity firms, mutual funds, and their counterparties and service providers who process billions of transactions each day. Exchanges and networks have also developed to issue, hold, and transmit cryptocurrencies. These systems all have millions of users, employees, and outside providers that process their data, monitor systems, and provide IT security. Each of those outside providers has their own bevy of providers who have their own providers.

You see where I'm going here. The financial services system in the United States is an interconnected web that is the financial equivalent of six degrees of separation from everyone on the planet, including the bad guys. Every user, provider, employee, computer, software application, seam, junction, and handoff point in every one of these systems and subsystems is a vulnerability point in terms of both physical and cyberspace security. Everyone shares the same internet. While many may use private or segmented networks to handle the storage and transmission of money and the most sensitive of data, unlike strategic military defenses, there is no single entity or mapping system that protects the U.S. economic infrastructure that is supported by the internet or determines how to rebuild it. Each day, the digitization of every square inch of commerce and the movement of money through new technological channels grows, as do the risks and vulnerabilities.

Moreover, the unique, interconnected nature of the financial industry means that if one major U.S. bank were upended by a cyberattack, the consequences could be felt globally. In 2020, the staff of the Federal Reserve Bank of New York analyzed the impact of a cyberattack on the United States' five major banks. Their conclusion should have galvanized resources to begin to correct systemic financial deficiencies. It estimated that if just one of those five banks was to stop making payments because of a cyberattack, 6 percent of the country's banks would breach their end-of-day thresholds, and 38 percent of U.S. banks' assets would be affected (excluding the targeted bank).[3] A March 2022 staff working paper by the Federal Reserve reaffirmed the potential impact and consequences of a multiday cyberattack on payments systems.[4] That is only the beginning of the disintegration of confidence that could sweep over the nation like a tidal wave unless resolved very quickly. In a business where confidence determines viability, the suspension of money moving in the continuous manner that it does each day would be a devastating event. The staff underscored that "uncertainty regarding the nature and extent of the attack could prompt runs to occur in segments of banks' operations that are otherwise unaffected. . . . This reflects the high concentration of payments between large institutions, and the large liquidity imbalances that follow if even one large institution fails to remit payments to its counterparties."[5]

It should be no surprise that financial services are among the industries most targeted by cyberattacks—they are where the money and data are. Hackers create fake banking websites and lure consumers in through phishing emails at a dizzying rate. Fake bank websites come and go by the hour. In 2021, the French bank Crédit Agricole topped the list of most-impersonated web brands with 17,755 unique phishing URLs. PayPal Holdings had 2,601, JPMorgan Chase 2,537, Wells Fargo 1,564, and Square 786. The number of fake bank websites is estimated to be in the hundreds of thousands.[6] Is that system—one that includes a game of chance where we hope that we land on the right website—one that we want to use to invest and safeguard our money? Moreover, these attacks appear to be escalating. "America is grappling with a cyber insurgency and our financial sector is the number one target . . . facing the world's elite hackers, composed of organized crime syndicates and motivated nation-states," says Tom Kellermann.[7] He served on a cybersecurity commission ordered by President Obama. Analyses in 2015 and 2019 by Websense Security Labs and the Boston Consulting Group concluded that financial services firms experience many multiples of cyberattacks compared to other companies.[8]

The financial stakes are much different from those of the past and the impacts more irreversible today. When I started in financial services in 1976, the financial system was safer and simpler because it was impractical to steal enough paper—cash or securities—to bring it down. There was a limit to the number of boxes of documents or bars of gold a thief could load into a waiting van. That was generally the limit of the threat. Financial institutions relied on closed, proprietary networks for the transmittal of data and payments. Those systems created 180 degrees of separation. Today, there is no paper to steal. Nearly every important financial piece of data and representation of value has been digitized into a form that can fit on a thumb drive, changing the scale of the risks enormously. And much of it is connected to an internet that is open to anyone. That change in the scale of the threat should have resulted in a corresponding ramping up of offensive and defensive cyber capabilities. Unfortunately, the increase in vulnerabilities has been exponential and the creation of solutions arithmetic.

As a result, every dollar that every one of us has sweated to earn could someday disappear into an ethereal afterlife ruled by rogue humans or artificial life forms. I am not an alarmist, and there are a million scenarios that could unfold that might alter future events both positively and negatively. But I know how to evaluate the financial odds. Security today seems to the "good-enoughers" to be just good enough—there have not yet been enough instances of large digital heists from traditional financial institutions to warrant a do-over. When the Bank of Bangladesh saw $81 million disappear in 2016, that was not a direct online theft such as cryptocurrencies have experienced. The Federal Reserve was simply following the counterfeited instructions of hackers. But at the rate that things are changing, those who are not concerned about falling behind the pace at which threats are being created have their heads in the sand. Given the risks, it makes all the sense in the world to be prepared, even if they never materialize.

THE VALUE OF FINANCIAL DATA

Banks are natural targets because they have and move money and data. But so far money has not been the principal target, or it has been well enough protected. Like many online and social media interfaces, banks are flush with data about their customers, an almost as valuable commodity as money in the digital world. Unlike other companies, financial institutions are subject to a broad range of laws and regulations that limit their collection and use of that data. Consider the universe of customer data that banks have: addresses, Social Security numbers, account numbers, dates of birth, educational background, balances, employment history, purchase records, credit ratings, and bill paying and credit card histories. The combination of this information and some basic algorithms can deduce a person's lifestyle, health, religion, sexual proclivities, social and political awareness, health status, charitable inclinations, marital status, fidelity, number of children, automotive choices, and even the make of the computer they use.

Inside a regulated financial institution, that data is protected as much as any data is these days. But those protections are impacted by legal, operational, and security loopholes in a borderless world of contradictory and overlapping laws and regulations where consumers seem uninhibited about sharing every aspect of their life with strangers. Consider how the blending of this data and the endless crates of digital data that are farmed each day through social media, GPS tracking, cell phone calls, and internet searches could be used and abused. Financial institutions are not yet wired like other companies that beg for, borrow, or steal every scrap of information about their customers to be able to surveil and control them—yet. But they engage with and connect to firms that do.

Today there are increasing threats to the sanctity of financial data given its value. A company or data broker may collect, store, and sell personal data in ways that are inconsistent with applicable laws and consumer privacy. In addition, hackers will target and steal data to use for marketing or criminal purposes. Like digital alchemists, they can turn personal data into credit, cash, and merchandise. The problem is that businesses and legislators won't be overly concerned about consumers' data when consumers themselves send mixed messages about how much they care about it. Consumers won't invest significant chunks of time to read user agreements. Most will mindlessly click "Agree" or fail to opt out of whatever informational heist companies seek to impose on them. They relentlessly release vast streams of personal information about themselves on websites and social media networks seeking a better user experience. As Shoshana Zuboff artfully details in *The Age of Surveillance Capitalism: The Fight for a Human Future at the New Frontier of Power*, consumers are all too willing to click away their data so that they do not miss out on the online and social media experiences and euphoria that they consider vital to their personal, business, and emotional well-being.[9] It is not a fair game, however, since the penalty for opting out is a life sentence to social and financial isolation.

More problematic is the fact that collectors and repositories of this consumer data may simply lie about whether and how it is collected, and how it will be used. Alternatively, data hogs may disclose that they intend to scoop up every scrap of data and turn the consumer into a merchandizing lollapalooza, knowing that the consumer will either not read that disclosure, not care, or simply click "Agree" to whatever appears in the small type. Social media, screen scrappers, data miners, and information brokers see dollar signs attached to every digital breadcrumb a consumer creates. And they gobble them up whether consumers know it or not. Since banks sit on an informational treasure trove—a mother lode of behavioral data—it makes them valuable players in this new digital world, and even more important targets of unscrupulous actors.

Not enough is being done to prevent, distract, or slow down companies of all types from harvesting and using every imaginable type of personal information for their own economic benefit. Much of it is already out the door, which is particularly frightening when you consider that we really have very little idea how technology may evolve and use that information in the future. In 1990, no one could have imagined that we would be where we are today, and the same will be true in the future.

Even less is being done to educate consumers, a method that is always much cheaper and more effective than ubiquitous regulation. Amazon announced in 2020 that Ring and Echo customers would automatically be opted into the new "Sidewalk" feature. It claimed that this would enhance user experience by sharing some of the user's internet connections so that devices, even those owned by someone else, could work over a longer distance.[10] But who said Ring could do that? Did Amazon ask? Do the people who live on the street have protectable interests in this? This is a good example of the take first and explain (maybe) later approach of Big

Tech that many experts have exposed. It seems peculiar that relatively few have questioned whether the vast collection of data to date has been legal, ethical, or wise. It just happens.

Banks will be tempted to join this data orgy in more direct ways to gain access to more customers, provide them with more functionality, increase efficiencies, and generate profits. There is increasing economic and competitive pressure on financial institutions to share their platforms and customers, facilitate the bundling of apps, deploy digital wallets, offer real-time finance, participate in the decentralization of data, partner with fintechs that promote DeFi (finance without traditional intermediaries), and offer open banking.

Open banking platforms are being implemented in Europe and Asia, increasing the challenge of controlling the use and abuse of financial data.[11] Open banking enables consumers to share their bank and credit card transaction data "securely" with fintech firms, digital banks, data aggregators, credit bureaus, payment networks, third-party providers, telcos, health care services, and retail outlets to increase functionality and efficiency. While platforms that share data with such applications and services may save consumers time and money, the moment data is shared across companies or industries, the risk of execution failure, violations of laws and privacy protocols, and fraudulent third-party provider access to data increases. Open banking naturally magnifies the risks of using an already insecure internet. By increasing the number of providers that may touch or be unaccountable for a user's data, it only multiplies the number of vulnerabilities. The UK has been a global leader in this area as half its small businesses and over 4 million consumers use services powered by open banking technology.[12] There is no doubt that the economies and conveniences gained from open banking are significant. But so are the increased security issues.

A January 2022 draft report from the Department of Commerce's National Institute of Standards and Technology (NIST) authored by a group of academics and technologists summarized the benefits that open banking could create. But it largely glossed over the security challenges open banking raises, concluding that "having an open platform *should* stimulate the means of securing financial systems, such as by enabling better methods for detecting and preventing fraud [emphasis added]."[13] Unspecific *shoulds* are generally worth very little in the real world. Several financial trade associations seemed to agree. The Bank Policy Institute (BPI), the Securities Industry and Financial Markets Association (SIFMA), and the American Bankers Association (ABA) filed a March 3 comment letter on the report concluding that the NIST report did not sufficiently address the complexities and risks that an open banking regime might introduce, or offer privacy or cybersecurity recommendations, contrary to both the title and purported purpose of the report.[14] That about says it all. Why should even more sensitive data be shared with more potentially untrusted third parties on networks that we cannot guarantee are or will remain secure? The euphoria for new technologies and the assumed time and cost savings are fatally blurring the risks to financial security.

THE DARK SIDE OF DATA

Consumer data has always been collected and analyzed to predict the buying patterns of consumers and maximize sales. Spotters sat in shopping mall parking lots and counted people, packages, and brands leaving the store. That seemed harmless enough and was perhaps why no one objected. Today the fact that spotters have been replaced by facial recognition devices and drones that can merge their data with GPS, phone, and social media data shows how things have evolved into much more sophisticated forms of behavioral analysis and control. Companies and governments have learned that far more power and profit can be amassed by conditioning consumers to take an action rather than merely predicting under what circumstances they might do so.

Technology has changed the business of data collection and the behavioral sciences that are linked to it. While many may think that the business of watching began with Facebook in 2004, this only proves the lack of historical perspective that many bring to these issues. Meta/Facebook, Google, and companies like them are the offspring of companies such as Simulmatics and the RAND Corporation, businesses created in the 1940s and 1950s to collect and harvest large amounts of data with monstrously large computers (physically more than powerfully) to roughly predict human behavior, particularly with regard to politics and elections. The problems we have today with the collection and protection of data began eighty years ago.

John F. Kennedy was reported to be the first beneficiary of the magical predictions that the "People Machine" developed by Simulmatics could make about how people would vote on certain issues. Simulmatics disappeared a little more than a decade after a tumultuous corporate life that ended with the introduction of its "Riot Machine" in the 1960s that predicted when cities might erupt because of racial rioting. The Simulmatics story told by Jill Lepore in her book *If Then*[15] underscores a point that I make in *200 Years of American Financial Panics*. Our present—whether viewed through political, financial, or social prisms—is always the culmination of decades of colliding public policies and corresponding market reactions that have fermented or rotted over long periods of time under circumstances that were often unanticipated. That is why history is so important to the future. Every financial crisis has been the result of a continuum of decades of related benign, neutral, or inept economic, social, and market events driven by the government, private industry, or both. Those who ignore history (more and more of us in this instant gratification digital society) will miss a valuable prism through which we can solve the question of why we keep having so many financial crises.

This data surveillance society driven by private companies, governments, and others (the "watchers") creates political and economic incentives with enormous direct and collateral consequences. Governments, consumers, and technology amateurs innocently stand by while it happens, mesmerized by what they see as the societal benefits of innovation. Those who gather the power of data to themselves, including governments, understand the control that comes with it—something that can be used for purposes far beyond the sale of products. So here we are in a world that most of us never voted for or wanted to create. Nevertheless, it is upon us.

People like Deborah are increasingly displeased about it. With her online buying and social media habits meticulously collected, cataloged, and analyzed by sophisticated supercomputers and algorithms; her generous sprinkling of personal information on social media sites; and GPS surveillance of her movements, data analysts can determine that Deborah is pregnant, unmarried, likely to develop diabetes in future years, and is, therefore, a questionable candidate for a new job, a new house, or health and life insurance. Fred and Ethel have also learned about the dark side of data. When they searched online for marriage counselors and new apartments, they were classified as poor credit risks given the likelihood that they were heading toward divorce. The increasing use of artificial intelligence does more than simply analyze all this data. It predicts the future, and to that extent it may create the future. That is a philosophical discussion for another day.

A hands-off approach to the legal and illegal collection of human-generated data has resulted in a world where consumers are convinced that the more information they provide, the better user experience—and perhaps the better life—they will have. Very often, that statement is false, but there is no doubt that governments' laissez-faire approaches to technical innovation as they regulate analog-world activities to death have been a potentially devastating policy lapse that may be irreversible. On the other hand, if they had taken a heavy hand in controlling the direction that technology and data collection took, we would probably be discussing the dampening impact they had on innovation, personal rights, and democracy.

These are extraordinarily complex issues largely because new technologies and malicious users of it continue to increase, reducing the margins for error and raising the stakes markedly. Until networks are more secure, there is no reason to continue to load more data and value onto them. The solutions require a new breed of global public/private/technology oversight that is so far removed from what we have ever used that it will take years to develop and accept. It is already too late to undo the watching damage that has been carefully camouflaged behind the seductive communication channels and illusory social benefits that technology provides. That toothpaste cannot be squeezed back into the tube.

This is the dark underbelly of the data society. What may create unprecedented quality-of-life enhancements may just as easily be used to cause societal and economic devastation. As every scrap of information on the planet is digitized, America's economic infrastructure is becoming increasingly efficient, but also vulnerable not just to attack but to collapse. No financial system will be immune, not the U.S. Treasury's electronic benefits being paid monthly to millions of Americans, the Federal Reserve's payments processing, clearinghouse transactions, market exchange trading, the mining and transference of cryptocurrencies, or the clearing of credit card transactions. While the government and businesses try to defend against cyberattacks, they know that there is an element of futility involved, much as lab rats run endlessly on spinning wheels.

DATA CAN'T BE PROTECTED FOREVER

A new generation of faster computing power is around the corner, so any security that has been trumpeted as effective today may eventually become anemic as new algorithms and quantum computers reduce the complexity of breaking encryption from trillions of years to seconds.[16] Peter Shor, a professor of applied mathematics at MIT, discovered in 1994 that a quantum computer could break most of the encryption that protects transactions on the internet. Breaking any defensive security system is basically a matter of mathematics, computing power, and time. Today's RSA 2048-bit encryption is adequate to protect a traditional binary computing system and the online and digital commerce it generates.[17] That explains in part why phishing and other insider vulnerabilities are used so frequently. They take advantage of human frailties rather than mathematical ones. But any system that is built on mathematics is vulnerable. In a recent survey of 75 million RSA certificates, it was determined that there is a 1 in 172 chance of being vulnerable to compromise due to shared cryptographic key factors.[18]

Quantum or some iteration of advanced computing is coming, and the question is, will our data be ready? Forms of security used today that rely on traditional forms of encryption could be susceptible to being penetrated, potentially creating a data free-for-all. The digital world of commerce is premised on the expectation that there will be enough lead time for current systems to prepare new defenses against the next generation of technological threats, which the good guys will get their hands on first. Those assumptions may no longer be true, making it the biggest bet ever made.

Virtual vulnerabilities grow exponentially every day as businesses, governments, and consumers are increasingly seduced by the thrill, profitability, and inevitability of technological innovations. Businesses continuously underestimate the need for secure passwords, too many organizations do not require multifactor authentication, and cyber thieves get more sophisticated by the hour. Each day we cede more of the economic underpinnings of the country to technology in the name of innovation and quality-of-life enhancements without fully appreciating the attendant vulnerabilities or building a system to better protect against them. Consider the facial recognition abuses in China and then the IRS announcement and subsequent about-face in early 2022 with respect to using facial recognition data to identify U.S. taxpayers. There is no doubt that facial recognition is both a valuable and dangerous tool that raises enormous issues.[19]

It is easy to imagine why some nations of the world are redirecting huge sums of money toward the development of artificial intelligence, high-speed digital communications systems, quantum computing, and cloud control. For its part, the U.S. government seems to be "driven by the unshakable truth that the world is ours," as Professor Brand sarcastically says in the movie *Interstellar*, where that unshakable truth turned out to be false. Are we sure that the United States will always be the economically strongest, most militarily powerful, and best technologically endowed nation in the world? If we act like it will, it is more likely that eventually it will not be. Other nations will run harder to catch up.

We should be questioning why legitimate entities should be traversing sensitive information highways that are available to everyone, including digital pirates and terrorists. In the analog world, everyone needs a government form of identification to travel domestically and a passport to travel internationally. Nothing close to that exists in cyberspace. There have been endless discussions, analyses, and reports about the risks to national economic interests created by the insecurity of the internet. But the actions taken to systematically protect against the commission of potentially life-altering cybercrimes continue to lag behind the pace at which cybercriminals hone their skills.

ARE WE MAKING PROGRESS ON SECURITY?

Today, the U.S. Cyber Command focuses on military defense infrastructure, providing support to combatant commanders and strengthening the nation's ability to withstand and respond to cyberattacks.[20] The Cybersecurity and Infrastructure Security Agency (CISA) in the Department of Homeland Security is arguably the closest thing to an agency responsible for U.S. financial infrastructure security, a responsibility that it shares with the Treasury Department. CISA works with businesses, communities, and government at every level to make the nation's critical infrastructure resilient to cyber and physical threats.[21] Eight years ago in 2015, it issued the Financial Services Sector Specific Plan in conjunction with the Financial Services Sector Coordinating Council (FSSCC) for Critical Infrastructure Protection and Homeland Security, and the Financial and Banking Information Infrastructure Committee (FBIIC). The report touts the sharing of information between companies and the government (which is currently limited and ineffective), the development and use of common approaches and best practices (which is more of a goal than a reality), collaboration among an alphabet soup of federal agencies (which speaks for itself when it comes to effective management), and robust discussions about policy and regulatory initiatives (a lack of discussion has not been the problem over the last twenty-five years).[22]

In the face of the escalating power of technology, the rush toward artificial intelligence, and the ongoing reconstruction of money, banking, and commerce that will only be accelerated by blockchain applications and the arrival of quantum computers, there are too many chefs stirring this pot, and not enough cooking going on. There are not enough delegations of responsibility and coordination to be assured of reactions by the government and the private sector when the inevitable assaults on financial markets, payments systems, or the economy are launched. The question is whether they will be effective.

The Government Accountability Office (GAO) has criticized the government's lack of centralized cyber defense coordination. In September 2020, it warned that there is an urgent need "to clearly define a central leadership role to coordinate the government's efforts to overcome the nation's cyber-related threats and challenges," since it is unclear who *ultimately* maintains responsibility for coordinating execution of the plans and holding federal agencies accountable.[23] It appears that the best America has to offer at this moment is a loosely configured company-by-company, agency-by-agency detection-and-defense strategy. With some exceptions, it is essentially every person and company for themselves. This state of unpreparedness will lead to chaos if just one protagonist decides that the U.S. economy should be disrupted or terminated. Imagine the consequences if that were the way the United States constructed its strategic military defense. JPMorgan Chase would have to procure its own ballistic missiles to defend and protect its square block in Manhattan.

Data has now become the most valuable economic asset, as the many breaches in corporate security demonstrate. The rush toward artificial intelligence may be the most significant event in history. Vladimir Putin has said that the country that wins the race to artificial intelligence will be the "ruler of the world." China has laid out a plan and commensurate spending to reach global dominance in artificial intelligence by 2030. It has already deployed a form of algorithmic governance to monitor its own population through facial recognition, and in 2020 it began imposing sweeping social evaluation profiling tools to reward and punish citizens based on their social scores.[24] China is outspending the United States in these areas. The time for talk has long passed.

The goal is straightforward. We must stay ahead of data and cybersecurity issues that could alter or bring down financial institutions and the economy. They will always be prime targets for the harvesting of marketing and behavioral data and attacks on the country's economic infrastructure. The lyrics of Don McLean's "American Pie" bemoan "the day the music died." Without vigilance and concerted action to take control of our financial futures and devise the rules and enforcement mechanisms that will ensure that humans and not machines shape that future, we risk witnessing the day the money died.

A comprehensive, integrated global early-warning system to protect, detect, and remediate the improper or malicious use of technology against the financial infrastructure of sovereign nations is necessary. If it isn't feasible, we had better be considering alternative solutions. Unless the United States takes the lead and creates a coalition of democratic nations to implement preemptive offensive and defensive financial cyber strategies, there may be no counterweight short of military power to respond to those who would weaponize technology to collapse economies. Each day that passes, we are missing the opportunity to reverse gears and take control of digital access, communications, and finances. Nothing less than the control of global economies is up for grabs.

Part III

THE LONG AND WINDING ROAD
TO CYBER INSECURITY

9

THE EVE OF DESTRUCTION

When Barry McGuire sang about the "Eve of Destruction"[1] in 1965, he was referring to the growing threat of nuclear war. That threat exists to this very day, although it has never ripened into the use of atomic weapons in an armed conflict. Perhaps a similar scenario will unfold with regard to the threats to critical infrastructures posed by the malicious use of technology. But it is hard to miss the risks that are brewing given the increasing frequency and size of cyberattacks on government agencies and private industry.

On December 15, 2020, the government disclosed that the computer systems of at least nine federal agencies, including the Treasury, Homeland Security, and Commerce Departments, had been breached in a global cyber espionage campaign believed to have been orchestrated by the Russian government as far back as early 2020.[2] The fact that the country was in a presidential election year may have played some role in the timing of the attacks, but there is evidence that planning began years before.[3] Hackers were reported to have infiltrated government systems through updates released by SolarWinds, a software company based in Austin, Texas. It unknowingly pushed malicious code to 18,000 customers, which included the White House, the Pentagon, and NASA and some of the country's leading telecommunications providers, as well as more than 425 of the U.S. Fortune 500.[4] The next day, SolarWinds posted on its website that it had been made aware of a cyberattack: "We have been advised that this incident was likely the result of a highly sophisticated, targeted, and manual supply chain attack by an outside nation state, but we have not independently verified the identity of the attacker."[5] These events were again treated as inevitable and unpreventable by many. The impact of the attack dominated the news for only a few days.

While malicious users of technology that are well financed may be able to use brute digital force to take down financial, electrical, transportation, and military defense grids, the SolarWinds hackers were more discreet. They embedded malicious code into a natural part of everyone's digital lives—software updates for the Microsoft Office 365 platform—to surreptitiously monitor email traffic.[6] Many updates are automatically downloaded and installed pursuant to user instructions that effectively put computers on autopilot. What could be more invisible and disarming? The attack was "executed with a scope and sophistication that surprised security experts and exposed a potentially critical vulnerability in America's technology infrastructure."[7] The fact that such a vulnerability was there after so many years of analysis and promises speaks for itself. The Russian embassy in Washington, DC, denied any involvement, saying that Russia "does not conduct offensive operations in the cyber domain."[8]

By using a flaw that was difficult to detect, the attackers were able to gain access and insert counterfeit "tokens."[9] Even if Russia was not responsible, the scope of what was done should have been a screeching warning. CISA, part of the Department of Homeland Security, labeled the incident as a "grave threat" to critical infrastructure entities and private sector companies, noting that the hackers also broke into computer networks using bugs other than the SolarWinds software. Of the Microsoft customers identified as victims of the SolarWinds hack, 44 percent were IT services companies. The irony of this attack was that cybersecurity experts had been used to circulate the malicious code. Eighty percent of the victim companies were based in the United States, but targets were also hit in the UK, Canada, Mexico, Belgium, Spain, Israel, and the United Arab Emirates.[10] A full recovery from the SolarWinds hack was estimated to take the U.S. government from a year to as long as eighteen months. CISA said that it would be well into 2022 before officials could fully secure the government networks compromised by the attackers.

The first response to these kinds of attacks is usually a remediation effort to remove the code from the network. That may be a complex effort that requires shutting down applications, limiting access, and monitoring entry points used to access networks. The longer hackers live inside systems, the longer a "strategic recovery" will take.[11] In those cases, hackers can establish a trusted identity such as an administrator and become extremely difficult to detect and remove as they authorize themselves to take the actions they take. The thought of hackers leaving logic bombs and other apps behind that could assume control of networks or systems at any time in the future is completely unnerving.

The SolarWinds attack underscored that there are no internet police who are solely charged with the responsibility for defending cyberspace and apprehending digital thieves. CISA has stepped forward, but it needs more resources, authority, and personnel to carry out the mission. Congress continues to misperceive and underestimate the problem. And this is all just the tip of the iceberg. Consider what this attack could have done if it intended to collapse the Departments of Homeland Security, Treasury, and Commerce. In 2016, something similar but more severe happened in Brazil when all of Banco Banrisul's thirty-six domain name system registrations were taken over, giving hackers control of a good portion of the bank's operations.[12]

Attackers remain three steps ahead of defenses as the Willie Sutton philosophy is being adopted by thieves across cyberspace. They go where the money is. Financial services are the fastest-growing and largest market for cybersecurity products and services, with data security spending by global financial institutions growing 67 percent from 2013 to 2016.[13] The biggest problem for banks is that once they are online, there are so many soft defensive points to be probed by hackers. That is why cyber thieves often target bank customers—the weakest link in bank IT security. It is cheaper and easier for malicious users of technology to use customer access rather than targeting the banks themselves. Customers tend to log into their accounts from multiple devices, including laptops they are using in coffee shops, or less protected mobile phones and public computers. In the meantime, criminals continue to get more sophisticated and are able to take advantage of an increasing number of new products and delivery systems. Nation-state hackers are increasingly focusing on the financial sector because it is a critical choke point that can exert pressure on adversaries. And what were once exclusively nation-state level capabilities are moving into the hands of criminal groups, as the lines between government hacking groups and criminal organizations seem to blur as cybercriminals get involved directly in politics.[14]

Most troubling perhaps is the inability of cybersecurity experts to corral the hackers, whoever they may be. A SWIFT Institute working paper in 2017 noted that "the most sophisticated nation states and organized crime groups have begun targeting the 'seams' between these well-defended networks, exploiting weaker institutions in the global financial system to pull off massive heists." In this way, attackers take advantage of multiple banks in multiple countries. The challenge for regulators, law enforcement, and government authorities has become overwhelming, even though they may not ever say it. How can all the seams, intersections, and junctions in our six-degree-of-separation networked society be protected?

On January 22, 2021, a vendor used by the $31 billion Flagstar Bank in Troy, Michigan, for its file-sharing platform had to inform the bank that a vulnerability had been exploited by an unauthorized party that was able to access some of Flagstar's information. As if it mattered to Flagstar's customers, the bank announced that it was just one of several of the vendor's clients that had been impacted.[15] According to an investigation, the attacks conducted in December and January were focused on vulnerabilities in the vendor's firewall software, which was reported to be twenty years old and due to be retired at the end of April 2021. The vendor issued patches within four days of the first attack. A ransomware gang ended up with the stolen data and used it to extort people, but it was not clear if it had stolen the information or purchased it after it was stolen.[16] Just as messages communicated online may not be encrypted end-to-end in their journey, the fact that banks are highly regulated is undercut by their use of vendors that are not. And much of banking and most of fintech technology is provided and managed by outside vendors. Unless there is comprehensive door-to-door security, including vendors, the standards imposed periodically along the way will not protect networks and their communications.

The chain of bad news never seems to end. On March 2, 2021, Microsoft disclosed ten-year-old vulnerabilities in its Exchange Server mail and calendar software for corporate and government data centers that had been exploited by Chinese hackers at least since January 2021. The group, which Microsoft dubbed "Hafnium," reportedly sought to gain information from defense contractors, schools, and other entities in the United States. Experts speculated that because Exchange Server is so widely used around the world, the hack would probably stand out as "one of the top cybersecurity events of the year."[17] These types of hacking events may simply be dress rehearsals for something much larger, and yet whatever attention they receive quickly fades. While SolarWinds allegedly patched the security issue within a week and shut down access points, it did not necessarily remove the intruders, which could have embedded themselves in as many places as needed to conceal and assure their continued presence. Consider the possibility that every important computer system in the country may be infected with logic bombs and "cuckoo's eggs," all waiting to hatch or be called into action to spy on, steal from, alter, or destroy their hosts.

Russian cyber theft of critical data from the U.S. government now stretches over more than two decades and continues to become stealthier and more sophisticated every year. It is what led to the creation of U.S. Cyber Command, the Pentagon's elite cyber warfare group established in 2008 to defend the military intelligence network, provide support to combatant commanders, and strengthen the nation's ability to withstand and respond to cyberattacks. While this is important, it is difficult to put too fine a point on the risks created by the potential foreign infiltration of the Federal Reserve's and U.S. Treasury's systems. It is the Treasury that is responsible for the nation's debt management, cash production, Social Security and other benefit payments, loans to other federal agencies, tax collection, payments to Treasury note holders around the globe, and economic policy formulation. The scale of its financial operations is massive. In the second week of February 2022, the Treasury had about $675 billion[18] in its general account at the Federal Reserve Bank of New York, which directly impacts the funding of U.S. markets. The movement of funds from that account back to reserves through the shrinkage in the T-bill supply and the reserves held in banks reduces the most highly sought-after collateral—T-bills—and systemic liquidity. That in turn impacts the ability of banks and other financial intermediaries to hedge or finance risk. Even a short hiatus in the Treasury's ability to access those funds, issue debt, and transfer funds could throw mortgage and finance markets into chaos.[19]

Richard A. Clarke, the highly regarded cyberspace advisor to several presidents has written extensively on the growing threat of cyberattacks. He describes a briefing that President Bush received a generation ago in 2002 about how vulnerable the financial industry and the U.S. economy were to cyberattacks. Stunned by what he heard, the president jumped up and said, "I want this fixed." Clarke noted that the president had assumed what everyone seemed to—"information technology is supposed to be our advantage, not our weakness."[20] This led to the Comprehensive National Cybersecurity Initiative (CNCI) and National Security Presidential Decision 54, calling for a twelve-step plan to secure government networks. It did not address the vulnerability of the financial sector to cyber war.[21] These initiatives did not become public for some time.

In 2009, President Obama ordered a review of federal efforts to defend the U.S. information and communications infrastructure and the development of a comprehensive approach to securing America's digital infrastructure. He appointed an executive branch cybersecurity coordinator who had his ear and could work closely with state and local governments and the private sector to coordinate responses to future cyber incidents and strengthen public-private communication and cybersecurity awareness.[22]

THREAT CAPABILITY RATCHETS UP

In 2007, Apple introduced the iPhone, and San Francisco Giants slugger Barry Bonds hit his 756th career home run to break Hank Aaron's home-run record. Tucked away behind the major headlines of that year was a cyber event that most people never noticed but that clearly changed the world. Russian entities allegedly took revenge on the tiny country of Estonia in April for moving a bronze statue of a soldier wearing a World War II Red Army uniform in the Estonian capital of Tallinn. Lesser transgressions have led to wars. Tallinn saw two nights of riots and looting in reaction to the move. Major cyberattacks consisting of blizzards of spam were sent by botnets, overwhelming servers and disrupting online services throughout the country. Cash machines and online banking services worked only sporadically, government employees were unable to communicate with each other, and the media was temporarily silenced. According to an anonymous Estonian official, the Russians initiated the attack, but malicious hacker gangs anxious to do their part chimed in. If the last decade has taught the countries of the world anything, it is that there are numerous malicious gangs and cyber activist groups within their own borders willing to cause the economic infrastructures to collapse. Estonia learned its lesson. It established an effective voluntary Cyber Defense Unit that has become a model for the world.[23] But the game was on.

Nine years later, a Department of Justice (DOJ) complaint filed in March 2016 charged a group of Iranian individuals employed by two Iranian computer companies that performed work on behalf of the Iranian government with launching DDoS attacks on forty-six companies, including Bank of America, the New York Stock Exchange, BB&T, and Capital One, between 2011 and 2013.[24] The attractiveness of DDoS attacks to hackers should not be surprising given the economics of such actions—the impact is great and the cost is almost nothing. At that time, it was reported that customers at six of the largest U.S. banks were unable to access their accounts online or make payments. The DOJ alleged that the targets of the attacks incurred tens of millions of dollars in remediation costs.

That same month, the Bank of Bangladesh reported $81 million stolen from its account at the Federal Reserve Bank of New York when hackers gained access to the bank's systems and sent messages through the SWIFT system—the Society for Worldwide Interbank Financial Telecommunication—making thirty-five requests to transfer funds to accounts in the Philippines and Sri Lanka.[25] Four of the requests were fulfilled, transferring $81 million to hackers. Other similar attempted transfers were discovered. The charges filed by the DOJ alleged that over a three-year period, the conspirators attempted to steal about $1 billion from financial institutions.

In May 2017, Equifax's computer systems were allegedly breached by four members of the Chinese People's Liberation Army who stole sensitive consumer data and Equifax's trade secrets. They obtained information on nearly half of all U.S. citizens. In February 2020, a federal grand jury indicted the four who were charged with, among other things, conspiracy to commit computer fraud, conspiracy to commit economic espionage, and conspiracy to commit wire fraud.[26] In July 2019, a former employee of a cloud computing services provider was arrested for alleged wire fraud and for violations of the Computer Fraud and Abuse Act in the form of exploiting misconfigured web application firewalls on the servers of the service provider to access and copy personal identifying information from approximately 100 million customers who had applied for credit cards from Capital One.[27]

These kinds of attacks have continued to increase, perhaps even more than we have been able to document. According to a report by VMware, cyberattacks against banks spiked 238 percent between February and April 2020.[28] "There has been a significant evolution in cyber threats facing the global financial industry over the last 18 months as adversaries have advanced their knowledge," a 2017 study commissioned by SWIFT found. "They have deployed increasingly sophisticated means of circumventing individual controls within users' local environments and probed further into their systems to execute well-planned and finely orchestrated attacks."[29]

The impact has been predictable. JPMorgan Chase & Co. says that it spends about $600 million and employs 3,000 people annually to address cyber defense. Emphasizing the vulnerabilities created by an integrated global financial system and the constant attacks by adversaries that are smart and relentless, JPMorgan's chairman and CEO Jamie Dimon wrote in his 2019 letter to shareholders, "The threat of cyber security may very well be the biggest threat to the U.S. financial system."[30] As if on cue, on October 28, 2019, the Russian General Staff Main Intelligence Directorate's (GRU) Main Center for Special Technologies carried out a widespread cyberattack against the country of Georgia disrupting the operations of thousands of Georgian government and privately run websites and interrupting the broadcast of at least two major television stations.[31] Previous GRU attacks reportedly include a December 2015 attack on part of Ukraine's electricity grid which left 230,000 people without power for between one and six hours; a December 2016 distribution of malware designed specifically to disrupt electricity grids that caused a fifth of the Ukrainian capital of Kyiv to lose power for an hour; a cyberattack on Ukrainian financial, energy, and government sectors in June 2017; and the disruption of the Kyiv metro and Odessa airport in October 2017.[32]

The list of such cyberattacks grows as countries like China, Iran, Russia, and North Korea that are all too eager to add their versions of cyber mayhem enter the fray. Banks have been targeted by groups such as Carbanak and the Lazarus Group that are thought to have links to North Korea.[33] The Defense Advanced Research Projects Agency (DARPA) within the Department of Defense has identified areas of concern in the financial services sector, among them the risk of flash crashes caused by manipulated sell orders that cause a rapid decline in the stock market, and attacks on order matching on stock exchanges.[34] In 2020, more than 100 financial services firms were targets of a wave of DDoS extortion attacks. They were conducted by the same threat actor that had sent extortion notes threatening to disrupt the websites and digital services of firms in Europe, North America, Latin America, and Asia-Pacific, hitting dozens of institutions within weeks.[35]

This is the new world we live in, one in which parties that control technology and artificial intelligence are not all democratic or benevolent. Many are hostile nations, terrorists, or fanatics. Indictments of hackers that will never appear in the country for arraignment or arrests of cyber thieves here and there by law enforcement will never make a dent in the frequency and ferocity of these attacks. Moreover, new technologies such as artificial intelligence and quantum computing may be all that a country needs to dominate a region or the world militarily and economically. These dangers are being discussed "quietly" by U.S. national security experts who are not satisfied with the resources being devoted by the U.S. government.[36] Less resources and coordination have been made available by the government to protect the financial services infrastructure, but the ball is starting to roll in the right direction. However, it is not moving fast enough or targeting the fundamental issues that are creating vulnerabilities and threats. These growing attacks will eventually begin to disrupt your daily routines in ways that you never imagined.

10

THE GENESIS POINT
OF VIRTUAL FINANCE

1994–2000

To understand how the United States has arrived where it is today financially in cyberspace, we must go back to the beginning. Fortunately, that is not very long ago.

The news of 1994 was marked by the dubious accomplishments of people like Howard Stern, Lorena Bobbitt, Tonya Harding, Dr. Kevorkian, and Marla Maples. But it was at that time that the web and online transactions were beginning to go mainstream. Unbeknownst to most, a new breed of potential threats to national financial security would be launched in October 1994 when Stanford Federal Credit Union became the first federally insured financial institution to offer its customers online banking. That one action was in retrospect a sea change for an industry that had never done anything outside closed, proprietary networks that limited users and boasted the highest security protocols available. A year later, commercial banks followed suit, beginning a revolution in financial services that has lasted to this day and opened a myriad of soft defensive targets in the country's financial services infrastructure—customers, service providers, and other third parties who connect to it.

I was at the Office of the Comptroller of the Currency when it allowed national banks in 1977 to establish nationwide ATMs. Later in private practice, I worked with banks, credit card companies, and tech firms on a variety of transactions that created new online products, presences, and delivery channels. Throughout that time, we focused on delivering financial innovation, naively believing that the challenges of transacting business in an open architecture were addressed by "those guys behind the curtain" who provided cryptography, encryption, digital signatures, internal procedures, and a host of secure interfaces. It took years to realize that what was being built would be handicapped in its ability to defend against the potential for wide-scale malicious attacks as the financial sector transformed from paper to digits.

If we knew in 1994 what we know today, the financial sector would not have launched headlong in the same way into the internet without first establishing stronger and standardized rules of engagement to ensure higher levels of security. There was simply not enough understanding about where it was all going or what the threats could be. The competitive impetus was enormous, and the expectation was that deficiencies could always be repaired along the way, as it still is today. But in 2023, we find ourselves in a difficult position given our myopic focus on the enhancements that technology can provide. The year 1994 was significant for financial institutions, most of whom ran past that decisive moment with one sole purpose—to match what their competitors were doing and provide their customers with the fastest, sexiest, and most up-to-date financial products and delivery channels. It was a good thing that Stanford Federal Credit Union's executives hadn't bungee-jumped off the top of their headquarters building.

Once online banking arrived, it became clear that the nature of money might also change. New forms of money were constantly being beta-tested. In early 1996, at the invitation of its creators, I visited the rollout of a new smartcard and electronic wallet by Mondex in Swindon, England. Mondex was created by Tim Jones and Graham Higgins of the UK's National Westminster Bank. The system had run internal trials with approximately 6,000 London-based NatWest staff beginning in 1992, leading up to its public trial in Swindon. It required extensive development and retooling of bank and merchant point-of-sale hardware throughout the city. British Telecom, NatWest, and Midland Bank sponsored and installed retail terminals for public transport and parking as well as for 700 merchants throughout the city.

Mondex smartcards were issued to the residents with some initial spending "money" on them. The Mondex card relied on an open system that permitted individual transfers between cards through electronic Mondex wallets without having to go through the central system of the bank. Both held cash information on their chips, allowing the data to be used like money and reconciled and balanced when the card was swiped at a terminal connected to the bank's network. Peer-to-peer transfers of Mondex money could be enabled by one person simply inserting her smartcard into the wallet of the person she was transferring funds to. Today, such sharing of "digital fluids" is understood to create significant security issues. I still have my Mondex smartcard with something less than £10 on it, and a Mondex electronic wallet.

This experiment was followed by a tidal wave of electronic money, smartcard, and authentication products boasting of their ability to transmit value electronically in a secure form. Some innovators such as DigiCash claimed to provide anonymous currency, a development that federal regulators and the Financial Crimes Enforcement Network (FinCEN) immediately frowned on. But it was as if someone had turned on a switch—banks, credit card companies, savings institutions, and credit unions began running toward technological pots of gold at the other end of the rainbow. Much of this history is captured in a book I coauthored in 1998, *21st Century Money, Banking & Commerce*.

The explosion of the internet forced the government to drag itself into the twentieth century. Twenty-five years later, progress has been made toward the creation of a safe and secure internet, but not nearly enough. The adoption of technology has outpaced the development of secure technology. Appendix B catalogs some of the significant government and private sector reports from around the world that identify and offer recommendations to deal with growing cyber threats and vulnerabilities related to the financial services sector over the last twenty-five years. It demonstrates the abundance of words that have been written during this period as the internet has morphed into an alternative world, in comparison to the more limited action and progress that has occurred. It reveals that we are effectively making a choice to remain insecure by our inaction, and that is a choice that we will regret.

1996–1998: IDENTIFYING THE PROBLEM

Executive Order 13010 issued by President Bill Clinton on July 15, 1996, first documented the increasing government concern for effective cybersecurity.[1] It was both the first and perhaps the most insightful of all the government mumbo jumbo that has been produced on the subject of cybersecurity. It identified nine national infrastructures including banking and finance, the only one that we will focus on, as being so vital that their incapacity or destruction could have a debilitating impact on the defense and economic security of the United States. The term "economic security" was not one that had often been used prior to that point when it came to discussing the internet or online commerce. EO 13010 established a Commission on Critical Infrastructure Protection (CCIP) to include the heads of ten federal departments and agencies to identify and consult with public and private sector owners and operators of critical infrastructures, assess vulnerabilities, and recommend a comprehensive national policy and implementation strategy to defend critical infrastructures from physical and cyber threats. It also established an Infrastructure Protection Task Force (IPTF) within the DOJ and chaired by the FBI. In 1996, it appeared that the country was on its way to addressing the growing problem of internet insecurity before it got out of hand. Later that year, Congress passed the Economic Espionage Act.[2] It made it a crime to steal or appropriate trade secrets to benefit any foreign entity.

At the time, there were other real-world events occurring that distracted attention and resources from cybersecurity. The Taliban took over Afghanistan in 1996, as it did once again in 2021, underscoring the illusive progress the United States has made over this period not only in its Afghanistan adventures but also in creating effective cybersecurity defense strategies. In 1997, Newt Gingrich was reelected as the United States' Speaker of the House of Representatives, the United Kingdom returned Hong Kong and the New Territories to China, and Microsoft released its Internet Explorer 4 beta version. In June, the CCIP issued a thirty-page report,[3] much of which occupied itself with process, focusing on cyber threats from nation-states, groups, organizations, individuals, and the interdependencies between systems. It offered a set of recommendations that have been steadfastly repeated without embarrassment since 1996. They included the need for trusted information sharing, risk management, research and development, and clear definitions of the role of government to be established. Perhaps most prophetically, that 1997 report offered the following warning:

> "Who's in charge?" of responding to a cyber attack on the U.S. is not a rhetorical question. Initial investigations reveal ambiguity in the alignment of responsibilities among law enforcement, intelligence, and national defense communities, particularly if an attack comes from or passes through another country. Ambiguities exist within and among levels of government and between government and the private sector.[4]

That statement is as relevant today as it was in 1997. Interestingly, the words "bank" and "financial services" were mentioned only a handful of times in the most generic of contexts in the report. Several months later, a more fulsome report was issued devoting five pages to banking and finance infrastructure.[5] The primary risks at that moment were viewed as physical in nature, "consisting either of natural disasters or a direct coordinated attack on the system's more vulnerable points, as well as the threat created by insiders abusing confidential information."[6]

While hinting at future problems, the report underscored the strength and resiliency of the modern U.S. financial system. It did discuss the growing availability of the internet as an aggravating factor in terms of the transmission of the kind of information needed to plan attacks.[7] It also pointed to an evolving threat of a larger-scale cyberattack by a sovereign adversary or terrorists, noting however that the current probability of such threat was "low."[8] Referencing the "extensive system redundancy, alternate power sources, and diverse communication links" embedded in the country's payments systems and security exchanges, there was a striking absence of any sense of urgency, except with regard to military, power grid, and transportation system defenses.[9] Indeed, it noted that the modern U.S. financial system had never suffered a "debilitating catastrophe," suggesting that some observers characterize it as shock proof and well prepared to confront a broad range of threats to its operations and integrity.[10]

The report missed the fact that up until that point, most financial services systems had been built around closed proprietary networks that were by their very nature more secure. That would not remain the case in the future. It did, however, state a common theme that we hear repeated to this day: financial institutions would benefit from better access to "reliable current information from government and from across the industry."[11] It also underscored a perennial problem—the difficulty of encouraging governments and industry to spend the dollars necessary to invest in the kinds of systems that will prevent attacks that have never occurred.[12]

Several principal findings and recommendations were included[13] that would be repeated in different ways by many of the reports and studies that would follow. Information sharing must be improved, contingency planning must be made a priority, the insider threat must be reduced as much as possible, education is critical, and backup facilities must be made a priority. What was missing, however, and what continues to be missing from subsequent analysis up to this day, are warnings that reports tend to contain "stationary" analysis. In other words, they focus on solving the last big problem, ignoring the fact that as the words of the report are being written, the next big problem is forming and will likely be front and center by the time the report is published. In a technological environment where change occurs exponentially, solving a past or current problem with methods that may be obsolete within months is not a useful problem-solving methodology.

The work of the CCIP was not left to gather dust as most other government reports are. It was followed by President's Clinton's Presidential

Decision Directive (PDD) 63—Protecting America's Critical Infrastructure.[14] It established a goal of creating reliable, interconnected, and secure information system infrastructures by 2003. Yes, 2003. This report established a national center to warn of and respond to attacks, ordered federal agencies to serve as models for how infrastructure protection should be attained, and encouraged voluntary participation of private industry. PDD 63 also set up a new bureaucracy to deal with the new challenge—a national coordinator within the White House, supported by a Critical Infrastructure Assurance Office to oversee critical infrastructure protection, foreign terrorism, and threats of domestic mass destruction (including biological weapons). Finally, PDD 63 launched the National Infrastructure Protection Center (NIPC) to coordinate information sharing efforts among security and law enforcement agencies, laying the groundwork for the future creation of information sharing and analysis centers (ISACs).

President Clinton and PDD 63 were remarkably prescient. If PDD 63 had been implemented and followed through with the necessary energy, resources, and focused sense of urgency that both the government and private sector should have exerted, it would have been a pin in the country's more successful chronology of cybersecurity protection. That did not happen. The urgency of this issue was not quite visible or immediate enough for Congress to appreciate it and appropriate the money. Private industry seemed equally unwilling to invest the kinds of dollars that would have elevated U.S. cybersecurity until it was sure that the problem posed an existential threat. But PDD 63 did push the private sector to begin to act as the government would become distracted by other shiny objects. One of them emerged in August 1998 when President Clinton was forced to admit to an "improper physical relationship" with a White House intern, Monica Lewinsky, and that he had "misled people" about the nature of the relationship. In October, impeachment proceedings were launched, sucking up vast amounts of important government attention and resources.

I became the chairman of the American Bar Association's Committee on the Law of Cyberspace in the summer of 1998 and worked to implement an aggressive agenda begun by my visionary predecessor Jeffrey Ritter, carving new paths into the unknown territory of cyberspace commerce. My biggest challenge was completing a twenty-country effort conceptualized by Jeff to analyze and harmonize the laws of cyberspace so that businesses would understand the new rules of the road, allowing online commerce to occur across local, state, and national borders as legislators around the world took up the issue. The Committee on the Law of Cyberspace thrust itself into the middle of what was happening in online commerce.

1999–2000: RESPONSES TAKE SHAPE

In 1999, the Financial Services Information Sharing and Analysis Center (FS-ISAC) was established in response to the White House's request to coordinate sector responses to incidents such as major financial services attacks. This was an important step forward in the effort to protect the country's economic infrastructure. FS-ISAC, a consortium of banks, brokerage and securities firms, credit unions, financial trade associations, insurance companies, investment firms, bank service providers, and payment processors, would "help ensure the resilience and continuity of the global financial services infrastructure and individual firms against acts that could significantly impact the sector's ability to provide services critical to the orderly function of the global economy."[15]

Today, 7,000 FS-ISAC members receive cyber intelligence, alerts, and threat assessments that can provide them with early warnings of attacks, and they have access to a global community of experts, cybersecurity exercises, best practices, and up-to-date training materials. FS-ISAC has been a critical force behind the launch of the Hamilton Series of cybersecurity exercises that have been conducted and the creation of Sheltered Harbor in 2019, a not-for-profit, industry-led initiative created to protect customers and financial institutions from a catastrophic cyberattack that could cause critical systems—including backups—to fail and customers to lose access to their money and accounts.[16] We will return to these positive steps later.

On January 7, 1999, the impeachment trial of President Bill Clinton began. I worked in the same building on Pennsylvania Avenue in Washington, DC, where independent counsel Ken Starr had taken space to conduct his investigation of the president. The building was surrounded twelve hours a day by press that literally camped out there waiting for the smallest hint of breaking news about the scandal. More than a few times, as I pulled into the underground garage in the building, members of the press armed with TV cameras and microphones rushed to my car to see if Monica was in the backseat or hidden on the floor. Later that year, Vladimir Putin became the acting president of Russia after Boris Yeltsin resigned. Though impeached by the House, President Clinton was acquitted by the Senate in the summer of 1999.

E-commerce was beginning to get traction as online shopping began to soar. The growing presence of the internet began to highlight issues near and dear to people's hearts, such as privacy. In November 1999, Congress passed the Gramm–Leach–Bliley Act, which required that financial regulators mandate that their regulated institutions adopt "administrative, technical, and physical safeguards" for the security of "nonpublic personal information."[17] This endowed the federal financial regulators with significant power to deal with online financial issues and risks, which they have never quite fully used to maximize virtual systemic stability. But two significant events were about to occur that would leave an indelible imprint on the future of cybersecurity.

The first was the "Millennium Bug," an event nicknamed "Y2K." It was thought that every computer in the world would malfunction due to programmers enabling computers to record only six-digit dates— xx/xx/xx—meaning that on January 1, 2000, computers would read the date to be January 1, 1900. That unleashed a fury of activity as businesses and individuals began to take corrective actions and prepare for what some expected to be a cataclysmic global computer systems failure. Whether the problem had been extraordinarily hyped or the millions of hours of research, preparation, and rewriting of billions of lines of code had saved the day, Y2K came and went without even a whimper.

The Y2K fire drill underscored, however, that the United States was the most wired and technology-reliant country in the world. Secondly, it taught us that understanding and mapping every piece of hardware, software, and application of the networks that businesses create or purchase is mandatory. Third, Y2K was a stark example of a future world in which humans might not continue to be in control unless they are proactive. Companies learned valuable lessons about the need for information sharing, the benefits of joint ventures between governments and private entities, the value of continuity and contingency plans, the absolute need for vendor management plans, and the impact of effective horizontal and vertical communication. Y2K was excellent training for what was to come in 2001 and thereafter as the internet grew in scale and functionality and terrorism became a fact of life.

By 2000, the Clinton White House was back in business, though it was pressed to get its agenda completed in the few remaining months it had. In January 2000, it released a 199-page report that unlike the 1997 reports was an *immediate* call to action.[18] It included extraordinarily prophetic warnings given that the country was less than a decade into the digital age and was still twenty months from the defining impact of the event of 9/11. In his summary remarks in the report, the president said that in the next war, America's infrastructure would be a target of computer-generated attacks on critical networks and systems that had been built quickly without adequate concern for security. He underscored that *now* was the time to fix it.[19]

The report's introductory message from the highly respected national coordinator, Richard A. Clarke, emphasized that the country needed a plan to build a defense of cyberspace "relying on new security standards, multi-layered defensive technologies, new research, and trained people."[20] The plan forcefully underscored the need for the government and private sector to protect the stability of the nation by identifying and defending critical infrastructures, particularly the financial sector and support activities, from growing cyber threats from hackers, nation-states, criminals, insiders, and terrorists.[21] The threats and actions required could not been laid out any better at that early date. There were ten action programs created that included seventy-eight milestones to be achieved. Unfortunately, these warnings are almost equally applicable today.

In January 2000, Bill Gates stepped aside as chief executive of Microsoft, and in February, Windows 2000 was released. Also in February, the GAO released a report commenting on the president's January 2000 plan, noting with regard to the numerous federal agencies involved with concurrent jurisdictions that "it is important that a federal strategy delineate the roles and responsibilities of the numerous federal entities involved in information security and related aspects of critical infrastructure protection."[22] It also emphasized that agencies need the appropriate funding and staff training to make their plans work.[23] This continues to be a serious issue today as demonstrated by the fact that most every presidential directive and executive order on cyberspace issued since then has recited these warnings and the need for more attention and resources.

At a press conference in the summer of 2000 in London, I announced the publication of the twenty-country report on the harmonization of the rules of jurisdiction in cyberspace completed by the ABA's Committee on the Law of Cyberspace.[24] It created a blueprint for how towns, cities, states, and nations could go about deciding what laws applied to transactions that took place in a virtual world that had no borders. As these rules of the road would evolve over the coming years, it would make it easier for online commerce to grow and prosper. The insecurity of the infrastructure being built remained an afterthought.

11

AN INSECURE
WORLDVIEW EMERGES

2001–2008

George W. Bush was elected forty-third president of the United States in November 2000. Ten months later, the world exploded. I was attending a board of directors meeting at a bank client on Long Island with my partner, Howard Goldstein, on September 11, 2001. The meeting was supposed to have been in our New York office at the Battery in downtown Manhattan but had been moved to the bank's Long Island headquarters the day before. During the meeting, the CEO's assistant burst into the boardroom in a panic, motioning everyone toward the television which she had just turned on. Within fifteen minutes, the first World Trade tower collapsed. We saw the second one smoldering but assumed that the fires would be extinguished. When it collapsed twenty-nine minutes later, we knew that the world would never be the same. The meeting ended abruptly in a fog of confusion. If it had been in our firm's offices about a quarter of a mile from the Trade Towers, the ensuing chaos would have been many multiples worse.

Normally I would have headed to LaGuardia Airport after the meeting to catch a plane back to Washington, DC. That could not happen. All forms of ground transportation westward toward Manhattan were immediately shut down. All airports were closed. Everything that we take for granted every day of our lives stopped. Howard and I got a ride from one of the bank's directors east to Hempstead and rented the last car the Hertz rental office had. The New York license plate hung by one screw off our subcompact, but it had wheels and moved. We drove north to Port Washington, congratulating ourselves on the way there for being among the few who would have figured out that the only way to travel west or south was to take the ferry across Long Island Sound to Bridgeport, Connecticut.

When we arrived at the ferry at around 3 p.m., we no longer felt so smart. The line of cars was endless, extending several miles beyond the Port Jefferson City limits. After waiting for five hours in our car, we paid our $49 and drove onto the ferry at around 8 p.m. I still have the ticket that we purchased. When we arrived in Bridgeport, we headed west. I dropped Howard off at his home in Westchester around midnight. I vividly remember driving south through New Jersey looking left at the skies over Manhattan that were lit up and still spewing smoke.

At about 4 a.m., I reached McLean, Virginia, just a half-mile from the CIA. The entrance to my cul-de-sac was blocked by police. I had forgotten that the normal security on the block would be significantly beefed up given that my neighbor directly across the street was then–secretary of state Colin Powell. Showing up at 4 a.m. in a car with New York plates was not the least suspicious method of arrival at a moment when everyone and everything was suspect.

"Sir, what are you doing here?" asked a Fairfax County police officer as I pulled up to the barrier blocking the way and rolled down my window.

"I live here," I replied. A second officer walked around the car and starred at the license plate.

"Why are you in a New York vehicle? Do you have identification?"

I handed over my license as he typed into the computer in his vehicle. Thirty seconds passed as they talked to each other. Several other police cars were strategically positioned behind them. Normally Secretary Powell had about a dozen State Department agents guarding him, his home, and the block. Security was seriously beefed up that night. Once I was allowed to pass and drive down the block, there were more cars of agents in front of his house and mine.

For the next five years, life on the block was different. Everyone had to be precleared through State Department security. Stray cars that entered once the barriers were removed a few weeks later were pulled over and questioned. It was not unusual to see the driver of a cab or plumbing van spread against his vehicle being frisked. One night, a car full of kids parked at the end of the block emitting the earthy fragrance of marijuana resulted in a bust initiated by the State Department security agents sitting fifty feet away. One young boy being removed from the car by the Fairfax police was heard to say, "My dad is going to kill me." Ironically, although the world had gotten much more dangerous, we did not need to lock the doors of our house for the next several years.

2001–2002: THINGS CHANGED OVERNIGHT

Everyone's concept of security, safety, and privacy changed on September 11, 2001. I saw that shift as clients called with an array of unprecedented issues created by the stoppage of almost every aspect of life and business. Planes didn't fly, trains didn't run, people didn't go to work, and supplies didn't get to where they needed to be. In such situations where commerce stops and people lose money, there is always a process to determine who if anyone is financially liable. I remember wondering what we would have done if the Federal Reserve and Treasury had been destroyed. What if every means of moving money electronically had been attacked? We would have been completely unprepared.

After the initial shock of an attack on U.S. soil, there was an expectation of retaliation. A new sense of caution and increased levels of physical security were installed everywhere. It became quite tedious to get through security in many buildings in New York and Washington. I would leave twenty minutes early to clear security at my client Merrill Lynch's offices near Ground Zero once they reopened. Getting cleared into the Mariner S. Eccles Building at the Federal Reserve Board on Constitution Avenue in Washington, DC, could take even longer. Suddenly identification became important, and our movements were subject to scrutiny. We were less safe, and with that, a part of our freedom had been snatched away. Life became less convenient.

With the explosion of online commerce happening simultaneously, the irony of verifying identities in the analog world while permitting rampant anonymity in cyberspace became obvious to me. It did not take much imagination to determine that it would become a breeding ground for terrorist activity. No one needed a license or passport to travel online. A renewed importance was attached to the protection of critical infrastructures in the country. The possibility of a bomb in a tunnel, under a bridge, or at a power station now became real. Experts began spinning out evil scenarios, including the placement of dirty bombs and computer-generated attacks on the country's power grids and transportation and banking systems, all in the hopes that it would help the country get prepared for the next attack.

The work begun in the Clinton administration on protecting critical infrastructures suddenly became much more real. Beyond President Bush jumping onto a fire truck wearing a fire hat at Ground Zero after the towers collapsed and warning that those who were responsible would feel the country's fury, the government seemed to appreciate that securing the walls of cyberspace would be a substantial part of the defenses against new terrorist challenges. Thirty-five days after 9/11, on October 16, 2001, the president issued Executive Order 13231 titled "Critical Infrastructure Protection in the Information Age."[1] Focused largely on physical security and defense infrastructures, it ordered the usual collection of federal agencies responsible for defense and security to develop policies and guidelines to protect national security information systems.[2] It also ordered the federal government to work with industry, state and local governments, and nongovernmental organizations to develop systems to share threat warning analysis and recovery information among government network operation centers and information sharing and analysis centers such as FS-ISAC. It was a comprehensive battle plan of action to begin the process of ensuring that the country was never so surprised or unprepared again.

The Computer Fraud and Abuse Act (CFAA),[3] originally enacted in 1984, was amended in 2001 by the USA Patriot Act,[4] criminalizing improper access to computers and networks, including any financial record of a financial institution.[5] In addition to criminal penalties, the CFAA allowed victims of computer hacking who had suffered damage or loss to sue under certain circumstances.

The first international treaty to combat cybercrime—known as the Budapest Convention—was adopted on November 23, 2001, and signed by more than sixty countries who agreed to adopt laws and measures to prevent, combat, and punish cybercrimes.[6] Russia was the only member of the Council of Europe *not* to sign the treaty.[7] Not until 2021 would the prospect of a global treaty reemerge when, surprisingly enough, it was proposed by Russia, which began to see it as a way to move digital power from Big Tech companies to governments.

On January 29, 2002, President Bush delivered his State of the Union address, describing North Korea, Iran, and Iraq as an "axis of evil." Two days later, a large section of the Antarctic Larsen Ice Shelf began to disintegrate, consuming about 1,254 square miles over the next thirty-five days, making climate change yet another global challenge that needed to be dealt with. On May 30, 2002, the final piece of debris from the World Trade Center was removed from Ground Zero, and five months later on October 2, Congress passed a joint resolution authorizing the president to use the United States Armed Forces, as necessary and appropriate, against Iraq. The country would begin more than two decades of war in the Middle East. That war and the internet would become the most significant factors to impact American culture for the foreseeable future. In some respects, the internet became the vehicle by which the convenience and freedoms that we had lost on 9/11 were somehow reinjected back into our lives. To the extent that the internet became an emotional compensating balance for the things lost to terrorism, we ended up compounding our problems.

In 2002, the Financial Services Sector Coordinating Council (FSSCC) was created to work collaboratively with key government financial and security agencies to strengthen the resilience of the financial services sector against attacks by proactively identifying threats and promoting protection, preparedness, and collaboration.[8] Its Financial Services Cybersecurity Profile, launched on October 25, 2018, provides data, education, cybersecurity risk management materials, cyberattack drills, and training for the financial services to be prepared to deflect as many attacks as possible and quickly recapture data to maintain continued operations. It does not represent a comprehensive defense structure. In December 2002, Congress enacted the Federal Information Security Management Act (FISMA)[9] to oversee the security of the federal government's information systems and establish and enforce information security standards for federal agencies and their contractors.

2003–2004: A NATIONAL STRATEGY

Sensing a new urgency for security after 9/11, the Bush White House released a National Strategy to Secure Cyberspace[10] in February 2003 containing just shy of fifty recommendations and action items focused on protecting a somewhat expanded list of U.S. critical infrastructures.[11] It was largely repetitive of concepts and goals discussed by the Clinton administration, but it created a renewed awareness of the threat that cyberspace could present.[12] This was not the first and it would not be the last time these recommendations and goals would be put forward.

It was becoming clear at this time that the challenge of internet security was a unique one that did not align well with traditional military strategy. Most online commerce occurs on networks and components that are largely privately owned, making the private sector best equipped and structured to respond.[13] But they are not permitted to under the law, nor do most have the requisite technological firepower to do so. Businesses generally don't want the government overseeing or regulating their internet commerce, and the government doesn't want private companies defending themselves. In the analog world, military and police forces are responsible for defending U.S. coasts and airspace. With no physical territory to defend in cyberspace, military, security, and intelligence agencies find themselves in a new world when dealing with cyberattacks. The defense of the computer networks of the five largest banks, for example, falls primarily on those banks and requires coordinated private sector actions which raise sensitive information sharing issues that must be interwoven with government actions that will inevitably be seen by some as unacceptable intrusions.

President Bush's national strategy report identified these issues but stressed the importance of a government role in private sector cybersecurity. It proposed the development of a National Cyberspace Security Response System (NCSRS) responsible for facilitating rapid identification, information exchange, and remediation to mitigate the damage caused by malicious cyberspace activity.[14] It also established a National Cyberspace Security Threat and Vulnerability Reduction Program to enhance law enforcement, vulnerability assessments, software security, and the physical security of cyber systems and telecommunications.[15] There were, however, no specific discussions of the particular vulnerabilities, impacts, and protections needed by the banking and financial services industries or their regulators in this national strategy report. The words "banking and finance" appeared in it only eight times.

Also in February 2003, the White House released another report focused on reducing America's vulnerability to acts of terrorism by protecting critical infrastructures and key assets from *physical attack*.[16] It devoted only two pages specifically to the banking and finance sector. It noted the vulnerability of banks to the extent that they rely on critical infrastructures such as electric power, transportation, public safety services, computer networks, and telecommunications systems to assure the availability of services. It was surprisingly self-reflective and honest, however, when it said that "overlapping federal intelligence authorities involved in publicizing threat information cause confusion and duplication of effort for both industry and government." On that note, the Department of the Treasury organized the Financial and Banking Information Infrastructure Committee (FBIIC) as a standing committee comprised of representatives from thirteen federal and state financial regulatory agencies to work with the National Infrastructure Protection Center (NIPC), the Financial Services ISAC (FS-ISAC), and the White House Office of Homeland Security to improve the information dissemination and sharing processes.[17] The Department of Homeland Security (DHS) was directed to work with the Department of Treasury, the FBIIC, and the FS-ISAC to improve federal communications with sector members and streamline the mechanisms through which they exchange threat information daily as well as during an incident.[18] The report also required the Department of the Treasury and DHS to convene a working group of representatives from the telecommunications and financial services sectors to study and address the risks that arise from each sector's dependencies on electronic networks and telecommunications services. A similar meeting for a similar purpose would be convened by President Biden eighteen years later.

Terror on the U.S. homeland had impacted policy makers' thoughts and analyses, forcing them to transition from focusing on what they hoped would go right to what might go wrong. That was a worthwhile shift in security and the treatment of technology. Federal banking regulators were no exception. While not experts in technology, cyberspace, and cybersecurity, they were naturally getting increasingly concerned about the implications for the country's critical financial infrastructure. In April 2003, the federal financial regulators issued an interagency paper that identified continuity objectives in the post-9/11 environment.[19] While it remained at a high level and was nontechnical in nature, it did focus attention on the fact that the reduction of potential systemic risk would require the rapid recovery of clearing and settlement activities for the purpose of completing material pending transactions on their scheduled settlement dates.[20]

From a logistical point of view, the federal banking agencies, including their joint venture in the Federal Financial Institutions Examination Council (FFIEC), have always been good at creating statements of what they expect financial institutions to do: secure, protect, and back up data and house it in geographically dispersed areas to maximize recovery in the event of a cyberattack. They are indeed experts in underwriting, lending, credit analysis, and risk management. But they have less resources devoted to the technical aspects of cyber deterrence and defense and normally don't suggest how they should be implemented. That is to be expected and is not a criticism. They are not cryptographic, networking, or cybersecurity experts, so often they can only produce cybersecurity frameworks that require adherence to a broad set of commonsense principles that institutions must put the operational flesh on. A concise list of these policies, procedures, and manuals is maintained by the FFIEC on its website.[21]

On April 9, 2003, Baghdad fell to coalition forces. A week later, ten new member states were admitted to the European Union. By May 1, the Iraq War was over. The year seemed to end with an exclamation mark as Saddam Hussein was captured by American troops on December 13. On February 3, 2004, the CIA revealed that despite the massive amount of global intelligence to the contrary, there had been no imminent threat from weapons of mass destruction before the 2003 invasion of Iraq.

President Bush was picking up political and popular support for his positions on both physical and cyberspace defenses. On July 7, 2004, he issued a secret National Security Presidential Directive 38 (NSPD 38) to secure cyberspace. National Security Presidential Directive 54 (NSPD 54) followed in January 2008.[22] Together they authorized DHS and the Office of Management and Budget (OMB) to set minimum operational standards for federal executive branch civilian networks. DHS was also empowered to lead and coordinate the national cybersecurity effort to protect cyberspace and the computers connected to it. Finally made public in June 2014, NSPD 38 also included a Comprehensive National Cybersecurity Initiative (CNCI), which contained twelve projects to deploy and connect an intrusion detection system of sensors and a cyber counterintelligence plan across the federal enterprise to increase the security of networks and enhance global supply chain risk management.

On October 29, 2004, a videotape of Osama bin Laden emerged threatening terrorist attacks on the United States and taunting President Bush. On November 2, George W. Bush was reelected to a second term as president of the United States. But as Americans became weary over the next few years with the national fetal position they had been in, they also began to turn on the wars in the Middle East and began to let their guard down.

2007–2008: FIGHTING ON TWO FRONTS

In April 2007, following a contentious relocation of a Soviet-era statue in Tallinn, Estonia, Russian hackers unleashed a series of coordinated DDoS attacks against government, banking, university, and newspaper websites that lasted three weeks and virtually shut down the most wired country in Europe.[23] This should have been a wake-up call for the world, particularly since every indication was that Russian hackers were responsible. It certainly was for Estonia. It established a voluntary Cyber Defense Unit, which to this day has been among the factors that have sensitized Estonia to the risks in cyberspace and hardened its defenses.[24]

Between July and August 2008, Georgia became the victim of a coordinated defacement and DDoS campaign that disrupted government and bank websites. The attack was directed using a strain of malware frequently used in Russia that flooded websites with traffic that included the phrase "Win love in Russia." The National Bank of Georgia's website was defaced. Georgia attributed the attack to the Russian government, which denied the allegations.[25]

On June 29, 2007, the first iPhone was released for sale in the United States, beginning a new technological era. In 2008, the Identity Theft Enforcement and Restitution Act (ITERA)[26] again amended the Computer Fraud and Abuse Act of 1984 to, among other things, expand identity theft and aggravated identity theft crimes, authorize criminal restitution orders in identity theft cases to compensate victims, and expand the definition of "cyber extortion" to include a demand for money in relation to damage to a protected computer. But cybersecurity was not the only problem facing the U.S. government in 2007, and it would have to take a backseat for several years to efforts to save the U.S. economy from a building financial crisis triggered by a binge of subprime mortgages that had been securitized and sold throughout financial markets.

The economy had shifted into high gear thanks in large part to a blistering real estate market driven by new buyers entering the market by virtue of reduced standards that were focused on providing credit to subprime borrowers. On March 16, 2008, Bear Sterns collapsed and was followed by the failure of a stunning pantheon of other U.S. financial companies that were either acquired or went into bankruptcy. Economically, the country was quickly slipping into what would become the most terrifying financial crisis since the Great Depression.

In the summer of 2008, Candidate Obama gave a speech at Purdue University earmarking U.S. cyber infrastructure as a "strategic asset."[27] But once elected, his administration was distracted by the country's financial crisis, and while it continued to fly the banner of cybersecurity,[28] it seemed to back away somewhat from the cybersecurity path that the country had been on in favor of being more business friendly and not dictating security standards for private companies.[29] For at least a year, there was no one in the White House overseeing the cybersecurity efforts of the U.S. government.[30] Richard A. Clarke, who had served in the White Houses of Presidents Reagan, Bush I and II, and Clinton, concluded that because cyberattacks left no physical remembrance of destruction as 9/11 did, it made it easier to ignore them. He cleverly cited Kevin Spacey in *The Usual Suspects*: "The greatest trick the devil ever pulled off was convincing the world he didn't exist."[31]

12

DISTRACTIONS, DIPLOMACY, AND CYBER THREATS

2009–2016

On January 3, 2009, the Bitcoin network was created when the first block of the digital currency was mined by whoever is Satoshi Nakamoto. On January 20, President Obama was inaugurated. On July 4, 2009, still reeling from the financial panics that had played out around the world, financial institutions in the United States and South Korea were among several targets of three waves of DDoS attacks spanning six days when 65,000 botnets clogged computers and slowed government and commercial websites such as the New York Stock Exchange, the Nasdaq, the White House, and the *Washington Post* for several hours at a time. Days later, Shinhan Bank in South Korea was impacted. The malware spread through emails that were encoded to overwrite hard drives with the string "Memory of the Independence Day." South Korean intelligence suspects it was the work of a specific criminal or state-sponsored organization.[1] On October 22, Microsoft released Windows 7.

2010–2012: BACK TO THE VIRTUAL CHALLENGES

By 2010, the country was returning to some sense of financial normalcy, which meant that other problems once again found themselves on the agenda. In February, DHS issued a report on network security and privacy.[2] It reiterated that the nation's electronic information infrastructure, which is under perpetual attack, is vital to the nation's economy and national security. It warned that such attacks would likely expand over time "as more of our critical infrastructure is connected to the Internet, and malicious cyber activity grows in volume and becomes more sophisticated and targeted."[3] DHS announced that it was upgrading its cybersecurity capabilities to secure and defend against such threats, including the use of "EINSTEIN," a network flow monitor designed in 2003 to surveil incoming and outgoing traffic for malicious activity directed toward the federal executive branch's unclassified civilian computer networks and systems. DHS also emphasized the need for the federal government to work with interagency partners to ensure an abundant, qualified cybersecurity workforce, and for private industry to become "more involved in developing comprehensive, game-changing, innovative solutions that improve and expand upon our current capabilities to protect, detect, and respond to cyber incidents."[4] This admonition came fourteen years after these risks and similar recommendations were first articulated by the Clinton administration.

The U.S. Cyber Command, the nation's unified combatant command for the cyberspace domain, began operating on May 21, 2010, when several military programs were consolidated. The country's financial infrastructure was not specifically mentioned as a target to be protected beyond Cyber Command's generic role in the cybersecurity landscape. In July 2010, the Dodd-Frank Act was signed into law by President Obama in response to the greatest financial crisis since the Great Depression and the factors that led up to it. It did not contain any provisions directed at correcting the liquidity deficiencies that had so greatly contributed to the financial crisis or the increasing cybersecurity threats hanging over the financial sector.

In November 2011, the Securities Industry and Financial Markets Association (SIFMA) began coordinating a series of industry-wide resiliency exercises called Quantum Dawn. They were intended to provide a forum for financial firms, regulatory bodies, government agencies, and trade associations to share notes regarding simulated cyber or physical attacks. SIFMA is now up to the fifth exercise, Quantum Dawn V. We will return to talk about the role and efficacy of Quantum Dawn in a later chapter.

Real-world terror and catastrophes dominated 2011. Egyptian president Hosni Mubarak resigned in February, shifting control of Egypt to the military until a general election could be held. On March 11, a 9.0 magnitude earthquake and subsequent tsunami hit the east of Japan, killing 15,840 people and leaving another 3,926 missing. In May, President Obama announced the killing of Osama bin Laden, the founder and leader of Al-Qaeda and the person responsible for 9/11. On July 31, over 12.8 million people in Thailand were affected by severe flooding, causing 815 deaths and $45 billion in damage in fifty-eight of the country's seventy-seven provinces. In September, Occupy Wall Street protests began in the United States, thereafter spreading to eighty-two countries over the next month.

In December and stretching into mid-2013, DDoS attacks were conducted against over forty-six U.S. financial institutions on behalf of the Iranian government and Islamic Revolutionary Guard Corps. The men indicted for these attacks in March 2016 worked for two private computer security companies in Iran that allegedly worked for the government. The U.S. Treasury Department responded in September 2017 with sanctions against eleven individuals and organizations accused of participating in the attacks.[5] Certainly this should have been an event that would attract the government's and the private sector's attention and resources. It did not.

In October 2012, President Obama issued Presidential Policy Directive (PPD) 20.[6] It was classified as top secret until October 16, 2017, but it was leaked in 2013. PPD 20 detailed government policies regarding offensive and defensive cyber actions and established warlike authorizations and rules of engagement. It reserved the use of defensive measures to protect national interests to situations where network defense or law enforcement measures were insufficient. In those cases, it required the *least intrusive* methods to mitigate a threat. But it also emphasized the development of offensive capabilities, acknowledging that it would require considerable time. It referenced the complicated and intertwined roles of more than a dozen federal agencies, all of whom were required to work together and establish criteria and procedures to be approved by the president.[7]

The subject of cybersecurity began to be a topic in the annual reports issued by the Financial Stability Oversight Council (FSOC), a super agency established by Congress in 2010 in the Dodd-Frank Act to oversee and regulate financial systemic stability.[8] FSOC consists of the country's federal financial regulators who, supposedly sitting as a group, will have more insights into the overall systemic threats the economy faces than they do sitting in their individual offices. Remarkably, its 2011 annual report had just one single reference to financial cybersecurity concerns. In 2012, the report was only slightly more instructive on the spectrum of growing risks, noting that cyber threats pose a "potentially significant risk to the stability of financial markets through the disruption of critical payment, clearing, and settlement systems for key financial institutions."[9] It noted that the evolving cyber-threat environment required strengthening the ongoing collaboration and coordination among financial regulators and private entities in the financial sector, and it recommended that mechanisms for sharing information related to cyber threats and vulnerabilities should continue to be explored.[10]

2013–2014: GETTING SERIOUS—PERHAPS

On February 12, 2013, in White House Executive Order 13636, President Obama summed up, repeated, and repackaged the comprehensive cybersecurity approaches, tools, and defenses that had been developing and recommended over the previous two decades.[11] He instructed the government (again) to establish a cybersecurity framework to increase the volume, timeliness, and quality of cyber-threat information shared with U.S. private sector entities and to build better defensive systems and cybersecurity services for critical infrastructure sectors. DHS and the Treasury were directed in Section 9 to again identify "critical infrastructure where a cybersecurity incident could reasonably result in catastrophic regional or national effects on public health or safety, economic security, or national security." The finance and banking areas were designated as "Section 9 Infrastructures," as they have come to be referred to since then.

That same day, the White House issued "Presidential Policy Directive/PDD-21: Critical Infrastructure Security and Resilience."[12] It repeated the need to strengthen the security and resilience of its critical infrastructure against both physical and cyber threats. It also directed the federal government to work with critical infrastructure owners and operators and international partners to take proactive steps to reduce vulnerabilities, minimize consequences, identify and disrupt threats, and hasten responses and recovery efforts to strengthen the security and resilience of the nation's critical infrastructures. Toward these ends, it directed the clarification of relationships across the federal government to effectively oversee the country's sixteen critical infrastructures, but then it seemingly undercut itself by delegating that task to about two dozen different federal agencies.[13] Treasury was assigned responsibility yet again for coordinating efforts in the financial services sector with seven other agencies.[14]

The country was reminded of the ever-present terrorist threat when two pressure cookers exploded near the finish line of the Boston Marathon on April 15, killing three people and injuring more than 260 others. In June, extraordinary surveillance at home and abroad by the NSA was disclosed in a leak of classified intelligence by NSA contractor Edward Snowden. Later in the year, FSOC's annual report included references to cybersecurity attacks during 2012 in which more than a dozen financial institutions sustained DDoS cyberattacks, concluding that, "with this recent experience, the financial sector has become increasingly adept in identifying, assessing, preventing, and mitigating cyber risks."[15] This report could not have been more understated or misinformed.

Several months later in August 2014, reports circulated that the account information of more than 75 million customers had been the subject of a data breach at JPMorgan Chase. The hackers appear to have entered the network through a single-factor authentication server that allowed them to gain access to more than ninety bank servers.[16] The breach was only discovered when the group was noticed on a website for a charity race that JPMorgan sponsored. Other financial companies were targeted in the attacks, including Dow Jones, Fidelity, E-Trade, and Scottrade. Nine people were charged; at least one was Russian. All too often the people charged in these kinds of crimes are in foreign countries beyond U.S. jurisdiction and are never brought to justice. Hardly a weapon to be feared.

In Executive Order 13681[17] issued on October 17, 2014, President Obama directed federal agencies to improve the cybersecurity of their consumer financial transactions by requiring that official government payment cards use more secure chip and personal identification number (PIN) payment terminals and payment cards. It also directed the delivery of a plan to the president within ninety days to ensure that all agencies making personal data accessible to citizens through digital applications required the use of multiple factors of authentication. The discovery of the massive SolarWinds hack in December 2020 would suggest that the directives issued to this date to protect federal networks had either not been fully implemented, were ineffective in reducing incursions, or were just no match for the increasingly creative and well-equipped hackers and the inherent insecurity of cyber networks.

In November, the infamous attack on Sony Pictures' email and computer systems became front-page news as company emails, personal information about employees, and copies of then-unreleased Sony films, plans, and scripts were made public. North Korea was blamed, and several North Koreans were indicted for the attack, which was generally viewed as retaliation for Sony's intended release of *The Interview*, a satirical movie about North Korea's leader.[18]

In 2014, FSOC's annual report finally devoted more attention to and concern for growing cyber threats to financial institutions. It mentioned the December 2013 data breach at Target Corp and the financial disruption that cyber incidents can have on critical financial infrastructure, particularly given the financial sector's interconnectedness and the likelihood of economic contagion. It underscored the recurring theme that there should be a close partnership between the public and private sectors, citing the work being done by the Financial Services Sector Coordinating Council (FSSCC) and the Financial Services Information Sharing and Analysis

Center (FS-ISAC) in collaboration with the Treasury and members of regulatory, law enforcement, and intelligence communities.[19] FSOC recommended that financial regulators continue to inform institutions, market utilities, service providers, and other key stakeholders of the risks associated with cyber incidents; ensure that regulated entities are adhering to existing regulatory and state-of-the-art principles; assess cyber-related vulnerabilities facing their regulated entities; and maximize the sharing of information.[20] Given that it was 2014, these references to cyber threats that appeared in just a handful of the 147 pages of the report and the recommendation to inform institutions of the risks were a pedestrian approach to the growing problem that could devour financial institutions and the economy. The JPMorgan breach would occur several months after the report was released.

Finally, in 2014, the Hamilton Series of exercises were launched in collaboration with the Treasury Department and the FSSCC to simulate a variety of "plausible cybersecurity incidents or attacks to better prepare the financial sector and the public sector for cyberattacks." Starting in 2018, FS-ISAC added range-based cyber exercises to raise capability levels and resiliency across the sector.[21] We will return to the Hamilton Series and its impact in a later chapter.

2015: MORE ALARMS

On February 13, 2015, President Obama signed Executive Order 13691 to again encourage and promote sharing of cybersecurity threat information between the private sector through information sharing and analysis organizations (ISAOs) and government.[22] It also clarified the authority and operational framework of the National Cybersecurity and Communications Integration Center (NCCIC), a civilian agency in DHS tasked with coordinating information sharing within the federal government and with entities outside the government.

On April 14, 2015, the president issued Executive Order 13694, directing federal agencies to freeze and block the transfer of any property of persons in the United States who were engaged in malicious cyber-enabled activities.[23] CISA followed with a 2015 Financial Services Sector Specific Plan in conjunction with Homeland Security, the FSSCC, and the FBIIC. The plan underscored the importance of information sharing between companies and the government (which is still done on a limited basis), the development and use of common approaches and best practices (which is more of a goal than a reality), collaboration among an alphabet soup of federal agencies (which speaks for itself when it comes to effective management), and the need for robust discussions about policy and regulatory initiatives (a lack of discussion is not the problem).[24]

In June 2015, another loud alarm sounded when the Office of Personnel Management (OPM) discovered that the background investigation records of current, former, and prospective federal employees and contractors as well as the Social Security numbers of 21.5 million individuals had been stolen. Usernames and passwords that background investigation applicants used to fill out their background investigation forms were also stolen.[25] The magnitude of this breach was yet another wake-up call, and for a few weeks, everyone seemed to take it for the serious event it was. FSOC's annual report of 2015 included a two-page summary of cybersecurity vulnerabilities among its 150 pages.[26] It used the word "cyber" thirty times, ten times more than in its 2012 report. While it referenced "recent cyber attacks," it did not specifically mention JPMorgan and its breach. It noted the work of the private sector being done by the Financial Sector Cyber Intelligence Group (FSCIG) and FS-ISAC to increase the speed of information exchange, and it was replete with recommendations that *someone* in the government should take responsibility for and implement enhanced cybersecurity protections. It also encouraged continued efforts by the Federal Financial Institutions Examination Council (FFIEC), which were as instructive and effective as any the government had taken, to collaborate and coordinate on cybersecurity issues affecting the banking sector. Finally, it recommended the establishment of a national plan for cyber incident response, "coordinated" by the Treasury, to "identify and articulate the role of law enforcement, the Department of Homeland Security, and financial regulators."[27]

References to "cyber threats" began to slip into other parts of the report dealing with insurance, securities, and payments and processing markets and systems. FSOC also began to be more specific about the risks to financial stability that could arise from attacks on payments and clearing systems, interconnected network systems, and service providers operating within critical financial infrastructures and supply chains. It raised the issue of "recovery" procedures that provide financial firms subject to destructive attacks the capabilities and procedures to resume operations as quickly as possible.[28]

The Cybersecurity Information Sharing Act of 2015 (CISA 2015) was enacted into law on December 18, 2015, as a part of the Consolidated Appropriations Act of 2016.[29] It joined the ranks of laws such as the Foreign Intelligence Surveillance[30] and Electronic Communications Privacy Acts,[31] which authorized government monitoring of communications, raising Fourth Amendment and other privacy concerns. CISA 2015 addressed impediments to collaboration between the public and private sectors and authorized private entities to monitor information systems, use "defensive measures," and share information about cyber-threat indicators or defensive measures with other private entities and the federal government. But it expressly prohibited the private sector from undertaking offensive cyber-security efforts such as "hacking back." It did create a limited protection from liability for companies that monitor information systems or share or receive cyber-threat indicators, raising objections from those concerned about companies being protected from privacy intrusions.

2016: IT'S GETTING REAL

In 2016, Britain voted to leave the European Union, while North Korea conducted nuclear and ballistic missile tests. But if there is any year that put a pin in the threatening effects that the internet could have for the viewer at home, it was 2016. The entrails of the election of Donald Trump that were examined indicated that Russia had used social media and other electronic means to attempt to influence the election. That seemed like announcing that there was gambling in Las Vegas casinos—why would we not expect Russia or other adversaries to do that? Why weren't we prepared? The Democratic National Committee and high-ranking Hilary Clinton officials had been hacked, and embarrassing emails had been dumped into the public domain. Facebook and other social media darlings began to be seen in a more sinister light as the use of their data became a focus of the controversy. This would be a story that would be fleshed out over the coming years, adding to our understanding about the vulnerabilities of elections and the complexity of determining the truth in a world where it was becoming increasingly difficult to distinguish true from fake news. This was the first clear linkage in the minds of most between cybersecurity and democracy.

In February 2016, the Bangladesh central bank network was breached, resulting in the transfer of $81 million to accounts in the Philippines. This was one of the largest bank thefts in history, noticed only due to a misspelling of the word "Foundation."[32] North Korean–affiliated actors were alleged to be behind the attack.[33] In April, a significant step forward was taken when FS-ISAC established Sheltered Harbor, the not-for-profit, industry-led ecosystem of financial institutions that is supported by a group of financial services trade associations[34] and prepares participating members for catastrophic cyberattacks that could cause critical systems to fail.[35] Sheltered Harbor allows participating institutions to back up critical customer account data each night through their own or outside providers' secure data vaults. Stressing its "resiliency planning guides and an expanding network of assurance and advisory firms," Sheltered Harbor advertises itself as helping institutions create the business and technical processes necessary to restore critical systems.[36] The effectiveness of Sheltered Harbor has not yet been tested by a catastrophic event.

The Financial Systemic Analysis and Resilience Center (FSARC) was established in 2016 and was later renamed the Analysis and Resilience Center (ARC) for Systemic Risk in 2020.[37] ARC is a nonprofit, cross-sector organization established by leaders from the financial and energy sectors designed to mitigate systemic risk to the nation's financial and energy sectors and facilitate operational collaboration between members, the government, and other key sector partners.[38] ARC analyzes systemically important critical systems, enhances capabilities to monitor threats to those systems, and develops collaborative resilience measures.

In June 2016, the Committee on Payments and Market Infrastructures and the International Organization of Securities Commissions (CPMI-IOSCO) issued cyber resiliency guidance for financial market infrastructures[39] classified as essential to "maintaining and promoting *financial stability* [emphasis added]."[40] The report was largely repetitive of many other pronouncements to that point, but it repackaged them for financial market infrastructures (FMIs) which process financial and payments data. "Financial stability" was mentioned nine times in the 32-page report. The guidance focused on cyber resilience based on the ability to detect a potential cyber incident and resume critical operations and the settlement of obligations as rapidly as possible. It set within two hours of a disruption as the goal to enable systems to be back online to complete settlement by the end of the day of the disruption. This guidance is a useful compendium of corporate policies and procedures to address a world of increasing cyber threats that places the responsibilities of cybersecurity squarely on the shoulders of FMIs. It does little to address the issue of systematic defense of the FMIs in light of the increasing threats that future technologies and the super sciences will create.

In October 2016, the Federal Reserve, the Office of the Comptroller of the Currency, and the FDIC issued a proposed rule to establish enhanced cybersecurity standards for large financial institutions.[41] Jumping on the operational resilience bandwagon to develop relevant new standards, they addressed cyber risk, governance, internal and external dependency management, incident response, and situational awareness. This proposed rule appears never to have been finalized in the form that it first took, having been opposed by the banking industry. Most importantly, as quantum computing and the potential for it to undo current online security became a more likely next generation of technology, the National Institute of Standards and Technology (NIST) launched a competition to develop "quantum-resistant" algorithms. Six years later, in July 2022, shortly after the G7 nations announced the intention to cooperate in the deployment of quantum-resistant cryptography, NIST announced four algorithms that will become part of its post-quantum standard.[42] NIST is hoping to have a quantum-resistant encryption system by 2024.

The 2016 financial stability report by the Treasury's Office of Financial Research (OFR) contains a ten-page section titled "Cybersecurity Incidents Affecting Financial Firms."[43] This marked a turning point in the threat level evidenced by federal financial overseers, noting the increasing incidence of attacks on financial firms because of their deep reliance on technology and relative vulnerability, and identifying vulnerability to malicious cyber activity as "a top threat with substantial potential impact." The report discussed the increasing attempts to disrupt, steal, alter, or destroy data through hidden weaknesses in widely used software (called zero-day vulnerabilities); target email accounts to steal passwords (spear phishing); infect websites and users with malicious software (malware); and implanting software that locks companies out of their own IT systems (ransomware).

OFR's report qualified as the most alarming statement that the U.S. government had made to that point concerning the threats that cyberattacks posed to the nation's financial institutions, adding that the magnitude of the threats and the resilience of institutions were difficult to measure.[44] It highlighted the redundancy of financial regulators and recommended that they develop a broader approach to resilience that focuses on "key links" among financial institutions and more collaboration among regulators to develop a common lexicon and a shared risk-based approach between financial networks, third-party vendors, overseas counterparties, and service providers.[45] Shortly thereafter, FSOC released its 2016 annual report. Unlike OFR's most recent annual report, FSOC devoted only a few of its 160 pages to nine references to cyber threats. It recited the need for increased information sharing, making networks more secure, supervising the exposure created by outside service providers, and recognizing the importance of incidence response and recovery. It provided very little in the way of practical deterrence.

As the Obama administration came to an end, on October 11, 2016, the G20's Financial Stability Board published a statement on the fundamental elements of cybersecurity in the financial sector. Candidly, it underscored that "mature entities recognize that it is impossible to guarantee a zero-failure environment."[46] Its recommendations included the oft-repeated list of cybersecurity strategies, including risk control, monitoring, exercise response, recovery procedures, information sharing, and education.

The year closed with the Commission on Enhancing National Cyberspace Security established by President Obama issuing a report on the digital economy on December 1.[47] Underscoring the advantages that attackers have and the increasing vulnerabilities in digital systems, it noted in a half-dozen passing references the continuing concern for critical infrastructures such as the financial system. It contained sixteen recommendations and fifty-three related actions to enhance U.S. cybersecurity. The half of the report that focused on cybersecurity repeated many of the same themes, including the need for public-private collaboration to harmonize standards and defenses to protect critical infrastructures. It contained penetrating insights into the obvious, stating that no technology comes without societal consequences and that the challenge is to make sure that the positive impacts far outweigh the negative ones as cybersecurity risks are lowered and privacy and other civil liberties are protected.[48]

13

RUNNING IN CYBER CIRCLES

2017–2020

In 2017, the G20 finance ministers and central bank governors warned that malicious uses of information and communication technologies could undermine security and confidence and endanger financial stability, hardly a revelation at that point.[1] In contrast, earth-shattering news—literally—did occur on August 17 when two observatories detected gravitational ripples created by the merger of two neutron stars to form one huge black hole. The event had happened 130 million years ago, but the light detailing its effects had just reached earth.

The newly elected Trump administration sent mixed signals from the start on growing cybersecurity problems and defenses, issuing strong statements that the United States was at war in that space as it eliminated the position in the National Security Council (NSC) responsible for developing policy to defend against cyber weapons. The White House argued that the post was no longer considered necessary because lower-level officials had already made cybersecurity issues a "core function" of the president's national security team.[2] Certainly, a position in the White House does not determine the level of resources and attention an issue deserves or receives. The inattention and lack of progress that cybersecurity had received over the prior two decades when there were White House officials in charge was clear evidence of that.

Federal banking regulators started to move faster to balance the conflicting benefits and threats of technology. Over the next several years, they and the Federal Financial Institutions Examination Council (FFIEC) would issue multiple cybersecurity warnings and bulletins about what was expected of financial institutions. Many of them were generic warnings and broad governance suggestions that were long on record keeping, process, and procedures and short on preventative details.[3] But they had clearly gotten the message, and the level of concern was being ratcheted up.

The private financial sector was also beginning to shift into hyperdrive as the attacks on financial companies increased in frequency and severity, highlighting the fragility of America's financial infrastructure. It had literally finally caught cybersecurity fever, with about two dozen reports issuing over the next five years regarding the increasing threats being posed to the economic security of countries around the world. Everyone was beginning to realize that this was a real problem and that *something* had to be done. All too often, the publication of more reports hashing and rehashing the same preventative and restorative remedies had substituted for real progress. A common theme that neither the private sector nor the government could be effective alone was now solidly established. It was unclear, however, whether they could be successful together at that point. An effective collaboration would naturally raise complex questions about politics, privacy, and the bounds of government oversight. Not many seemed to be zeroing in on one of the main culprits: inferior software and hardware and an absence of AGE—authentication, governance, and enforcement.

2017: MORE WAKE-UP CALLS HIT THE PRIVATE SECTOR

Immediate resistance to a Trump presidency produced polar-opposite cultural and political perspectives that seemed to intrude on everyone and everything. Meanwhile, the plodding analysis and development of cyberattack action plans continued, with no specific goals in sight. On February 15, 2017, the Treasury's Office of Financial Research (OFR) issued a follow-up to its 2016 financial stability report reiterating the obvious—that financial firms are vulnerable because they rely heavily on information technology and the interconnectedness of their activities and networks but "are primarily responsible for their own security."[4] On March 18, 2017, the finance ministers and central bank governors of the G20 issued a warning about the malicious use of technology "to disrupt financial services crucial to both national and international financial systems, undermine security and confidence, and endanger financial stability."[5] They promoted the resilience of financial services with the aim of enhancing cross-border cooperation and requested that their Financial Stability Board perform a "stock taking" of existing relevant released regulations and supervisory practices.

But more creative and constructive analyses and solutions were beginning to emerge. On March 27, the Carnegie Endowment for International

Peace issued a report that went several steps further than the Budapest Convention in 2001 and proposed prohibiting nations from knowingly supporting any activity that intentionally manipulated financial data and algorithms and requiring them to cooperate to mitigate such activities. This was an approach that a logical mind would think was necessary if there is any hope of achieving real internet security. Carnegie's intention was, among other things, to send a clear message about the fragile stability of the global financial system, build confidence, create momentum for greater collaboration, and enhance the impact of existing agreements.[6] It also urged nations to achieve a greater degree of definitional clarity about when cyberattacks would be considered crimes, espionage, or acts of war that could trigger retaliation.[7] From this point forward, Carnegie would be one of the more focused voices on the increasing threats to the financial services sector.

On May 11, 2017, President Trump issued Executive Order 13800 directing a gaggle of agencies to identify how they could support the cybersecurity efforts of Section 9.[8] In June, Facebook announced that it had reached 2 billion monthly users, while a Petya malware cyberattack affected organizations in more than sixty-four countries. In July, hackers revealed that they had stolen data from HBO, including episodes and scripts of *Games of Thrones.*

The International Monetary Fund (IMF) released a blunt study in August reciting the typical warnings that had long since become boilerplate in these reports, pointing to the Bangladesh Bank and other recent breaches.[9] It called attention to the fact that, compared to all other U.S. industries, the financial industry was experiencing a groundbreaking shift in web application attack patterns that could no longer be viewed as simply IT problems.[10] It labeled the situation "a textbook example of a systemic risk," noting that the risks related to cyberspace go well beyond the internal monitoring and risk management capacities of individual institutions.[11] The report reached a very frank conclusion: *security is an illusion, and cyber risks cannot be eliminated.*[12]

The Bank for International Settlements (BIS) released a report[13] in August 2017 debating whether cybersecurity requires a rules-based or principles-based approach.[14] Reading the report, one gets the feeling of rearranging deck chairs on the *Titanic* as it argues for the creation of regulatory and corporate governance records to create legal defenses given the unlikelihood of preventing cyberattacks by persistent and sophisticated attackers. It did however raise the question of whether there should be a different or second cyberspace. We will return to that potential solution later. Also, that month, the IMF released a report reciting the litany of high-profile cyberattacks on financial institutions, the complexities created by the anonymity of cyberspace interactions, and the potential impact of cybersecurity attacks on financial stability given the interconnected nature of financial services.[15] The report distinguished itself by criticizing the continuing deficiencies in software, underscoring what may be the original sin of cyberspace insecurity—coding. It noted that software is often rushed to market for competitive purposes and not compelled to emphasize security until it achieves market dominance—a point in time where it is usually too late.

The Institute of International Finance (IIF), an organization of 450 financial sector organizations from over seventy countries, followed with the publication of a report that laid out four scenarios of cyberattacks that could impact the stability of financial systems and networks.[16]

1. Payments systems' disruptions that may prevent or slow cross-border and domestic transactions.[17]
2. The theft or corruption of financial data that might halt purchases and sales or affect the pricing of goods, securities, and other financial instruments.
3. Disruptions from problems with utilities, telecoms, cable companies, and technology providers that spread through the interconnected nature of financial services networks.[18]
4. The incalculable damage that could be done from the loss of confidence in the financial system.[19]

This report suggested that regulatory agencies and the private sector focus more on deterring and preventing attacks and that law enforcement bodies in charge of policing cyberattacks (whoever they are) be given additional legal tools to prevent attacks. Included as examples of this were the authority to expedite the extradition of hackers and impose penalties on noncooperative countries or cyber havens to disrupt the cost-benefit ratio that encourages profit-seeking cybercriminals.[20]

In October 2017, Equifax, one of the big four credit-reporting agencies announced that more than 140 million customer records including birth dates and Social Security numbers had been compromised allegedly by four members of China's People's Liberation Army (PLA). The hack reportedly occurred because versions of the Apache Struts software had a vulnerability that could allow attackers to remotely execute code on a targeted web application. Apache had disclosed the problem on March 7, 2017, and offered a patch and instructions, but Equifax reportedly failed to immediately install the patch. Within a few weeks, the hackers were inside Equifax's systems gaining access to Equifax's web server and collecting credentials so that they had unfettered access to large databases. The DOJ indictment explained that so much data had been stolen that the hackers split it into many smaller 600-megabyte chunks so that their transmissions would avoid suspicion. They were so deep inside Equifax's network that they were able to use the company's existing encrypted communication channels, making their activity appear to be normal network activity.[21] The indictment also details how the PLA team allegedly set up thirty-four servers across twenty countries to infiltrate Equifax, making it difficult to pinpoint them as a potential problem.

Another step forward in the identification of the seriousness of the cybersecurity problem was the New York Cyber Task Force's remarkably well-done paper analyzing the defense of cyberspace.[22] Though not specifically directed at the threats to national financial security, its thirty-nine pages were as direct as any published statement to date, chastising cybersecurity professionals for "treading water" without making any real progress for nearly five decades.[23] Claiming to be "inspired by military tactics," the report argues for better cyber defenses by raising the cost to attackers at every level. The New York Cyber Task Force underscored that the task of dealing with challenges would only increase in complexity as technology evolves toward new plateaus such as an IoT that will connect almost everything seamlessly.[24] David E. Sanger, author of *The Perfect Weapon*, makes an important point in this regard in his brilliant analysis of the risks of cyberspace when he notes that the pace of technology is creating vulnerabilities far faster than it is solving them.[25]

Echoing the IMF report, the New York Cyber Task Force shined a light on the absence of any real consequences for writing poor-quality software, which then leads to the purchase of more insecure software to address the problem.[26] It summarized the frustration its drafters felt, noting that few studies or cyber strategies laid out an overall practical approach to bridge the competing priorities in cyberspace, defaulting to checklists of critical tasks with no underlying theory of how they could lead to success.[27] This is the fundamental point that has been evaded for the last quarter century. As much as any government or private sector effort had done to this point, the task force report was a call to arms—a loud and annoying alarm bell relentlessly left to ring all night in the forest where, perhaps, no one heard it.

The G20's Financial Stability Board (FSB) issued a 126-page report in October 2017 following up on its October 2016 statement on the fundamental elements of cybersecurity in the financial sector. It merely cataloged the actions that most regulators were taking, noting that FSB members are aware of the threats and the importance of the issue.[28] The SWIFT Institute, a think tank division of the Society for Worldwide Interbank Financial Telecommunication, followed with a report concerning the growing threats facing financial institutions. Its report focused on consumer fraud related to the increasing incidents of cybercrimes, as well as a rise in attacks on ATMs by criminal groups in Europe and Asia.[29] It stressed what should have been yet another alarm bell—that there had been an increase in nation-states that were participating in massive heists that took advantage of the "seams" between well-defended networks.[30] The 2017 annual report of FSOC devoted somewhat more time to cybersecurity but continued to approach it as just another challenge financial institutions face requiring Congress to adopt new laws to give the bank regulators new examination and enforcement authorities.[31]

2018–2019: BIG TECH, BIG PROBLEMS

While reports emerged from every direction, Big Tech was steaming along seemingly unabated in its efforts to gather data and control. Netflix underscored the dominance of the internet culture in 2018 by becoming the largest digital media and entertainment company in the world at a market value of $100 billion. In March, as Facebook continued to grow, its CEO, Mark Zuckerberg, made the first of several apologies for the company, this time concerning mistakes made after it was learned that data on 50 million of its users had been gathered from private Facebook accounts by Cambridge Analytica to develop psychological profiles of American voters to be sold to political campaigns for use in the 2016 presidential election.[32] In April, streaming music services overtook worldwide sales of CDs and vinyl.

On May 14, 2018, the Supreme Court struck down a federal law prohibiting sports gambling, formally opening another new internet business. In July, the DOJ charged twelve Russian intelligence officers with cyberattacks against Democratic officials during the 2016 U.S. election, and Facebook removed a Russian-linked network of sites that had interfered in the 2016 election. Zuckerberg again testified before Congress a little more than two weeks later. On August 2, Apple became the first American public company with a market value of $1 trillion, followed a month later by Amazon.

"Rise of the Machines, Artificial Intelligence and Its Growing Impact on U.S. Policy" was released in September 2018 by the U.S. House of Representatives' Committee on Oversight and Government Reform's Subcommittee on Information Technology.[33] It made no reference to the impact of artificial intelligence on financial services. But it did warn that U.S. leadership in the area was no longer assured given the pace of advancements and spending by countries like China. It also highlighted the massive amounts of data required by artificial intelligence that could "invade privacy or perpetuate bias, even when using data for good purposes." The report emphasized that artificial intelligence has "the potential to disrupt every sector of society in both anticipated and unanticipated ways" and called for the government to address these challenges.[34] The government's track record suggests that its ability to resolve yet another technology challenge is doubtful.

That same month, the Carnegie Endowment joined the voices warning that the threat landscape had evolved to encompass not only profit-motivated actors but state and nonstate actors targeting financial institutions.[35] The building theme of American subservience caused by its failure to maintain technological superiority was unmistakable. Carnegie zeroed in on the fundamental challenges that the financial sector and the U.S. economy faced, noting the asymmetry between kinetic and digital defense strategies that raises questions about the sufficiency of existing plans and capabilities for defending critical infrastructures against foreign adversaries.[36] The report noted the increasing calls for corporate and government counterattacks—hack backs—and recommended prioritizing the national collection and sharing of cybersecurity data and the formation of collaborative playbooks between and among the public and private sector. As an example, it referenced the All Hazards Crisis Response Coordination Playbook developed following the thirteen 2014–2016 Hamilton Series exercises involving representatives from several U.S. government agencies and the financial sector.[37] Those exercises led to the development of industry standards to increase the resilience of payments systems and banks if systems go down.[38]

The next month, the Brookings Institute issued a relatively short but important paper on financial stability and cyber risks.[39] It focused on how cyber risks differ from traditional financial shocks and the contagion that they could create in financial networks. It recommended that international regulations be harmonized, better models be designed to measure cyber risks, better mapping be constructed to evaluate how cyberattacks could impact the plumbing of markets and institutions, and more exercises be conducted at the domestic and cross-border levels to bridge cybersecurity communities and financial stakeholders, including C-level executives.[40] These recommendations represented an effort to shift cyber defense to being more proactive and offensively minded. FSOC's 2018 annual report referenced increased incidents of cyberattacks on banks but otherwise matter-of-factly repeated its message that a cybersecurity event could have severe negative consequences impacting the stability of the financial system.[41] Progress was seen in some areas in 2018. Barbra Streisand revealed that she had cloned her dog twice.

The Cyber Risk Institute (CRI) was established in 2018 as a subsidiary of the Bank Policy Institute (BPI) to support the cybersecurity and resiliency of global financial institutions.[42] It was led by a coalition of trade associations and more than thirty financial firms seeking to develop better cybersecurity and resiliency strategies and standards. Its cybersecurity risk assessment product, known as the Financial Services Cybersecurity Profile (Profile), is a risk assessment tool that consolidates over 2,400 regulatory questions into 277 diagnostic statements to make it easier to meet regulatory compliance requirements.

The Profile is a terrific product, but it demonstrates how the goals and resources of financial institutions can be hijacked by the need to comply with laws, rules, and regulatory standards. Complying with those requirements does not necessarily ensure high levels of internet security. It only ensures regulatory compliance so that the institutions will not be cited for violations. Those are two different and sometimes contradictory goals.

As a preelection year, 2019 began with a very long list of Democratic candidates announcing their intention to run for president in 2020 to unseat Donald Trump. In March, the political distractions in Washington exploded into a lathered frenzy as Democrats announced a wide-ranging set of investigations into President Trump even after Special Counsel Robert S. Mueller announced that he had found no evidence that President Trump colluded with Russia in the 2016 election. The White House, Congress, the courts, and the press were perpetually obsessed with Donald Trump. At least in the case of the press, we knew why. Any story or headline about him equated to online clicks and eyeballs, which in the current economy create the revenue that companies need to stay in business.

In February 2019, Christine Lagarde, the president of the European Central Bank (ECB) and former head of the IMF warned that a cyberattack could trigger a serious financial crisis.[43] In March 2019, the Carnegie Endowment released a report on cyber threats to the financial system,[44] sounding more private sector alarms over the continued evidence of cyber escalation in the financial sector and the growing number of adversarial nation-states that had significant resources to invest in disruptive cyberattack technologies. It underscored that cyberattackers were building advanced capabilities to target core banking systems, becoming more aggressive in disrupting defenses, and creating new ways to link organized criminal gang activity.[45] Carnegie argued for even greater leadership in the financial community that could work closely with law enforcement, something that would require a shift in mind-set from traditional "compliance-based approaches" that financial institutions live with every day.[46]

In March 2019, a former tech company software engineer for Amazon Web Services (AWS) allegedly hacked into Capital One's networks compromising the credit card applications of approximately 100 million individuals as far back as 2005, accessing 140,000 Social Security numbers; an undisclosed number of people's names, addresses, credit scores, credit limits, and balances; and other proprietary information.[47] The Comptroller of the Currency and the Federal Reserve Board sanctioned the bank and fined it $80 million for ineffective technology, cloud, and network security controls.[48]

In May 2019, the Trump administration drew battle lines with China over the hardware and software it was supplying worldwide given the growing threat that those products could be used to spy on and control networks. Executive Order 13873[49] limited the movement of commerce through any information or communications technology or service that posed an undue risk of sabotage or an undue risk to national security. The continuing conflict between the U.S. government and Chinese telecommunications company Huawei ended up in court that month as the company sued over the resulting federal ban on its products. The White House would further prohibit transactions with TikTok a year later through Executive Order 13942.[50] On August 11, 2020, through Executive Order 13943,[51] it prohibited transactions with WeChat. Two weeks before leaving office in 2021, the president issued Executive Order 13971[52] limiting the use of Chinese software by restricting the products and services of Alipay, CamScanner, QQ Wallet, SHAREit, Tencent QQ, VMate, WeChat Pay, and WPS Office.

Congress would act later in 2019 and President Trump would sign into law in 2020 the Secure and Trusted Communications Networks Act of 2019.[53] This required the Federal Communications Commission (FCC) to publish on its website any communications equipment or service that posed an unacceptable risk to the national security of the United States. The act also prohibited the use of any federal subsidies or grants with regard to communications equipment or services from these companies. The FCC list of suspect companies currently includes five Chinese companies: Huawei Technologies Company, ZTE Corporation, Hytera Communications

Corporation, Hangzhou Hikvision Digital Technology Company, and Dahua Technology Company.[54]

In addition to the FCC, the Department of Commerce and the Office of Foreign Assets Control (OFAC) were required under various laws to create lists of "covered companies" that would be subject to varying requirements, limitations, and prohibitions. These were powerful new tools in the hands of the U.S. government that could arguably help protect financial services companies when it came to the integrity of equipment and services ranging from branch cameras to IT networks and computer systems. How these tools will be used, whether they will reach the full spectrum of products that could be corrupted by geopolitical espionage, and what the international consequences are remain unanswered questions. The FCC is evaluating the extent to which companies use Chinese telecommunications equipment, but the scope of the challenge keeps getting larger.

As an example, drones are increasingly being used by individuals, companies, political operatives, militaries, and law enforcement to bolster data gathering, surveillance, espionage, and control efforts. The largest manufacturer of commercial drones, however, is a Chinese company, DJI, which in December 2019 was one of eight companies blacklisted by the Treasury for actively supporting the "surveillance and tracking" of Uyghurs and other ethnic minorities in China.[55] The Biden administration, however, reversed fields on many Trump administration actions with regard to China as only politics can do, issuing Executive Order 14034 on June 9, 2021,[56] rescinding President Trump's executive orders on TikTok (13942); WeChat (13943); and Alipay, CamScanner, QQ Wallet, SHAREit, Tencent QQ, VMate, WeChat Pay, and WPS Office (13971). This effectively started the process all over again, requiring Biden cabinet members to reassess the situation and submit a report recommending executive and legislative actions. Two steps forward and three steps back yet again.

In May 2019, Apple introduced its new TV streaming platform, Apple TV+; a news service, Apple News+; and an Apple credit card. On April 25, Microsoft became the third U.S. public company to reach a $1 trillion market value. Days later, hackers seized control of the computer system of the City of Baltimore, demanding a Bitcoin payment in return for unlocking it. A week later, San Francisco voted to ban the city's use of facial recognition. President Trump declared a national emergency due to cyber threats, broadly prohibiting U.S. companies from using foreign technology without a license, a prohibition that would be nearly impossible to monitor and enforce.

About this time, the debate over the advantages and disadvantages of decentralization of the delivery of financial services began to move to the fore. Would the financial system be safer with more fragmented and diverse peer-to-peer validations of transactions through technologies such as block-chain? Or would the disintermediation of trusted financial intermediaries (i.e., banks) in payments, clearing and settlement systems, and networks cause authorities to lose control and feel less able or inclined to support them when they got in trouble? A June 2019 report by the G20's Financial Stability Board appeared to side with fintech advocates, concluding that decentralized financial technologies could lessen some of the financial stability risks associated with traditional financial institutions and intermediaries and be more resilient to cyber risk than highly centralized systems.[57] In effect, decentralized financial systems were offered as an antidote for malicious technologies, which would only be able to disrupt a slice of the system at a time. But that is only half the analysis.

Cybersecurity issues were beginning to get high-profile traction as it became apparent that new products and delivery systems could potentially rearrange the status quo that financial institutions had become accustomed to. The head of Japan's central bank had predicted that cybersecurity could become the financial system's most serious risk, and Jamie Dimon, CEO of JPMorgan Chase, said that cyberattacks "may very well be the biggest threat to the U.S. financial system."[58] In July 2019, the Carnegie Endowment weighed in again with a 117-page report reiterating that global financial systems needed to be vigilant in the face of increased cyber threats, particularly since smaller, more vulnerable financial companies may be more exposed because they do not have the resources to adequately defend themselves.[59] That month, Facebook agreed to pay the Federal Trade Commission a $5 billion fine—the largest ever to date for violations of consumer privacy.

In September, the Carnegie Endowment presented a new conceptual framework for assessing systemic cyber risk, preparedness, and the buffers available to absorb cyber risk–induced shocks,[60] underscoring the interconnected nature of cyber threats in the financial sector which required an integrated approach to defense. Importantly, it argued that cyber risks are in fact threats to the stability of the world's financial infrastructure.[61] It referenced the NotPetya attack in Ukraine in late June 2017, a self-replicating virus using nation-state-grade technology that disrupted networks for several weeks and cost between $2 and $10 billion.[62]

On September 16, personal data about every Ecuadorean citizen was found in an unsecured cloud server. In October 2019, Carnegie released a valuable analysis reemphasizing the need for an integrated approach for financial market cyber resilience consistent with those released by the

Committee on Payments and Market Infrastructures–International Organization of Securities Commissions (CPMI-IOSCO) in June 2016.[63] It focused on the importance of teamwork and intelligence sharing through the implementation of a Threat Intelligence–Based Ethical Red Teaming (TIBER) framework.

On December 18, the House of Representatives voted to impeach President Trump for abuse of power (230–197) and obstruction of Congress (229–198). The year closed with the issuance of FSOC's 2019 annual report. It read much as if the pages from prior years had been cut and pasted into the report.

It was hard to escape the feeling that so many of the studies, reports, and publications that were being released reflected an immense frustration with an internet in which security threats seemed irresolvable. Many seemed like they were reporting from afar on a burning fire, recommending that someone call the fire department. Few concrete solutions to the growing problems were ever proposed in any of these reports or studies that could stop or prevent attacks. Most essentially recommended that enough water be kept on hand to douse the flames and that sufficient lumber be available to rebuild the destroyed structures.

2020: CLOUDS, QUANTUMS, AND CHINA

As 2020 dawned, an Iranian general was killed by a drone outside the Baghdad airport. Iran pulled out of the 2015 nuclear deal limiting its uranium enrichment in response to the United States having done the same. On February 5, the Senate voted to acquit President Trump, ending that impeachment saga. In February, the possibility of COVID-19 becoming a global pandemic became real, and on February 27, the Dow Jones Index lost 1190.95 points, its largest one-day drop in history as fear of the pandemic shutting down the world became real. To avert an economic disaster, Congress, the Treasury, and the Federal Reserve acted over the coming months to cut interest rates, make loans to increase economic liquidity, and roll out an arsenal of other economic stimulus and pandemic relief. Much of the year would be dominated by the spread of and fight against COVID-19 as cases and deaths increased to pandemic levels. The World Bank predicted that the COVID-19 pandemic would shrink the global economy by 5.2 percent in 2020. In November, after a disastrous year of COVID-19 and its financial aftermath, Joe Biden was elected president of the United States.

Financial cybersecurity concerns continued to grow. The staff of the New York Federal Reserve released an important report in January analyzing the impact that cyber risks could have on the U.S. financial system, particularly given the increasing use of artificial intelligence and Big Data.[64] It reinforced the message that payment and settlement systems are a natural high-value target for a malicious attacker, noting that the impact of a cyberattack on the largest five participants in the network could be "very large." The Bank for International Settlements documented the growing potential for losses from cybersecurity,[65] concluding that cyber value at risk may account for up to a third of total operational value at risk.[66] That was a startling number for a threat that seems largely uncontrollable. A companion piece released in May by BIS analyzed a unique database of more than 100,000 cyber events and determined that the financial sector was exposed to a larger number of cyberattacks, but handled them at lower average cost because of a greater relative investment in information technology security.[67] It noted that the increasing use of cloud services reduced cyber costs, especially for relatively small cyber incidents, but concluded that "as cloud providers become systemically important, cloud dependence is likely to increase tail risks."[68] It raised the likelihood that crypto-related activities, which are largely unregulated, would be particularly vulnerable to cyberattacks.

CLOUDY DAYS AHEAD

Enter the "cloud" as another new source of risk, adhering to the rule that technology creates new vulnerabilities faster than it solves old ones. The cloud is a combination of hard drives, computer servers, signal routers, and fiber-optic cables that store data on redundant, geographically dispersed hardware owned by private companies that are made accessible to users via the internet. There is no ethereal cloud. It all comes down to hardware and software physically located somewhere in the analog world.

Cloud computing became popular in the last twenty years. In March 2006, Amazon launched its first cloud-based service through Amazon Web Services, renting out server space and database tools that allowed users to launch and maintain applications more cheaply.

Cloud services now generally impact software, platforms, and infrastructure. Cloud-based software are applications such as Dropbox that make it possible for users to operate at higher speeds and efficiencies without having to download anything. Platforms provide digital environments that software can run on to work more efficiently. Cloud infrastructure provides server space that customers can manage remotely. There are internet-connected public clouds run by Apple, Amazon, Microsoft, and Google; consumer clouds that hold pictures and social media posts or store music and email; and company private clouds for employee and customer access.[69]

A wide array of operational and economic benefits stem from the availability of cloud computing resources, from improved functionality and reliability to lower costs and increased efficiencies. Concentrating digital communications and commerce onto a few super servers is more efficient. Cloud users are less likely to overload a computer, can enjoy greater capacity to store data, and may use more computational power than their own computer can produce.[70] Clouds often use software that shares data among different machines so that it may not have to move among many pieces of hardware and not have a permanent home on a specific chip in the cloud. Cloud technology also allows financial institutions to segregate sensitive data and benefit from economies of scale and the ability to automatically scale up and provide levels of security implementation and operational resiliency that many smaller institutions could not afford on their own. The cloud makes services more affordable and globally accessible, helps businesses update products for their customers, and enables remote work across industries. But it also opens us up to new risks and creates a new shared responsibility model that requires all involved to engage in data protection and integrity.[71]

The technical risks created by the cloud are similar to the risks associated with traditional technology infrastructure providers. A machine—perhaps one that can be trusted today but not tomorrow—controls the data. But as Vic Winkler lays out in his book on cloud security, the cloud also creates new challenges for users, such as capacity planning failures, incomplete data deletion, insecure connections, exposure to unauthorized parties, network availability, multi-tenancy or hypervisor vulnerabilities related to risks that may arise from multiple users sharing the same physical infrastructure, disaster recovery, the security of backup clouds to the cloud, transparency, disclosure of incidents, control of data, and legal and regulatory compliance of cloud providers.[72] Users provide endless streams of data to companies trying to maximize profits subject to security protocols that they may not understand or control. This increases the ease of behavior tracking, further entrusting our privacy to tech giants.[73] The FFIEC released a statement on cloud computing security in August 2020[74] noting the new operational and technical issues for banks, including the risk of "lock-in" as financial institutions become too dependent on a particular service provider, the concentration of cloud providers and data, and the increased impact of their failure or disruption on financial stability.[75]

Most problematic perhaps is the fact that cloud services aggregate data in one place, acting as a Fort Knox of data that attracts hackers. If clouds ever fail or succumb to cyberattacks, the damage can be enormous, measured only by the maliciousness and creativity of the hacker and the redundancy and resilience of the defenses in place. Banks have learned that their cloud providers can be breached and personally identifiable customer data exposed, if not technically, through insider actions. At Capital One, the attacker allegedly used several programs to obfuscate her identity and discover a misconfigured firewall in the company's cloud services. Clouds may blur a hacker's visibility, but they and all their data and backups are not impenetrable.

THE CYBERSPACE SOLARIUM COMMISSION

In March 2020, a seminal report issued by the prestigious Cyberspace Solarium Commission confirmed the relatively unprotected state of the economic infrastructure of the United States, essentially concluding that to a large degree, the country had failed in its overall efforts to defend against cyberattacks and had defaulted to a "norm of inaction." After almost twenty-five years of analyzing, writing reports, and ringing alarm bells, the commission underscored the absence of and need for a plan "to ensure that we can reconstitute in the aftermath of a national-level cyberattack" and be able "to ensure that our economy continues to run."[76] The commission developed a familiar strategy built around layered cyber deterrence, economic resilience, government reform, and infrastructure fortification.[77] It recommended a strategic approach that imposed costs on problem actors. The report called out China, Russia, Iran, and North Korea and argued for stronger offensive and defensive enforcement efforts.[78]

Among the Cyberspace Solarium Commission's dozens of important recommendations were repetitive calls for more legislation, bureaucracies, information sharing, and congressional oversight. Among other things, the commission smartly recommended (1) the establishment of a National Cybersecurity Certification and Labeling Authority (NCCLA) to manage a program on security certifications and labeling of technology products; (2) standards for final goods assemblers of software, hardware, and firmware liability; and (3) the development of a cloud security certification program.[79] It also emphasized that advances in artificial intelligence and machine learning are creating both opportunities and challenges in cyberspace, which are likely to create a new technological arms race that adversaries like China and extremist organizations would use to attack

U.S. interests. The commission warned, "The economic and military advantages of the future will not necessarily go to the most powerful among nations, but to the actors that field the best algorithms or technologies."[80]

Finally, the commission recognized that quantum technologies present significant risks to U.S. national security systems, a reason why nations across the globe, including China, are investing significant resources to develop a viable quantum computer and "quantum-resistant" encryption.[81] The commission suggested codification of the concept of a "systemically important critical infrastructure" such as banking and finance so that they have access to special assistance from the government.[82] Congress responded within the year, drawing more than two dozen provisions in the 2021 National Defense Authorization Act directly from the report's recommendations, an action supplemented by two executive orders issued by President Biden in May 2022.[83]

The clear message that ran throughout the Cyberspace Solarium report was the need for more and better resources, data, and metrics to analyze the problems. It focused on deterrence as a means of shaping behavior, as well as denying benefits to and imposing costs on violators.[84] One disappointing aspect, however, was the relative absence among the commissioners, staff, researchers, and drafters of experts with deep experience in understanding financial institutions and global payments systems. That is probably why the words "financial services infrastructure" do not appear in the report, and there are less than a half-dozen references to financial services and the systems that they operate within. That is a serious defect that has plagued many reports issued to that point.

THE CONCERN GROWS

In April 2020, the Carnegie Endowment released another paper which presented a qualitative analysis focused on systemic risks created by two different networks: the financial network and the cyber network.[85] Using "cyber mapping," it analyzed risks that could potentially threaten financial stability, noting that the identity of systemically important actors and components of the financial network have not yet been comprehensively determined. The report wisely recommended the creation of a more technologically enabled regulatory system that can accurately measure systemically important cyber shocks and integrate macro-prudential analysis with the related financial stability risks.[86]

The Financial Stability Board cautioned that same month that "cyber incidents pose a threat to the stability of the global financial system," noting that "a major cyber incident, if not properly contained, could seriously disrupt financial systems, including critical financial infrastructure, leading to broader financial stability implications."[87] On April 15, the Federal Bureau of Investigation and the Departments of State, the Treasury, and Homeland Security issued an advisory highlighting the cyber threat posed by North Korea, noting that its malicious cyber activities have threatened the United States and the broader international community and "pose a significant threat to the integrity and stability of the international financial system."[88] Those recent events included the following:

- the attack on Sony Pictures Entertainment (SPE) in retaliation for the 2014 film *The Interview*;[89]
- a hack that resulted in $81 million being transferred from the Bangladesh Bank's account at the Federal Reserve Bank of New York through unauthorized transactions on the SWIFT network;[90]
- the WannaCry 2.0 ransomware attack, which infected hundreds of thousands of computers in hospitals, schools, businesses, and homes in over 150 countries;[91]
- the DPRK state-sponsored fraudulent ATM cash withdrawal scheme known as FASTCash, stealing tens of millions of dollars from ATMs;[92] and
- the April 2018 DPRK state-sponsored hack of digital currency exchanges making off with nearly $250 million worth of digital currency that were laundered through hundreds of automated digital currency transactions.[93]

The Bank for International Settlements followed in May with a report underscoring the increasing vulnerability of computer systems, networks, and cloud services to cyberattacks, concluding that the average cost of cyber events had significantly increased over the last decade.[94] The report warned that as cloud connectivity increased and cloud providers became systemically important, cloud dependence would increase cyber risks. Not surprisingly, it found a positive correlation between the price of Bitcoin and the intensity of crypto-related cyber events, making stronger regulation of the activities of intermediaries in crypto asset markets necessary.

The most intriguing development of 2020 was the emerging belief that the internet was broken and may not be fixable. During a July 23, 2020, press conference, the U.S. Department of Energy (DOE) announced that it was working on a prototype of an unhackable quantum internet expected to be completed within the next ten years under the National Quantum Initiative Act that had been enacted in December 2018. The new internet would run in parallel to the current one as a supplementary network for the banking and health industries, as well as serving the national security interest.[95] Scientists from the DOE Argonne National Laboratory in Lemont, Illinois, along with others from the University of Chicago, created a fifty-two-mile "quantum loop" in the Chicago suburbs, hoping to add the DOE Fermilab in Batavia, Illinois, to establish an eighty-mile test bed. All seventeen DOE national laboratories will eventually be connected to form the quantum internet backbone to bring the country "one step closer to a completely secure internet."[96]

In the midst of these cyber gyrations and virtual word salads, the GAO entered the fray again in September 2020 to criticize the country's critical infrastructure protection.[97] The report reads like an auditor's punch list based on the GAO's prior reports of 2011, 2015, and 2018. It expressed frustration that the financial services industry faced increasing cybersecurity-related risks created by (1) financial data insecurities attributable to service providers and supply chain partners; (2) the increasing sophistication of malware; and (3) the interconnectivity of networks, the cloud, and mobile applications.[98] Foremost on its mind was the fact that in February 2016, hackers were able to install malware on the Bangladesh Central Bank's system through a service provider, which then directed the Federal Reserve Bank of New York to transfer money to accounts in other Asian countries, resulting in the theft of approximately $81 million. It also criticized Treasury for not tracking or prioritizing the government's efforts to implement risk mitigation efforts laid out in 2019 and concluded that without more prioritized efforts and resources, the financial sector would be "insufficiently prepared to deal with cyber-related risks."[99] At the same time, the European Commission issued a proposal on digital operational resilience for the financial sector that announced a goal of making the most out of the current digital world by creating uniform, consistent standards of resiliency as well as developing a regulatory scheme for crypto assets and distributed ledger technology (DLT) infrastructures.[100]

The critical importance of the U.S. financial infrastructure and its exposure to cyberattacks was again underscored in November 2020 when the Carnegie Endowment issued a well researched and documented report by Carnegie's FinCyber Project. Its opening pages laid out a developing but now common theme:

> Malicious actors are taking advantage of this digital transformation and pose a growing threat to the global financial system, financial stability, and confidence in the integrity of the financial system. Malign actors are using cyber capabilities to steal from, disrupt, or otherwise threaten financial

institutions, investors, and the public. These actors include not only increasingly daring criminals, but also states and state-sponsored attackers.[101]

The report seems to default to an overreliance on governments doing and fixing things, a reliance that should by this point be seen as illusory at best without the buy-in and leadership of the private sector on a global basis. We must solve the dilemma created by private financial companies being unwilling to invest in universal cybersecurity and reticent to share cyber-threat incident data with competitors and regulators who may turn and use it against them. Carnegie released another report in November 2020 highlighting the growing concerns with cloud computing and the dawning of quantum computing.[102] It concluded that defending networks would continue to require significant resources and attention to detail.[103]

The G7 released another report in November 2020, reminding public and private financial sector entities that they should regularly exercise their cyber incident response and recovery plans to rehearse and prepare to effectively respond to and recover from cyber incidents.[104] Given the history of analysis that preceded it, the report's analysis was rudimentary in proposing elements of preparedness. In December, Carnegie released a similar paper containing a more sophisticated toolbox of processes and procedures that financial institutions should use and assistance they could seek.[105] A sense of inevitability oozes throughout these reports, though, as they are largely directed at what happens when the walls are breached and the castle begins to fall.

The Carnegie Endowment's FinCyber Project has spent significant resources to document cybersecurity breaches of financial institutions' systems since 2007. It maintains a valuable catalog of incidents in its "Timeline of Cyber Incidents Involving Financial Institutions."[106] According to that catalog of hacks, approximately fifty attacks against financial institutions occurred around the world in 2020 and 2021. They included malware attacks on Brazilian bank account holders; DDoS attacks on Hungarian banking and telecommunication services through servers in Russia, China, and Vietnam; a ransomware attack that caused the suspension of nationwide operations by Banco Estado, one of the three largest banks in Chile; the shutdown of 143 cash machines of a Belgian savings bank; the infection of Android devices by malware hidden in a cryptocurrency converter app; and a ransomware attack on CNA Insurance. Reading through the report provides a dark picture of the world we are speeding toward, with little or no rules, standards, or bumper guards.

The SolarWinds hack occurred in December 2020. Described as a "worst nightmare cyberattack,"[107] it should have riveted public and private sector attention and triggered immediate action to defend the country from an increasing series of cyberattacks. It garnered attention for a week but did not ignite a firestorm of action to better fortify cyberspace. The event was catastrophic. Nation-state hackers believed to be directed by the Russian intelligence service, the SVR, gained unprecedented access to the networks, systems, and data of thousands of SolarWinds customers, which included more than 30,000 local, state, and federal agencies (such as Homeland Security, State, Commerce, and Treasury) and private corporations (such as FireEye, Microsoft, Intel, Cisco, and Deloitte), when SolarWinds inadvertently delivered the backdoor malware as an update to Orion software. This "supply chain" attack inserted malicious code, creating a backdoor for hackers to access and impersonate users and accounts of the targeted organizations.[108] The hackers modified code and mimicked Orion software communication protocols to hide in plain sight. If that were not effective enough, they then erased their tracks from the crime scenes so that investigators couldn't prove the identities of the perpetrators.[109] Insured losses from the attack were estimated to be $90 billion.[110] Reputational harm, downtime, legal expenses, litigation, and regulatory fines have added significantly to those numbers.

What is even more troubling is that it was eventually learned that the attackers had first gained access to the SolarWinds systems in September 2019, giving them fourteen or more months of unfettered access to surveil and infect the system that had been penetrated with time and other logic bombs and malicious code that could lay dormant and take years to find. How it could have taken so long to detect the SolarWinds attack has a lot to do with the sophistication of the underlying code and the hackers who executed the attack. As we will see later when the United States and the UK linked the Ukrainian cyberattacks in February 2022 to Russia within forty-eight hours, much of the failure in creating better cybersecurity is attributable to the nature of the internet and a failure of sufficient will to address its inadequacies.

The SolarWinds attack should have triggered a cavalcade of government and private sector actions to defend against further cyber intrusions, but the attention it garnered was fleeting as usual. That same month, FSOC's annual report contained fewer words devoted to cybersecurity than prior years, culminating in another matter-of-fact set of recommendations that threw the hot potato to Congress, the FBIIC, the FS-ISAC, and individual banking agencies.[111] The more global financial systems become reliant on technology, the more unclear it seems to be who is responsible for protecting them from cyberattacks.[112]

14

WAKE UP AND SMELL THE QUBITS

2021 and Beyond

From a technology perspective, 2021 provided a clear picture of what lies ahead. Cybersecurity attacks increased in frequency and severity, and glimpses of advancements in crypto technologies, quantum computing, biometrics, facial recognition, 5G, the IoT, and artificial intelligence were about to launch yet another generation of financial products, delivery mechanisms, and vulnerabilities. And with little to no paper left, the damage that could be done by illegal penetrations of digital data sets would be more expansive and damaging than ever imagined.

The prospects of quantum computing, the next level of computer technology that transcends traditional computer language and operating systems, promises to exponentially enlarge digital capabilities by massively increasing the speed at which data can be processed. In doing so, it could also magnify cybersecurity vulnerabilities. Google posted a premature statement that it had achieved "quantum supremacy" when its Sycamore solved a mathematical calculation in 200 seconds that was estimated to require 10,000 years for a current supercomputer to solve.[1] Much more will follow quicker than expected as qubit capacity is enlarged.

Unlike conventional warfare which revolves around the control of physical geography and usually advances using proportional responses over extended periods of time, cyber war may necessarily be structured in an all-or-nothing manner. The stakes of cyber war may demand the deployment of all weapons immediately in a kill-or-be-killed moment. Once a cyberattack is launched by a country, *if* the target country can identify the attacker, it may not immediately be able to determine the extent of the attack and may have to assume that all systems are at risk, thus responding in kind.[2] The stakes, weapons, and strategies are entirely different from conventional kinetic warfare. It is this possibility of instantaneous mutual assured destruction that in most cases creates the discipline and rules of engagement that prevent it from occurring. But that assumes that every protagonist is equally invested in avoiding such a destructive scenario, which of course would be a rash assumption these days. The barriers to entry in the world of traditional military power have been extraordinarily high and therefore limited to participation by industrialized countries. As terrorism has altered those dynamics, the chances of fanatics obtaining a malicious piece of software or vastly improved computing power dramatically changes the risk-reward scenario in a financial cyber war.

2021: A NEW SHERIFF IN TOWN

As President Biden was sworn in, the next technological frontier challenge was rapidly unfolding. It should have been painfully clear that the time for more talk and study about cyberspace insecurity had long since passed. The Biden administration claimed to have gotten that message, but based on the last twenty-five years, it will have to prove that it can do more than provide additional cosmetic fixes meant to create pablum for political speeches and excuses when attacks occur.

We began 2021 with concern over the continuing spread of the CO-VID-19 pandemic, sprinkled with hope that the proximity of a vaccine marked its end. America had recorded more than 3,000 deaths a day for the first time, reaching a total of 375,000. On March 25, President Biden announced a goal of 200 million vaccinations in his administration's first 100 days. His appointments suggested that his administration understood the urgency that cyber insecurity was creating. But a blizzard of other distractions exploded onto the scene.

On January 6, 2021, a riot at the U.S. Capitol in Washington breached the building, causing lawmakers and Vice President Pence to shelter in place or be evacuated in a breathtakingly sad day in American history that saw five people lose their lives. Two days later, Twitter banned President Trump permanently "due to the risk of further incitement of violence," only to be followed by a similar action by Facebook. If nothing else, these actions heightened the complaints that large tech companies were taking political sides in an increasingly acrimonious political landscape.

On January 11, House Democrats introduced new articles of impeachment against President Trump for "incitement of insurrection" and encouraging his supporters to riot at the Capitol. He was impeached by the House of Representatives once again two days later on January 13, a week before his term was to end and Joe Biden would be inaugurated as the forty-sixth president of the United States. It was the first time in history a U.S. president had been impeached twice. For his part, President Biden immediately focused on erasing the Trump legacy, signing fifteen executive orders on his first day in office, including orders to rejoin the World Health Organization (WHO) and the Paris Climate Agreement, revoke the Keystone XL Pipeline, and mandate masks on federal properties. On February 11, he rescinded President Trump's national emergency order to fund the border wall with Mexico. Two days later, the former president was again acquitted by the Senate. The reemergence of COVID-19 in the form of an Omicron variant later in the year, the impending invasion of Ukraine by Russia, and an obsession with enacting massive domestic spending and other feel-good legislation would sap enormous government resources.

The financial services business in the United States was at a crossroads created by both the innovation and threats brought about by new technologies, coupled with the failure of a long line of presidents and congresses to modernize its regulation. Deposit taking, the creation of money, lending, investments, advisory services, and trading had all been virtually copied and reintroduced in cyberspace in ways that defied traditional classification and regulation, creating a plethora of new unregulated competitors for consumer and institutional dollars. Financial institutions began to appreciate that they were tech companies, or at least had to compete with them. Central banks around the globe were scratching their metaphorical heads trying to determine how their roles were changing. The Federal Reserve Board launched a program to consider the adoption of a central bank digital currency (CBDC) that could expedite the transmission of everyone's personal and retail payments.[3] Other countries like Malaysia and China raced out ahead of the pack to deploy CBDCs.[4] By this time, North Korea had used cyberattacks to steal some $2 billion from at least thirty-eight countries across five continents in just five years,[5] more than three times the amount of money it was able to generate through analog counterfeiting activities over the previous four decades. While communist in name, they were capitalists when it came to crime. CBDC must have them salivating.

Protecting the global financial system had become a pressing organizational challenge, but there were too many silos of responsibility, including governments, financial authorities, industry participants, and consumers, to make it work efficiently or effectively. The overabundance of state and federal financial regulators are ill equipped to deal with widespread technological threats—that is not what they had been created to do—and had to nurture relationships with a group of unfamiliar national security agencies to be prepared to deal with the overwhelming threats coming their way.[6] Greater clarity about everyone's respective roles and responsibilities and vastly greater international collaboration became a common theme among the experts.[7] It was becoming clearer that the ability of the government and the private sector to safeguard critical infrastructures such as finance and banking could define their capacity to defend democracy itself.

The Cyberspace Solarium Commission issued a report to assist in the presidential transition in January 2021.[8] Once again, it was full of recommendations that someone (unclear who) would have to find the time and energy to implement. Its principal recommendations included establishing an Office of the National Cyber Director, developing a national cyber strategy to include a framework for deterring and disrupting cyberattacks against the United States, and improving federal cybersecurity efforts such as strengthening partnerships with the private sector. Those recommendations essentially reiterated the commission's recommendations from its March 2020 report, but with yet a new sense of urgency and perhaps hope.

In March 2021, concerned about the SolarWinds and Hafnium hacks by the Russians and Chinese, the Biden administration began "assessing" and "rethinking" options for overhauling the nation's cyber defenses.[9] As the White House briefed reporters on their current thinking, it suggested that a "new structure" that combined traditional intelligence collection with the talents of private sector firms be put in place, noting that it was FireEye that ultimately identified the SolarWinds breach.[10] That was hardly a novel idea at this point. At the same time, Microsoft warned that "cyber-criminals are using the back doors Chinese hackers left behind to deploy ransomware" and that Microsoft customers had "limited time measured in hours, not days, to patch their systems."[11]

The White House promised yet another presidential order on longer-range fixes. A *New York Times* article reiterated well-worn themes, including agency boundaries that prevent the NSA and CIA from operating within the country's borders, the inability after all these years to stop or even detect hacks, the unique roles that the government and the private sector must play, and the complex privacy and civil rights challenges raised when the government is given broader authority to snoop. It also reported that Microsoft was investigating whether one of its partners in the Microsoft Active Protections Program, which includes ten Chinese firms, may have leaked information to Chinese hackers or was itself hacked.[12]

The National Security Commission on Artificial Intelligence issued a 756-page report in March 2021 delivering a sobering message. It described in painstaking detail how artificial intelligence would be the most powerful tool to benefit humanity, but would also be used in the pursuit of power to deploy potent new weapons and become the tool of choice to perpetrate crime. It concluded that "America is not prepared to defend or compete in the AI era."[13]

> For the first time since World War II, America's technological predominance—the backbone of its economic and military power—is under threat. China possesses the might, talent, and ambition to surpass the United States as the world's leader in AI in the next decade if current trends do not change. Simultaneously, AI is deepening the threat posed by cyber attacks and disinformation campaigns that Russia, China, and others are using to infiltrate our society, steal our data, and interfere in our democracy. The limited uses of AI-enabled attacks to date represent the tip of the iceberg.[14]

In setting forth a blueprint to deal with the challenges and opportunities created by AI, this massive report contained little mention of the impact and use of AI with regard to financial institutions and payments systems, likely given the relative absence of deep expertise in those areas among its commissioners, staff, and contributors.

That same month, the National Counterintelligence Strategy highlighted the expanding and evolving nature of threats to U.S. critical infrastructures from foreign state and nonstate actors, hoping to "raise awareness of the human threat to critical infrastructure, provide information on how to incorporate this threat vector into organizational risk management, and offer best practices on how to mitigate insider threats."[15] Focusing on ever-present insider threats within organizations, the report concluded that there is an unprecedented imperative for public-private collaboration to maintain insider threat programs that identify suspicious behavior in real time and provide the resources to respond. Twenty-five years into the cybersecurity threat era, the report calmly concludes that government and private sector organizations will become "more successful in combating insider threats when they develop greater awareness of the consequences of these threats nationwide."[16] Federal Reserve Board chairman Jerome Powell said one month later that he had become more concerned about a cyber incident crippling financial institutions and the functioning of the payments systems than a collapse like the global financial crisis of 2008.[17] Well, that should have gotten our attention.

CYBERATTACKS GET NOTICED

CNA Financial suffered a ransomware attack in March 2021 impacting employee and customer services for three days. In addition to losses attributable to downtime, reputational damage, and all the other remedial actions CNA had to take, it reportedly paid $40 million in ransom to unlock its data.[18] Then in May 2021, two more massive shots were fired across the bow of the U.S. cyber defense system when Colonial Pipeline announced that a ransomware cyberattack had forced it to close operations and temporarily halt pipeline operations. Hackers had entered its networks on April 29 using a leaked password later found on the Dark Web to enter a virtual private network (VPN) account that didn't use multifactor authentication. On May 7, an employee had found a ransom note on his computer just before 5 a.m. demanding cryptocurrency. The entire pipeline was shut down by 6:10 a.m.[19] Colonial paid a ransom of $4.4 million in Bitcoin as supply lines dried up and gas lines formed due to East Coast stockpiles of gasoline dropping by about 4.6 million barrels.[20]

This was followed by a cyberattack on the largest beef supplier in the world by REvil, a Russian-speaking hacker gang. JBS S.A., a Brazilian-based meat-processing company, supplies approximately one-fifth of meat globally. JBS's American facilities in Utah, Texas, Wisconsin, Pennsylvania, and Nebraska were among those disrupted or rendered temporarily inoperative. This attack underscored the vulnerability of the world's food supplies. JBS paid the hackers an $11 million ransom in Bitcoin.[21]

On May 12, President Biden issued an executive order that resembled those issued by Presidents Clinton, Bush, Obama, and Trump. It began with penetrating insight into the obvious—the country faces persistent and increasingly sophisticated malicious cyber campaigns that threaten the public sector, the private sector, and ultimately the American people's security and privacy, and the federal government must improve its efforts to detect and respond against these actions and actors.[22] The order sought, among other things, to remove restrictions that might limit the sharing of threat or incident information among executive departments and agencies. While one would have thought that these things had been done years ago, it also sought to "modernize" the government's approach to cybersecurity through the adoption of best practices, zero-trust architectures, secure cloud services, and the centralization of cybersecurity data. It established yet another new bureaucracy, the Cyber Safety Review Board, to review and assess significant cyber incidents.

The order should be commended for focusing on software supply chain issues, including the need for the creation of a seal of approval for algorithmically based products that meet certain security standards. But it was silent on the equally complex security issues created by computer hardware. That would be indirectly confronted in an executive order issued on June 3, 2021, dealing with the threats posed by the military-industrial complex of the People's Republic of China (PRC), including Huawei Technologies.

CHINESE CHECKERS

Hardware used by every financial institution from the Federal Reserve Board to JPMorgan Chase, whether it is the surveillance camera on the wall, the timer on the vault, or components in the server that stores sensitive data, are just as susceptible as software that is manufactured and manipulated to spy and transmit information to places where it should not be. So it became fair game to ask what the Chinese were putting in the technology products the rest of the world was buying. Huawei became the poster company representing the threat of embedded chips given its huge role in the development and build-out of new technologies such as 5G.[23] Underscoring the remarkable ability of politics to send mixed messages, President Biden rescinded the orders that President Trump had imposed on a variety of Chinese companies in Executive Order 14034 on June 9, 2021,[24] relating to TikTok, WeChat, Alipay, CamScanner, QQ Wallet, SHAREit, Tencent QQ, VMate, WeChat Pay, and WPS Office. These actions effectively rebooted the question of what to do with Chinese technology companies.

By June 15, the U.S. COVID-19 death toll had topped 600,000 as 65 percent of adults in the United States had been vaccinated with at least one dose. On July 8, President Biden announced that U.S. troops would withdraw from Afghanistan by August 31 despite increasing Taliban gains throughout the country. That withdrawal did occur, leading to horrible pictures and narratives that revived memories of the retreat America had witnessed as troops and civilians were airlifted from Saigon at the end of the Vietnam War in 1975. The killing of thirteen American troops and thousands of civilians during the withdrawal only punctuated the disaster as America ended its longest war of twenty years in Afghanistan.

The European Banking Authority issued its "Analysis of RegTech in the EU Financial Sector" in June.[25] The report reiterated the increasing risks being created by the growing use of technology by financial institutions and concluded that regulators must deepen their knowledge and address any skill gaps while supporting the convergence and harmonization of supervisory practices. It was essentially a process-oriented manual, with only modest analysis and discussion of the enormous new threats to the stability of financial infrastructures.

Switching gears on China, the Biden NSA, CISA, and the FBI issued an advisory on July 19, 2021, titled "Chinese State-Sponsored Cyber Operations: Observed TTPs."[26] The thirty-one-page advisory laid out the Chinese dilemma by confirming that "Chinese state-sponsored cyber actors aggressively target U.S. and allied political, economic, military, educational, and critical infrastructure (CI) personnel and organizations to steal sensitive data, critical and emerging key technologies, intellectual property, and personally identifiable information (PII)," which "support China's long-term economic and military development objectives." Pulling no punches, it accused China of masking its activities by using a revolving series of virtual private servers (VPSs) and common open-source or commercial penetration tools, consistently scanning target networks for critical and high vulnerabilities. It identified forty tactics and techniques deployed by Chinese-sponsored entities with correlating detection, mitigation, and defensive recommendations.[27]

A LOT MORE OF THE SAME

President Biden followed his May 12 executive order with a national security memorandum on improving cybersecurity for critical infrastructures,[28] repeating the growing concern over cybersecurity threats posed to the systems that control and operate the country's critical infrastructures. It established (finally or yet again under a new name) an Industrial Control Systems Cybersecurity Initiative as a voluntary, collaborative effort between the federal government and the critical infrastructure community to defend and enhance cybersecurity in essential control systems and operational technology networks. It also directed the secretary of homeland security, in coordination with the secretary of commerce and other agencies, within a year to develop and issue cybersecurity performance goals for critical infrastructures. We can only hope that these latest repetitions of warnings and the reinvention of action items will be more effective than they were in 2002.

In the first week of August 2021, the United States passed the 35 million mark in COVID-19 cases as global cases surpassed 200 million with 4.2 million deaths. That same month, a new global cybercrime treaty presumably to replace the Budapest Convention of 2001 garnered enough votes to commence discussions in January 2022. Ironically, although Russia never signed the Budapest Convention in 2001, it was a main proponent of this measure, now seeing state controls of the internet and access to data as a surveillance tool that could serve its political purposes. The UN, the United States, the EU, and many parties to the Budapest Convention opposed the measure, indicating fundamental disagreements about "what constitutes cybercrime, how law enforcement should gain access to data for cross-border investigations, and more broadly the role of governments in regulating the internet."[29]

On September 7, El Salvador became the first country to adopt Bitcoin as legal tender. A week later, SpaceX launched the first all-civilian spaceflight from Cape Canaveral, Florida, for a three-day orbit around the earth. Another canary in the coal mine died when T-Mobile announced on August 18, 2021, that an attack had breached its computer network, pulling Social Security numbers and other personal information of more than 40 million current and prospective customers. The stolen data included first and last names, birth dates, and driver's license information from a subset of current and potential customers. The breach was among the larger thefts of Social Security numbers. In what passes these days for an apology about inferior security, T-Mobile established an online portal for victims. As if reading from the common script used in each of these occurrences, T-Mobile said "it found and closed an access point used to break into its servers," which was part of a "highly sophisticated cyberattack."[30] Several weeks later, a twenty-one-year-old American who had lived down the street from me in Virginia but was then living in Turkey announced that he was the highly sophisticated attacker of T-Mobile and that he did it to "generate noise" and gain attention. He said that he had discovered an unprotected router in a security system that he described as "awful."[31] He was not a canary—he was an elephant in the pantry!

On August 25, 2021, President Biden met with top executives from several of the largest companies in technology, insurance, energy, education, and financial services, as well as members from his cabinet and national security team, to discuss yet again how to combat cybersecurity threats. Financial services executives Brian Moynihan, CEO of Bank of America; Jamie Dimon, CEO of JPMorgan Chase; Thasunda Brown Duckett, CEO of TIAA Bank; Andrew Cecere, CEO of U.S. Bancorp; and Alan Schnitzer, CEO of Travelers, joined the CEOs of Google, Amazon, Apple, IBM, and Microsoft. They participated in breakout discussions led by cabinet members and White House staff to identify the "root causes" of the attacks and establish "concrete steps for beefing up cybersecurity practices."[32] That month, the Federal Financial Institutions Examination Council (FFIEC) released an update of its 2005 and 2011 guidance regarding effective authentication and access risk management principles and practices for customers, employees, and third parties, among other things, encouraging the adoption of layered security and multifactor authentication.[33]

Also in August, the Senate Committee on Homeland Security and Governmental Affairs issued a follow-up on a 2019 report by the Permanent Subcommittee on Investigations titled "Federal Cybersecurity: America's Data at Risk." The 2019 report highlighted systemic cybersecurity failures of eight key federal agencies, stating that *none* had "met basic cybersecurity standards and protocols, including properly protecting Americans' personally identifiable information (PII)." Many had not promptly installed security patches to remediate known vulnerabilities that hackers might exploit.[34] Two years later in 2021, according to the inspector general, seven of the agencies made minimal improvements, and only DHS managed to employ an effective cybersecurity regime for 2020. It described in detail a generally abysmal record of cyber defense, reporting, and remediation by federal agencies.

Prior to the release of iPhone 13 in 2021, a serious iPhone software vulnerability was traced to the Israeli firm NSO, which offered proprietary software called Pegasus. Researchers at Citizen Lab discovered a "zero-day, zero-click exploit" that could turn an iPhone into a spying device, transmitting geographical location, call logs, contact lists, and photos. Apple issued a fix on September 13. Citizen Lab accused NSO of marketing technology to governments that used it recklessly in violation of international human rights law, a continuing story that will no doubt generate its own books. With 1 billion iPhone users in the world—approximately 25 percent of the planet's mobile phone users[35]—any iPhone vulnerability that could allow someone or something to spy on or impact the activities and thoughts of 12 percent of the world's population is extraordinarily dangerous.

On September 21, the Biden administration gave notice that it was tackling cybersecurity issues head-on. It announced that as part of the "whole-of-government effort" to counter ransomware, the U.S. Department of the Treasury was taking actions focused on disrupting criminal networks and virtual currency exchanges responsible for laundering ransoms. The new rules "strongly discouraged" the payment of cyber ransoms and increased record-keeping and reporting requirements, which place a burden on companies that abide by the law.[36] Discouraging ransomware makes a nice sound bite, but it hardly solves the challenge of having faulty systems penetrated and files encrypted by malicious code. That same day, the Treasury's Office of Foreign Assets Control sanctioned Suex, a cryptocurrency exchange, for its alleged role in laundering ransoms for cyberattacks.

Three weeks later on October 7, Deputy Attorney General Lisa Monaco announced the DOJ's intention to impose fines on federal contractors that fail to meet "required cybersecurity standards," including the disclosure of cybersecurity breaches under the False Claims Act (FCA).[37] The FCA is generally used to sanction people or organizations that defraud the U.S. government by submitting false invoices, so while it may be a stretch to rely on the FCA, it is certainly not the first time the government has treated this and other statutes like Silly Putty, pulling and pushing it until it fit the government's needs.

These actions by the Biden administration were long overdue and may turn out to be effective steps, but only if they constitute parts of a comprehensive strategy intended to prevent cyberattacks. If they are not, hanging some companies out to dry that fail to disclose attacks in a timely fashion according to the government will end up being cosmetic, reactive actions premised on the defeatist presumption that cyberattacks cannot be stopped. Making companies accountable so that they will have to calculate more than just profits derived by new technology products is an important step forward. An even better start down that path would be to lever off the Budapest Convention of 2001 and hold countries similarly accountable and mandate the extradition of hackers and servers.

On October 14, 2021, the G7 published principles to guide the development of central bank digital currencies.[38] The report notes that the international use of CBDC could bring "important benefits," but could also pose "unintended consequences." Referencing the work done by an array of other governmental entities and the Bank for International Settlements, which had issued a similar set of principles a year earlier,[39] the G7 emphasized the need for CBDCs to include a governance framework, protect data privacy, counter illicit financing opportunities, be interoperable with other existing systems, and maximize operational resiliency and cybersecurity protection. The concern over the security of CBDC was just another principle that was matter-of-factly included. There was no suggestion that it should be a fundamental gating issue before any country proceeds to deploy it.[40] Once again, this report raised the question of whether global regulators were being captured by the euphoria of technology.

In the last week of October 2021, the Financial Action Task Force (FATF), the global money laundering and terrorist financing watchdog that sets international standards to combat illegal activities in more than 200 countries and jurisdictions, issued guidelines to impose greater money laundering requirements on cryptocurrencies and crypto exchanges.[41] It is yet another lengthy, consult-speak checklist of procedural, process, and substantive actions required to reduce money laundering through the uses of virtual products and entities. The same day, the White House national cyber director, Chris Inglis, the man who has likely been responsible for the rapid action the Biden administration has taken, put an op-ed in the *Wall Street Journal* announcing the release of "the first statement of strategic intent for the newly created role of National Cyber Director."[42] Calling the risks in cyberspace "grave" and requiring "multilateral action," Inglis noted that merely "focusing on disrupting the illicit activities of cyber transgressors won't create the digitally enabled world we want." He did not provide any specifics that might be interpreted as going beyond mere disruption but did spend much of the article extolling the benefits of technology.

On November 3, 2021, the Biden administration, again noting "persistent and increasingly sophisticated malicious cyber campaigns," issued a directive requiring federal agencies to reduce the significant risk of known exploited vulnerabilities compiled by CISA. The directive applied to all software and hardware found on federal information systems managed on agency premises or hosted by third parties on an agency's behalf and required a review within sixty days to establish a process for ongoing controls and remediation of vulnerabilities.[43] On November 8, 2021, Robinhood Markets, Inc., disclosed that an intruder gained access to its systems, made off with the personal information of millions of users, and demanded a ransom payment. The intruder was able to gain access simply by impersonating an authorized vendor when he called a customer representative of the company.[44] Apologize, rinse, and repeat.

On November 18, the federal bank regulatory agencies issued final rules to improve the sharing of information about cyber incidents that affect the U.S. banking system. The rules require banking organizations to notify their primary federal regulator of any significant computer security incident no later than thirty-six hours after the banking organization "determines that a cyber incident has occurred." There is no indication of what happens on the other side of that regulatory curtain once the information is received. This obligation applies to incidents that have affected or could "materially disrupt" the viability of the company's operations, its ability to deliver banking products and services, or the stability of the financial sector. The rules also require bank service providers to notify bank customers when they experience a similar computer security incident for four or more hours.[45]

Congress again whiffed on appreciating the severe cyber threats that were emerging in December 2021. As it wrapped up its work for the year on pending bills, Republicans and Democrats were not able to agree on expanded reporting requirements. The *Washington Post* wrote that "the failure of such a popular and bipartisan effort—which would have marked the largest expansion of government involvement in private-sector cyber-security in years—raises questions about whether Congress is up to the task of responding to a wave of ransomware and other attacks that have battered industry in recent years."[46]

In November, the Commerce Department proposed new rules related to the TikTok controversy that would effectively allow it to ban software applications deemed unacceptable security risks.[47] It would force social media platforms such as TikTok to submit to third-party auditing, source-code examination, and monitoring of logs that show user data.[48] To the extent that these applications do present a security threat, the country had just lost a year in figuring out the problem as the Biden administration reversed and then revised what the Trump administration had done. Setting up an FDA-like drug approval process for software is a controversial idea that could be expanded to apply to a broader swath of applications beyond the TikToks of the world. We will return to this when we consider how increased governance and regulation of the internet could bolster security.

In mid-December 2021, another frightening vulnerability was exposed when it was revealed that the open-source Apache logging library Log4j had been hacked. It exposed global applications and services and hundreds of millions of computers to a threat that the head of CISA described as "one of the most serious I've seen in my entire career, if not the most serious." Hackers linked to China, Iran, North Korea, Turkey, and other governments were reportedly lining up to exploit the vulnerability in what cybersecurity experts described as "one of the most dire cybersecurity threats to emerge in years," which could "enable devastating attacks, including ransomware, in both the immediate and distant future." Close to 600,000 attempts by malicious cybercriminals to exploit the Log4j bug were logged within days, impacting 44 percent of corporate networks worldwide.[49] Systems were reportedly being attacked by the introduction of strings of code in an email, or the establishment of account usernames that allowed hackers to load arbitrary code on the targeted server and install malware or launch other attacks. Amazon, Microsoft, Cisco, Google Cloud, and IBM all reported that at least some of their services were vulnerable. Experts were concerned that attackers could potentially develop a worm to exploit the flaws and affect computer systems for years to come.[50]

This should have been yet another hint of the three-alarm fire that is the internet. The world's largest tech firms and governments would have normally been expected to deploy hundreds of experts to patch a flaw in a critical piece of internet infrastructure software. But that is not what happened. Log4j is open-source software that runs for free and allows anyone to inspect, modify, and use it. So when something goes awry, volunteers jump in to solve the problem, underscoring the systemic risk that is embedded into the very fabric of an internet where mass surveillance and malicious behavior are already a problem.[51]

On December 17, 2021, citing a concern with the increasing growth of digital assets and incidences of ransomware, FSOC issued its annual report including its most comprehensive and in-depth report on the risks created by cybersecurity vulnerabilities.[52] The report mentioned "cybersecurity" twenty-four times, once again reiterating that a "destabilizing cybersecurity incident could potentially threaten the stability of the U.S. financial system." It recommended that federal and state agencies continue to monitor cybersecurity risks and conduct cybersecurity examinations and use public-private partnerships to identify and mitigate cybersecurity risks.[53] The alarms were modestly raised in this report, but the spectator rhetoric was all too familiar.

By December 2021, the world was beginning to experience the effects of COVID 3.0—the Omicron variant. As the year closed, countries once again closed their borders, and the number of infected people skyrocketed. Scientists identified the new variant as far more contagious than previous forms of the coronavirus and less impacted by vaccinations,[54] but also less likely to have dire symptoms or result in death.[55] By Christmas Day 2021, reported cases of the coronavirus had reached 52 million, causing 815,000 deaths in the United States in just twenty-three months.[56] The new year arrived, and Apple's market value briefly surged past $3 trillion on January 3, the first American company to scale that peak. Perhaps as significant an event was the fact that in 2021, Toyota surpassed General Motors as the best-selling automaker in the United States. That would have been unthinkable twenty-five years ago.

In 2022, more evidence was seen that countries like Russia would not hesitate to use cyber weapons to supplement their military might to achieve their geopolitical goals. On January 14, with Russian troops massed on the Ukrainian border, a warning in Ukrainian, Russian, and Polish appeared on the official website of the Ukrainian Ministry of Foreign Affairs after a massive cyberattack, warning everyone to "be afraid and expect the worst."[57] That website, along with several other government agencies' websites, was temporarily disabled.

It was not immediately known who was responsible, but Russian hackers had previously been blamed for similar attacks on Ukraine, Georgia, and Estonia. WhisperGate malware, something very close to the NotPetya Russian virus that did $10 billion in damage around the world in 2017, was discovered in Ukraine pretending to be ransomware but actually making machinery inoperable. Experts predicted that the world would see cyber operations from Russia's military intelligence agency GRU and a Russian FSB agency operative Berserk Bear both inside and outside Ukraine. On January 18, CISA warned critical infrastructure operators to take "urgent, near-term steps" against cyber threats, eventually issuing a "Shields Up" warning. This should have made it crystal clear how much more dangerous the internet had made the world.

SOME REAL PROGRESS AT LAST

On January 26, 2022, OMB issued a memorandum establishing a federal zero-trust architecture (ZTA) strategy for all federal departments and agencies. It created an assumption that no actor, system, network, or service operating outside or within the security perimeter could be trusted.[58] It required federal entities to meet specific cybersecurity standards and objectives by the end of fiscal year 2024 in order "to reinforce the Government's defenses against increasingly sophisticated and persistent threat campaigns." Departments and agencies were directed to consolidate identity systems, authenticate and encrypt all traffic, employ multifactor authentication, authorize users, tighten access controls, collect and analyze metadata about users, and require users to log into applications rather than networks. The memo admitted that transitioning to a ZTA would "not be a quick or easy task for an enterprise as complex and technologically diverse as the Federal Government," but it claimed that the transition would provide a "defensible architecture."

Boldly identifying itself as "a dramatic paradigm shift in the philosophy of securing infrastructures, networks, and data," the memo was a welcome step forward but already decades too late. Its thirty-nine pages were full of sound and effective cybersecurity policies. But the challenge once again would be implementing and managing them consistently throughout the largest bureaucratic infrastructure in the world while ensuring that the thousands of outside providers that connect to federal systems are equally secure. Based on past precedent and the insecure nature of the internet, our expectations should be modest.

In February, the unimaginable happened when Meta/Facebook reported that it lost daily users for the first time ever, stalling just shy of 2 billion log-ins a day. Its stock lost more value in one day than any other in history—$230 billion—as it released a less than sparkling quarterly report suggesting that it was straining to monetize Instagram's reels, a short-form video product; develop a competitive response to TikTok; and control expenses as it implemented efforts to optimize the metaverse.

Facing the possibility of increasing litigation and legislative challenges to its competitive dominance, the real culprit in Meta/Facebook's slow-down may have been Apple. In 2021, Apple introduced an "App Tracking Transparency" update giving iPhone owners the choice of whether they would let data hoarders like Meta/Facebook monitor their online activities. As users opted out, Meta/Facebook had less user data to harvest and therefore less product to sell. The *New York Times* even suggested that the company was "in trouble."[59] As a lawyer that has represented large companies, I know how some react when faced with challenges. Those that have a ruthless corporate ethos to begin with become even more ruthless.

Later that month, at a cybersecurity conference in Munich, U.S. deputy attorney general Lisa Monaco announced the formation of a new FBI unit—the Virtual Asset Exploitation Unit—dedicated to blockchain analysis and virtual asset seizure to crack down on cyberattacks involving demands for ransoms in cryptocurrency.[60] A week later, President Biden's cyberspace czar, Chris Inglis, published an article in *Foreign Affairs* about the need for a new social contract for the internet.[61] I was intrigued by the title, hoping that it would push for a new concept of the internet to stop attacks rather than just finding better ways to react to them. The article underscored that digital ecosystems had become places where "private information is easily accessible, predatory technology is inexpensive, and momentary lapses in vigilance can snowball into a continent-wide catastrophe." It lamented that responsibility for addressing these vulnerabilities had become diffused, requiring individuals, small businesses, and local governments to shoulder "absurd levels of risk."

Instead of breaking new ground, Inglis once again called for increasing collaboration within the U.S. government and across the public and private sectors to address these issues. While he did suggest that businesses and government need a different approach to the internet, he focused on strengthening defenses and patting President Biden on the back for having essentially paved the way by what I view as his repetition of twenty-five years of admonitions in his recent presidential directives and orders. If you weren't paying attention, you might actually be excited about this. The proof is yet to come. Also in February, the threat of a cyber war became a reality as Russia prepared to invade Ukraine and hackers launched attacks on dozens of government and business websites,[62] with nearly two dozen companies in the liquefied natural gas production industry, including Chevron Corporation and Cheniere Energy, affected. It appeared to mark the first stage of Russia's effort to destabilize the U.S. energy industry.[63]

Online services of Ukraine's armed forces and its two largest lenders, PrivatBank and Oschadbank, were disrupted by suspected cyberattacks on February 15, as the country suffered its largest denial-of-service attack in its history.[64] PrivatBank's ATMs were down for an hour, and it experienced "instability" on its mobile app.[65] On February 23, the invasion began, and a wiper malware—HermeticaWiper—was unleashed on Ukraine, Latvia, and Lithuania computers. That malware is believed to have been embedded in those systems months before. Websites at government agencies and banks were disrupted for several days. Lithuanian authorities had reportedly been impacted as far back as November 12, 2021.[66] Several days later, Anonymous allegedly disabled government websites and played Ukrainian music and displayed its flag on Russian state TV. This was the official start of a new world disorder.

The Biden administration sought to move the digital asset ball down the field on March 9, 2022, when it issued a long-anticipated executive order.[67] Unfortunately, it did not pack any substantive punch and added very little except once again requiring some two dozen different agencies to coordinate action and produce dozens of reports. As noted already, the order glossed over the critical gating issues of market reconstruction and insecurity while also ignoring the growing role of the states in crypto asset regulation.

Some of the issues created by cryptocurrencies are new and are arising in a very different kind of economy. But the basic issues are similar to the ones that were studied by many of these same agencies in the mid-1990s when what was then called electronic money was introduced by companies like Mondex and DigiCash. Asking many agencies to study the issues once again and ignoring the role of the states creates a perpetual jump ball that goes up but never comes down. It is the perfect political delaying tactic.

On March 15, the president signed the Consolidated Appropriations Act of 2022 to fund the government through September 2022.[68] It included the Cyber Incident Reporting for Critical Infrastructure Act of 2022 (CIRCIA). That act dealt with, among other things, the complexities involved in identifying, evaluating, and reporting cyber incidents at critical infrastructures like banking and finance. As someone who has advised financial institutions in such situations, I can attest to the difficulty of applying the requirements of banking, securities disclosure, and cybersecurity laws and rules to cyberattacks during the heat of the moment. The details of such incidents are often only revealed over time, and it is not always readily apparent what happened, who did it, whether it was malicious, and what the damage will be. Satisfying notice and disclosure requirements of multiple federal and state agencies under these circumstances with the regulatory clock ticking can be a tricky challenge, particularly since the reputation of the company may hang in the balance. Companies are often forced to weigh the risk of reporting insufficient information against the risk of not reporting soon enough.

The CIRCIA also increased the reporting requirements for cybersecurity incidents that must be filed with CISA, mandating that they be made by companies, with some exceptions, within seventy-two hours. Ransomware payments are required to be reported within twenty-four hours. CIRCIA also directed CISA to promulgate rules within three and a half years to clarify which entities must file reports, which cybersecurity incidents must be reported, and what information must be contained in such reports. By the time the rule is finalized, several new generations of technologies and types of cyberattacks will surely have unfolded.

Proving that Congress got the message about the intricacies of requiring companies to report themselves to the government, CIRCIA protects them from liability for providing such information to the federal government and allows such reports to be treated as proprietary information that may not be used to assert a waiver of their legal privileges. In the category of creating the illusion of progress, CIRCIA created more bureaucracies: a Cyber Incident Reporting Council, a Ransomware Vulnerability Warning Pilot Program, and a Joint Ransomware Task Force.

Events that further belied the supposed security of the internet and cryptocurrencies occurred at the end of March 2022. Two hundred thousand messages between 450 members of the Trickbot Group, a Russian-affiliated cybercriminal gang, were made public by a Ukrainian hacker. They revealed Trickbot's plans to attack more than 400 U.S. hospitals. The attacks were deemed plausible given the fact that Trickbot's Conti ransomware had been the malicious code of choice to attack and disable numerous private and government computer systems throughout the United States in 2021.[69] Only days later, it was announced that hackers had stolen $625 million in cryptocurrency from the video game company

Axie Infinity. Axie operated "play-to-earn" systems that sold creature-centric NFTs that could be bred, deployed, and purchased with crypto-currencies. The breach occurred when the supposedly secure underlying blockchain that powered the game was infiltrated.[70] This was followed by the theft of $182 million worth of digital assets when hackers attacked an algorithmic stablecoin called Beanstalk, wiping out all of the Ether it held and collapsing the value of its Beans.[71]

With the Ukrainian war as a backdrop, people were starting to get a sense of helplessness when it came to the security of the internet. It was likely no coincidence that same week that President Biden submitted a budget that diverged from those of recent vintage in calling for the largest military and law enforcement spending in history.[72] He cited "evolving intelligence" and warned that Russian cyberattacks on critical American infrastructures were "coming." [73] Recognizing the potential threats to current cryptographic security measures posed by the speed and capabilities of quantum computing and the progress being made by China to dominate the field, the Biden administration issued two new executive orders.[74] Together, they established a presidential advisory committee as well as a working group of critical infrastructure owners and operators to ensure U.S. dominance in the field of quantum computing. They also directed the federal government to catalog its cryptographic-reliant systems, analyze future risks, and deploy quantum-resistant cryptography by the end of 2023.[75]

In May 2022, the Federal Reserve Board issued a report on U.S. financial stability. It identified significant risks to the system, notably identifying money market funds and stablecoins as potentially creating systemic run risks.[76] Much of it was traditional Fed speak, but what caught my attention was the fact that in discussing financial stability, the report mentioned cybersecurity and the threats posed by internet commerce and communication only once. That is a shocking omission in light of the ever-increasing cyberattacks on people, governments, and society and the canary in the coal mine represented by the theft of $14 billion worth of cryptocurrencies in 2021 that were supposed to be secure because of the immutability of the blockchain networks they are built on.

Part IV

REALLY REAL VIRTUAL REALITY

15

VIRTUAL SHOOTOUTS

When so many financial leaders and government officials around the globe increasingly and consistently sound the alarm about how cybersecurity is adversely impacting financial stability, it deserves serious attention. The informative but somewhat painful history of cybersecurity presented above emphasizes three important observations.

First, not enough has happened in the last twenty-five years to prevent or address the threat and potential damage of cyberattacks. Second, relying on the government alone to cause a different result to occur in the future is foolish. Third, to continue to enjoy the benefits of cyberspace and at the same time protect the country's financial infrastructure, a more rational and safer internet is required. Let's move on to solutions.

First, I do not want to suggest that everyone has been sitting on their hands for the last twenty-five years. As we have seen, thousands of public and private sector people and organizations have been working thousands of hours on hundreds of ideas and projects to identify online vulnerabilities and build procedures to create the resiliency to bounce back quickly. Not nearly enough, however, has been done in an efficient, coordinated, and effective way to create an internet that can prevent cyberattacks. An approach that assumes defeat (the presumed penetration of a system, network, or computer) and focuses its resources on redundancy and rebuilding after the damage has occurred is one that is applied to no other aspect of life except natural disasters. Cybersecurity breaches are not natural disasters or acts of God. They could not be any more man-made.

Each industry and company is largely on its own when it comes to cybersecurity, with some doing more than others. After the military, the financial services industry has perhaps done the most to prepare. Spurred on by private organizations, trade associations, and concerned regulators, financial services companies (banks, savings institutions, credit unions, securities firms, exchanges, insurance companies, and funds) have been collaborating to share information and build systems that ensure that if an institution or payments system is taken down, there is enough support and redundancy to put it back online and become functional as soon as possible. In the United States, however, there are thousands of interconnected financial institutions, payments systems, markets, and exchanges that combine to create a nearly infinite number of points of vulnerability. They are increasingly under cyberattack by foreign nations and criminal groups seeking to steal money, damage companies, disrupt the financial system, and undermine the functioning of the economy. Creating complete institutional resiliency and redundancy with appropriate backup facilities to be able to rebound quickly after each cyber event is becoming a more difficult and complex challenge with each passing day. But it is one of the few U.S. critical infrastructure sectors that have had mandatory cybersecurity and incident reporting requirements codified into law and regulation for over twenty years.[1]

Under the Gramm–Leach–Bliley Act of 1999 and its implementing regulations,[2] cyber incident reporting to federal regulators is triggered as soon as an institution suspects misuse of customer data. Federal regulators examine compliance with these requirements, as well as institutions' operating and governance processes, data security, data-handling processes, and third-party risk management measures. On January 12, 2021, federal bank regulators issued a proposed rule on "Computer-Security Incident Notification Requirements for Banking Organizations and Their Bank Service Providers."[3] States and foreign jurisdictions have similar requirements that may apply simultaneously. They include the New York Department of Financial Services (NYDFS) cybersecurity regulations,[4] the European Union General Data Protection Regulation (GDPR), and the European Union NIS Directive 1.0.[5] Given the broad authority federal regulators have in these areas, it is surprising that they have not acted to mandate a more secure environment. This suggests that they may not think or understand that more needs to be done, or that from practical and political vantage points, more can be done.

War games are an important first step; they are the next best teacher after experiencing the real thing. Cybersecurity tabletop exercises simulating network penetrations are critical to the preparedness and systemic resiliency of the country's financial infrastructure. Such exercises around the world have assisted government regulators in identifying gaps in response plans, refining their crisis communication, and improving coordination. The most significant financial services tabletop exercise programs have included the Hamilton Series, Quantum Dawn, Cyber-Attack Against Payments Systems (CAPS), Exercise Cyber Star (Singapore), Exercise Raffles, Waking Shark (UK), SIMEX18 (UK), Resilient Shield (UK and U.S.), UNITAS (European Central Bank), and the G7 cybersecurity exercises. In 2019, the UK National Cyber Security Center published "Exercise in a Box," an online cyberattack tool for organizations that provides participants a summary report with results and recommendations to improve cyber resilience.[6]

THE U.S. HAMILTON SERIES

The Hamilton Series, one-day cybersecurity simulation exercises for banks, were first launched in 2014. Led by the Department of the Treasury in conjunction with several private industry organizations, their purpose was to improve responsiveness to financial cyber threats and attacks for participants from the public and the private sectors. Government agencies involved have included the Department of Homeland Security, Treasury, and federal and state bank regulators. Private sector organizations such as the Financial and Banking Information Infrastructure Committee (FBIIC), the Financial Services Sector Coordinating Council (FSSCC), and the Financial Services Information Sharing and Analysis Center (FS-ISAC) have also participated. Exercises have replicated actual cyberattacks[7] and have included cross-sector exercises with electrical, liquidity, futures, derivatives, third-party supplier/service providers, and system assurance sectors.[8] In 2018, FS-ISAC added range-based cyber exercises for more technical, hands-on-keyboard experiences to coordinate risk management and compliance with regulatory requirements.[9] At the time of this book's writing, there had been nineteen exercises in the series focusing on information sharing, early-warning notification, resiliency and redundancy, and backup facilities. While important and useful, some argue that these exercises cannot realistically include the full range of indirect attacks and contagion that financial institutions are exposed to by users or infiltration.

QUANTUM DAWN

Beginning in 2011, the U.S. Department of Homeland Security and the Depository Trust and Clearing Corporation, and then later the Securities Industry and Financial Markets Association (SIFMA), organized tabletop exercises for the securities industry through the Quantum Dawn program. The six exercises held since have seen participation grow to more than 180 global financial institutions in 2019. Each exercise simulated a set of cyber incidents which have provided a forum for participants to test incident response playbooks and protocols across equities trading, clearing processes, and market closure procedures in response to an ecosystem-wide attack on market infrastructure.

Quantum Dawn II tested procedures that would impact the decision to close equity markets. Quantum Dawn III (September 2015) simulated a large-scale cyberattack lasting three business days that disrupted trading and recreated a failure of the overnight settlement process at a major clearinghouse but did not result in market closures. Quantum Dawn IV (November 2017) simulated a "bad day" on Wall Street during which financial institution payment infrastructures were targeted to create rolling impacts on the sector and markets. The events were assumed to cause widespread consumer panic and market contagion after major news outlets were hacked, resulting in the dissemination of "fake news" stories. More than 800 representatives from over 150 financial firms as well as more than 50 regulatory authorities, central banks, government agencies, and trade associations across nineteen countries participated in Quantum Dawn V (November 2019). It simulated a targeted ransomware attack orchestrated by an insider starting in a U.S. institution that ultimately impacted major banks across Asia and the UK. The sequencing of the unfolding scenario highlighted the sector's interconnectedness. It also tested responses to cross-border cyberattacks and the benefits of information sharing.[10] Quantum Dawn VI (November 2021) allowed financial firms, central banks, regulatory authorities, trade associations, law enforcement, and information sharing organizations to rehearse response mechanisms against a broad range of ransomware attacks.[11]

OTHER EXERCISES

FS-ISAC members participate in the CAPS exercises, which target the vulnerabilities of financial institution payments systems and processes. Participants practice mobilization, working under pressure, analyzing information, and defending against attacks. The exercises focus on strengthening teams, understanding system vulnerabilities, and improving processes and response plans.[12] The North Atlantic Treaty Organization (NATO) ran exercises in

April 2021 and 2022 in conjunction with the Estonia-based Cooperative Cyber Defense Centre of Excellence, which included the Bank for International Settlements; NatWest Group PLC; Mastercard, Inc.; FS-ISAC; and SWITCH-CERT, a Swiss computer emergency readiness team. These simulations were "live-fire exercises" including actual attacks involving more than 2,000 participants from thirty nations against systems set up for the drill and cybersecurity specialists defending against them.[13]

HAVE THESE EXERCISES PUT US IN GOOD SHAPE?

These and exercises like them are critical to enhancing global cybersecurity. But there are still significant questions about how effective such measures will ultimately be in the face of wide-scale live fire taking advantage of insecure networks from many different angles. Whether the games are comprehensive enough and are building a knowledge and experience base that will work effectively in a massive financial cyberattack is unknown and will not be known until such an event happens. That is the unfortunate chicken-and-egg dilemma created by any challenge where the "Start" button may also be the "Game Over" button.

The process seems fraught with peril and dripping with a sense of defeat since it focuses on responding, redundancy, and rebuilding. Many attacks are deterred and defended against every day, but failure against the most sophisticated and well-armed persistent attackers almost seems to be expected. Imagine if the military approached the defense of the country that way. We would simply need an Army Corps of Engineers to rebuild everything after it was inevitably destroyed. There must be a better way to protect the economic infrastructures of the world.

More problematic perhaps is the fact that exercises are for the members of certain organizations and are often by invitation only. Suffice it to say that "invited guests" are not the only vulnerabilities in the country's financial infrastructure. There is no assurance that these exercises, no matter how professional, targeted, and successful, will help any but the largest institutions or lead to effective measures, particularly if an attack is systemic rather than institutionally focused. If one big bank is slated to backstop another big bank, the plan can work, unless both big banks are simultaneously attacked and disrupted. With about 6,000 banks, 6,000 credit unions, 6,000 insurance companies, 3,400 broker-dealers, 7,500 mutual funds, 8,000 private equity firms, and 3,600 hedge funds in the

United States alone, coordinating and securing the virtual relationship of such an enormous number of companies and users is an incredible challenge for an online world where failure is baked into the design. We won't know whether these defense mechanisms will be successful until an attack happens, and of course then it may be too late.

If the internet is going to support nearly all commerce and transfers of money, we need a system that can prevent and deflect 99 percent plus of the most serious incursions. If that goal can't be reached, why are we using this internet?

16

CALL THE INTERNET REPAIRMAN

We are not much further down the path of preventing cyberattacks and data breaches than we were twenty-five years ago, though we are better at reacting when they inevitably arrive. The recitation of history I have dragged you through gives you a clear idea of what to expect in the future unless something changes. In the 1980s, the commercials about the Maytag washing machine repairman sitting idly waiting for the call that never comes were popular. They underscored the reliability of Maytag machines. When it comes to the internet, we have neither reliability nor repairmen.

The internet's efficient and charming qualities accommodate both the advantages and the vulnerabilities that the virtual world offers. Users have developed a deep infatuation for the internet's ability to enhance the quality of human life and maximize profit. But we also know that it is a mysterious, interconnected spaghetti bowl of instant digital gratification that is littered with secret entrances and mysterious backdoors. No matter how secure a network may be intended to be, if it connects to insecure networks and millions of people, it is insecure. That means we either secure the entire internet, do a better job at constructing and using new secure private networks, focus on segregating and securing sensitive data, or do some combination of all three.

This internet cannot practically be repaired without turning back time. Simply enhancing its security to make it more trustworthy and reliable would be a complex endeavor that would cost a lot and create inconveniences. Its billions of users and thousands of miles of aboveground and underwater cables, satellite links, routers, servers, and Wi-Fi connections create an infinite number of junctions, intersections, and handoffs that are used by law-abiding users as well as instigators of virtual chaos. Implementation of 5G technology, quantum computing, and the IoT that will connect everything in a centralized network will only exacerbate our ever-increasing reliance on the internet and the exponential growth of cybersecurity challenges. Asking users to give up any of the measurable benefits that the internet provides in return for greater but unquantifiable security will be a tough sell—until they see what a digital Pearl Harbor looks like.

I learned about the fragility of cyberspace the hard way. In the 1990s, I was among those whose accounts were hacked in one of the first publicized online data breaches. I had shopped on the website of a professional sports league that was the subject of an attack and theft of customer data. At that time, the thought of customer information being stolen online was hardly a consideration in the euphoric response that users were having to the new capabilities they were provided by the Web. The email I received about the incident changed my views entirely. It described how the breach had been the fault of one of the seller's service providers and essentially let me know that I was on my own and should contact my credit bureau, bank, and credit card company and, oh yes, be vigilant. I was completely underwhelmed by its handling of the situation. Shortly thereafter, I started advising my clients that how they approached online security events could impact whether they continued to have customers—and perhaps a company. I had concluded that when breaches occurred—and they would—they should be treated as marketing opportunities to demonstrate to customers how important they were as opposed to simply throwing them a virtual life raft in choppy online waters.

The amazing thing is that not much has really changed since then. I received an email in September 2021 from a national department store reporting a potential breach of my data.

> Earlier this month, we learned that in May 2020 an unauthorized party obtained personal information associated with certain of our customers' online accounts. . . .
> Promptly after learning of the issue, we engaged a leading cybersecurity expert and notified law enforcement. Our investigation is ongoing, and we are working quickly to determine the nature and scope of the matter. . . . We will continue to take actions to enhance our system security and safeguard information.

What has changed is that these kinds of incidents have become more frequent and more serious given how much more data and value have been put online. The text of the September 2021 notice was like many that are sent these days. While the tone is quite conciliatory and helpful, it can be paraphrased as "we really don't know what happened, when it happened, or how to fix it, so good luck with this crazy internet stuff."

Consider the juxtaposition of the words "promptly" and "earlier this month" in the message. A casual reader may get the impression that the company had moved expeditiously. Well, it had, but only when it finally knew something had occurred eighteen months before. I am not blaming the company. That may be all it could have said. But what are customers supposed to do with the information that their credit card credentials may have been put on the Dark Web a year and a half ago? The notice certainly did not suggest that customers could hold the store or any of its technology vendors responsible for any damage they incurred. But thank goodness the company's "investigation is ongoing" and it is "working *quickly* to determine the nature and scope of the matter" so that it can "*continue* to take actions to enhance . . . system security and safeguard information." Unfortunately, customers have become numb to these types of notifications and the many risks that are accumulating in cyberspace, thinking that if this is the worst, they can bear it. Apologize, rinse, and repeat.

Most users do not appreciate the number and nature of vulnerabilities that can destroy significant aspects of their lives. If they did, they would not be so cavalier with their data online. Should they have been allowed over these twenty-five years to cram their analog lives into this virtual world without a complete understanding of the vulnerabilities and risks being created? Governments protect consumers from many less dangerous things. More importantly, now that we better understand the risks, why are attempts to rectify the situation so limited? Part of it is just like every other economic and societal collapse we have experienced. The risks are considered manageable when everyone is enjoying and profiting from the ride on the way up. The government, economy, and society are so invested in the status quo that nothing will change it until the worst happens. Bet on it.

There are ways to repair the internet as a network, or at least make the risks more manageable. First and foremost, the economic incentives of online communication and commerce would have to be realigned to force companies and users to value security as much as innovation. This would not directly impact the actions of malicious actors beyond making it a bit harder for them to act. Second, a wide range of structural, procedural, protocol, and engineering improvements would have to be initiated and universally adopted to achieve greater security. There would be a monetary and convenience surcharge for this that many may not be willing to pay. These improvements *could* impact the capabilities of malicious actors. Third, modularization of the internet would have to become more ubiquitous, with new and more private secured networks used not only for wholesale traffic (e.g., the "back office" networks used by payments systems and exchanges) but also for the retail interfaces used by consumers and businesses. This would require a new concept of cyberspace—a new internet—and a new way of handling data.

The other end of the cybersecurity spectrum requires controlling untrustworthy actors. That would involve a combination of adjusted economic incentives, real authentication, behavioral constraints, strong enforcement, and ways to simply circumvent them. The available tools to implement these controls include, among other things, laws, taxes, protocols, alliances, extradition treaties, engineering standards, policing, and fines.[1] As an example of the scale of the problem and how these solutions would work, let's focus on the complicated makeup of the country's financial infrastructures and what would have to be fixed to provide greater security that would *prevent* rather than simply react to cyberattacks and data breaches.

The greatest online challenge in protecting banking and finance is that it takes just one successful malicious act to penetrate a system's defenses and create massive havoc. But institutions, governments, financial networks, markets, and users must simultaneously and continuously solve for a universe of infinite attacks aimed at millions of connections, seams, and junctures from a virtually limitless set of directions. That is an expensive and nearly impossible task.[2] Moreover, any company that is attacked is not permitted to fire back, so why would an attacker be afraid to launch its virtual missiles? It is a unique situation. Private companies are attacked, and governments respond in their own time and way, consistent with a myriad of political priorities.

The internet stacks the deck against its benign users, particularly since there are limited technology cops on the beat that can encourage and enforce participation using both a carrot and a big stick to address aberrant technological behavior. Solutions proposed to this global enforcement problem range from deploying a master artificial intelligence regulator to relying on forms of algorithmic accountability that incentivize businesses to verify that their systems act as intended and identify and rectify harmful outcomes.[3] It seems only fitting for machine-driven society to eventually deploy machine cops—algorithms regulating algorithms!

New rules, structures, and protocols can make the internet and cyberspace more secure environments.[4] All of them are logical and well thought out. The European Commission has embarked upon a several-year project to consider a Digital Services Act[5] that would create uniform EU standards to make the internet more transparent and secure and develop standards to impose liability for unlawful behavior. Nevertheless, remaking the internet piece by piece will be a remarkably complex task requiring so many compromises along the way that by the time it got done, it would be ineffective, obsolete, or both.

In the financial services world, the list of financial participants, network junctions, connecting points, and intersections creates so many systemic vulnerabilities that it is hard to imagine all of them ever being adequately secured one by one on an unsupervised basis. Banks connect to the Federal Reserve and the Department of Treasury for various payments, fiscal, monetary, and other financial transactions, as well as to various payments system, ACH, credit card, and electronic funds transfer networks. Hundreds of millions of customers, vendors, lawyers, financial advisors, auditors, and state and federal regulators are also connected to those banks and networks. It is a virtual six degrees of separation between and among each and every individual, company, regulator, and government authority in the financial sector, and the entity or person with the most vulnerable security determines how secure every other system can be. This connected world is represented in figure 16.1.

In addition to these connections, no matter how secure or offline a network is, the ever-present human frailties in the system can undo them at any time. Any employee can be bribed for their network password, and some employee somewhere will always click on a phishing email that plays

POINTS OF FINANCIAL INSECURITY

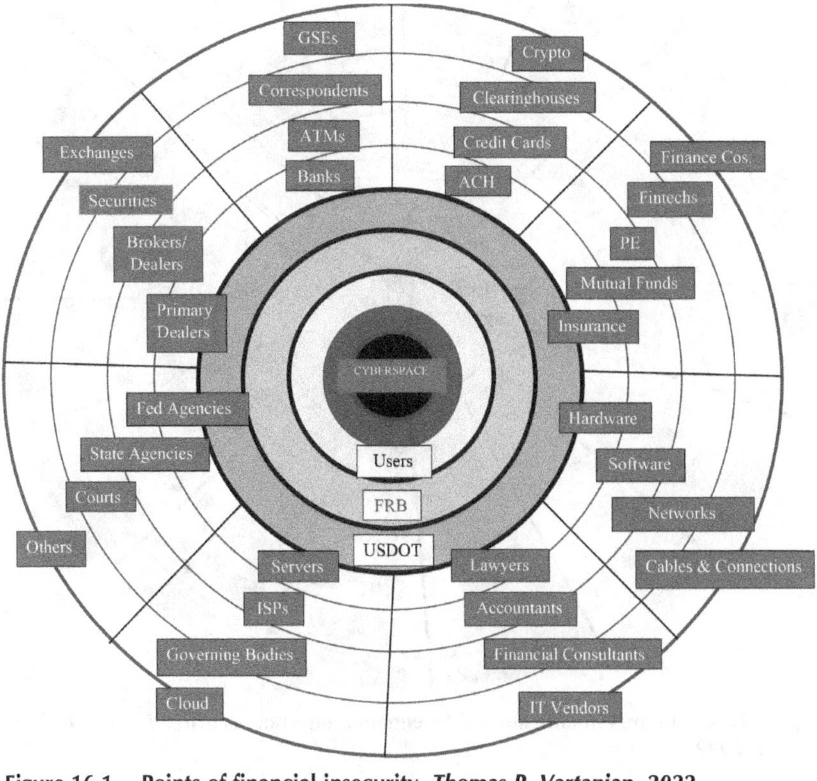

Figure 16.1. Points of financial insecurity. ***Thomas P. Vartanian, 2022***

on the most vulnerable of human emotions—"HAVE YOU SEEN THIS PIC-TURE OF YOUR SPOUSE!!" One click, and the hacker is in, while the clicker is sorely disappointed.

Behind the effort to realign economic incentives, there are more than 100 different fixes in more than two dozen categories that would make the internet more secure. The sheer volume of solutions and the enormity of the challenges are represented in figure 16.2.

While networks are much more resilient than they were twenty-five years ago, they still have a long way to go to achieve acceptable levels of security before we should feel comfortable entrusting all the data and economic value on the planet to them. The following list of potential security enhancements makes that point—it is long and ambitious. It is largely only what is done in the analog world to secure communications and transactions, so should it really be controversial?

IMPROVEMENTS NEEDED TO ENHANCE INTERNET SECURITY

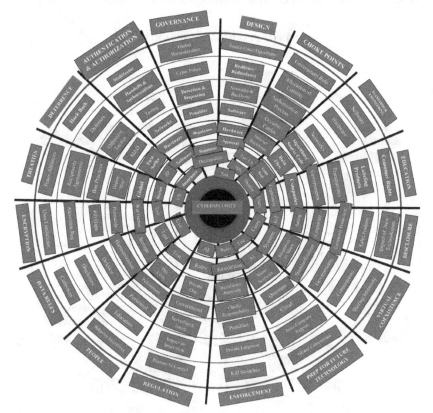

Figure 16.2.   Improvements needed to enhance internet security. *Thomas P. Vartanian, 2022*

17

THE INGREDIENTS
OF A SAFER INTERNET

L et's consider some of the improvements that would have to be made
if it was decided that the current internet was to be made significantly
more secure. You may consider this a fable, but as we proceed, ask yourself
whether you are willing to accept each and how they would change your
online experience.

AUTHENTICATION AND AUTHORIZATION

Anonymity as it is flaunted in cyberspace seems to include within it the
seeds of societal disintegration. Adversarial nation-states, hackers, terrorists, fanatics, pranksters, and criminals cannot be trusted in an environment
that provides them the option of acting anonymously. Walter Isaacson,
professor at Tulane University and CEO of the Aspen Institute, succinctly
describes the technological problem encoded into the current internet as
reflective of a brilliant structure that allows the earth's computers to be
webbed together and navigated through hyperlinks that travel only one
way. "You knew where the links took you. But if you had a web page
or piece of content, you didn't exactly know who was linking to you or
coming to use your content."[1] The internet has foolishly enshrined anonymity as an online human right.

Authentication is a reality that we all live with every day. We have
driver's licenses, passports, employee IDs, and marriage licenses. My dog
even has a license as well as a chip in his neck. Knowing who we are dealing
with may seem like an obvious expectation, but it is precisely the opposite of
the current internet's hallmark of ghostly anonymity. No one lets strangers
who won't identify themselves into their homes, and no one purposefully

sends money through or to parties they do not know or trust. So why do users voluntarily or tacitly allow strangers to rummage around their professional records, social lives, and financial valuables? Is it the amorphous promise of greater prosperity and a more fulfilling social and financial life? Each step toward eliminating anonymity that we take will make cyberspace a little bit more civil, secure, and more distant from the one we have.

Systems currently impose loosely configured authentication processes to identify a user when an application is launched.[2] Identifying online credentials often takes the form of an alphanumeric password, facial recognition, token, or some other biometric feature. Authentication methods are rapidly improving as more sophisticated applications of confirmatory email links, push notifications, and biometric scans are adopted. But given how far below the perfection curve that security is, we will still be running a losing race until standards that require, for example, an individual rather than a computer, program, network, application, or IP address to be authenticated are in place. The security of any system can only be as good as the level of seriousness devoted to identification, authorization, governance, and enforcement. Since the public internet does not generally confirm the identity of a specific human user, it is ineffective when it comes to detecting, preventing, or prohibiting illicit online activity. That is why botnets are easily created by the millions and can create havoc using DDoS attacks, circulating fake news, and influencing national elections. Nevertheless, Twitter users were repelled by the thought of any online identification linking a human to an account when Elon Musk suggested that once he owned Twitter, the company would impose such a requirement.[3]

When I log onto my bank's online banking network, the system attempts to authenticate me (my computer) as the registered owner of the account. Even if two-factor authentication is used, it does not rule out the possibility that someone has replicated or is using my computer or my cell phone. Biometric and other identifying technologies are available to assist in the authentication process, but they require additional software and the willingness of the customer to install and use them. Even when such authentication applications are being used, there is no guarantee that I am at the site I intended to visit or that some interloper has not hacked my computer and is not piggybacking on my identity access. Similar kinds of fraud and cons exist in the real world. The difference is that the scale of the crime that can be committed online is nearly limitless. I can lose cash and jewelry if my house is burgled, but if my online banking and other financial networks are penetrated, I can lose everything. That makes the need for security in the virtual world greater than in the analog world, a scaling that has not adequately occurred.

Private networks such as company email networks normally require higher levels of authentication to identify the bona fides of each user as an authorized user. Additional procedures can be used to authorize an authenticated entity to have access to online resources at each stage of the transmission. Zero-trust architecture (ZTA) evaluates each request for access, requiring the ability to continuously monitor any active session no matter how deeply or broadly it may extend. The White House finally directed in a 2021 memorandum that the entire federal government, including its employees, contractors, and business partners, adopt and integrate multifactor authentication (MFA) standards at the application layer for access to a network.[4] MFA involves levels of identifying users, whether through passwords, confirming messages, facial recognition, or other biometric measures, with each level providing just a little more assurance that the parties are who they say they are.

Highly secure ZTA networks may require users to authenticate into applications as opposed to the underlying networks to better guard against hackers guessing weak network passwords or reusing passwords obtained from a data breach.[5] Phishing-resistant approaches to MFA that can defend against such attacks have been developed. For example, the World Wide Web Consortium's (W3C) open "Web Authentication" standard supported by consumer devices and popular cloud services enables the creation and use of public key–based credentials to authenticate users to ensure that no operation is performed without user consent.[6] The federal government's Personal Identity Verification (PIV) standard is another such approach.

In addition, blockchain applications may resolve some authentication issues. When a digital ledger is distributed among members of the network, it is more difficult for someone to maliciously alter it. As each new "transaction" or block of data is added, a majority of the network must verify its validity. Adding the use of public key encryption also adds a way to handle the verification of identity.[7] However, this does not mean that blockchain is impenetrable to cyberattacks and security fraud. A June 2022 from DARPA warns about the vulnerabilities of blockchains.[8] The theft of $73 million in Bitcoins from a Hong Kong–based cryptocurrency exchange, the $60 million robbery of Ether from the Decentralized Autonomous Organization fund, the disappearance of $16 million from an indexed crypto-fund, and the "51% attacks" on Ethereum Classic, Zen-Cash, and Verge in 2018 that resulted in the reversal of transactions and double spending of digital coins all suggest that blockchains are indeed mutable. They are also susceptible to endpoint vulnerabilities, software flaws, malware, routing and phishing attacks, and transaction privacy leakage where hackers determine if a single user is receiving cryptocurrency in different transactions.[9] The inception of the IoT should encourage greater and more widespread use of more secure blockchain systems. Experts like Bruce Schneier seem to doubt it.[10]

Semi-decentralized versions of blockchain applications may provide enhanced authentication and verification for financial institutions, but tough choices will have to be made. First, businesses and consumers will have to agree that they are willing to endure the costs and inconveniences that accompany greater authentication. Second, someone must do the authenticating. That could be a government entity, private organization, machine intelligence, or a combination of the three. This is often the point at which improvements get tabled for want of consensus. If it is a governmental entity doing the authenticating and it is a repressive government, the result may be even more dangerous than an anonymous internet. Look no further than China for an example of how technology can be used to censor, control, and manipulate a population. Kai Strittmatter hammers this point home in his terrifying book *We Have Been Harmonized: Life in China's Surveillance State.*[11]

Third, standards would also have to be developed to determine what to authenticate and how often. Users, software, hardware, applications, networks, systems, handoffs, junctions, and everything in between could be authenticated, creating bureaucratic complexities and delays. Encryption could be applied to packets of data at each step in its movement, depending on how much cost and time users are willing to tolerate, allowing identities to be disclosed only to the two communicating endpoints.[12] Once installed, high levels of authentication would require constant testing, which would increase costs, personnel, and general aggravation. Each time an authentication protocol was launched, like virus protection, it could slow down the user's online experience, which always seems to be considered a superior goal to security. To make the entire internet secure, every user would have to commit to higher authentication and identification protocols and enforcement. Otherwise, hackers would simply break into the less secure network and crawl from them into more secure ones as they do today. Balancing the alternatives, costs, and benefits as the stakes increase will be an enormous challenge.

Banks around the world are hoping to encourage the creation of national digital identity schemes as a means of both reducing online friction and improving security for consumers and companies. A report by the International Banking Federation and the consulting firm Oliver Wyman provides an excellent overview of this fledging effort.[13] It underscores the benefits to consumers as well as banks of lowering operating expenses, reducing fraud risks, and creating new potential revenue streams as banks assume the role of digital identity providers. A short list of countries around the world are already offering digital identities. The United States is not one of them.

ENHANCED DETERRENCE STRATEGIES AND ECONOMIC DISINCENTIVES

Much of today's online defenses revolve around deterrence—using layered security to make it harder and more costly for hackers to penetrate a computer, network, or system. Deterrence is the virtual barbed wire that companies can use to discourage attacks. The best way to achieve maximum deterrence is to envision the work and expense that a malicious party must devote to the task to understand how its workload and costs could be increased to decrease the likelihood and number of successful attacks.[14] Deterrence is an important tool in the cybersecurity arsenal, and it works in many instances, particularly against less talented and armed attackers. The use of cards and keys, multifactor authentication, access filtering applications, passwords, and firewalls, however, will not impede well-resourced persistent attackers, particularly as cyber defenses are increasingly exposed to the abilities of state-sponsored and state-run attacks.[15]

Many of the reports that we have discussed have emphasized the importance of this kind of security—encryption, firewalls, multifactor authentication, zero-trust architectures—to create technical and economic hurdles to deter would-be attackers. But deterrence is not nearly enough given how much of our lives have shifted to the internet. Moreover, it will always be ineffective to the extent that the punishment, if any, for not pouring enough resources into engineering and compliance standards is less than the economic gain of moving fast and breaking things. And where too many people have too much access to too many networks, there are bound to be some that will make mistakes or are willing to commit crimes to provide access to a hacker.

Mutual assured destruction (MAD) has become a fundamental part of online security deterrence strategies. We would not expect Russia or China to take down the U.S. power grid or financial system; the United States would eventually retaliate as soon as it could get back online and identify the culprit or its affiliation. Moreover, the destruction of the largest economy in the world would not be good for the economies of many other countries, particularly China. But countries do stockpile zero-day exploits and less potent malware to infect the systems of other nation-states and deploy applications to spy or to simply lie dormant to be detonated in the future. This suggests that the internet is becoming one huge virtual improvised explosive device (IED) ready to be detonated. But unlike nuclear war, it is not clear who has their fingers on these virtual IEDs, what they want, whether they are rational or even sane, or how they can be controlled. This creates a complex set of dynamics to navigate that would befuddle even Sun Tzu, the fifty-century BC author of *The Art of War*.

The art of cyber war makes first-strike capability extraordinarily important. Some of us grew up with the threat of nuclear war hanging over our heads and can remember being instructed by teachers to duck under our desks during drills. Even as a third-grader, I could never quite figure out how those wooden desks were going to shield me from being incinerated in a nuclear disaster. In such an attack, the assumption was that the country would have between fifteen and twenty minutes to counterattack and launch its own missiles before another country's nuclear warheads would reach the U.S. mainland. The resulting annihilation of both sides was considered a significant deterrent to any rational, economically driven nation ever attempting to go down that road. In cyberspace, all bets are off. The first cyber strike may turn off the power, disable weapons, make computers inoperable, and destroy financial value simultaneously, leaving the target country unable to respond. What happened in Estonia and the Ukraine were good examples of first-strike cyber war potency.

At the other end of the spectrum, the question is who can or should respond to a cyberattack and how the response should be calibrated to the attack. That requires some agreement as to how such cyber aggression should be classified as either a cybercrime, an act of war, or a nuisance. Many such acts are carried out by state-sponsored or state-affiliated organizations, with the target often being one or more private companies. That complicates the ability to counterattack. After Sony Pictures announced the release of a movie about the assassination of North Korea's leader, it was digitally disemboweled by North Korean hackers who spilled its secrets and dirty laundry over time throughout the internet. Sony was defenseless to respond or to prevent the damage from growing. U.S. law prohibited Sony from "hacking back" to protect itself or retaliate, even if it could identify the perpetrators. It and other companies in its position are forced to rely on the government to retaliate *in a proportionate manner at the appropriate time.*

The issue of hacking back is a very complicated one. So, consequently, the response falls into the political realm, where people who focus on re-election and secret geopolitical rules, relationships, and tradeoffs make the decisions. Under these very imperfect conditions, making deterrence an effective defense against cyberattacks will be a daunting challenge.

Misaligned economic incentives play a significant role in cyberspace insecurity and how effective deterrence can be. Coders and software manu-factures are incentivized today to get to market first and worry about security later. Companies modify software programs primarily to achieve efficiencies and save money. Users have no interest in anything that slows down their online experience, and malicious actors see mostly upside from what they do, whether it be ransomware profits or their ability to create havoc for which the chances of any kind of reprimand are low. When bad actors don't see a significant financial cost to launching attacks, it is not an economic formula that can deter the kinds of malicious activity we most fear.

Realigning the economic disincentives to launch cyberattacks or for maintaining systems that are susceptible to them is the most efficient way to reduce vulnerabilities and attacks. When markets stop rewarding in-novation and begin punishing insecurity, networks will naturally become more secure as coders test more diligently, companies scrutinize software, and hackers learn that substantial fines may be imposed. As we will discuss later, the combination of better internet governance, rules of conduct, quality controls, enforcement, and kill switches could realign some of those economic incentives.

BETTER ENGINEERING: ZERO-TRUST ARCHITECTURE

To be more secure, the next generation of the internet would have to be based on more secure technologies that are better engineered. While those technologies are available, the challenge is transforming an internet that was not conceived as a secure environment into one. The coding and creation of software, the components from around the world that make up computer networks, and the potential that hidden features in each could be used for surveillance or espionage are all the result of a marketplace that has focused on innovation rather than security. It will be difficult to reverse that cycle, particularly since better engineering is directly related to the budgets that are available.

Some networks use "air gaps" or applications to segregate users based on the level of security required. One of my clients blocked all communications from foreign countries. While not infallible, these types of filters can add security. And now that many governments and organizations are realizing that increased security is needed, the corresponding economic reward for producing more secure software and hardware, and the costs attributable to creating and maintaining insecure applications and networks, may eventually incentivize the development of more secure online experiences. Companies are developing quantum-resistant cybersecurity and anti-ransomware data protection platforms for military and corporate clients through zero-trust data protection architectures and the increased fragmentation of data into many encrypted packets that can only be reconstructed and decrypted by authorized users.[16] As new products to reengineer the security of the internet emerge, the challenge will be the acceptance, uniformity, and integration of such new solutions relative to the costs and inconvenience to users.

Legacy methods of network security are proving each day to be increasingly insufficient. In too many networks, once attackers can breach the perimeter, their lateral movement may be largely unhindered. That has led to better-engineered security solutions such as zero-trust architecture. ZTA assumes that everything inside an organization's network is considered distrusted.[17] The term has its origins in the work published by the Defense Information Systems Agency (DISA) and the Department of Defense, which argued for moving security from network perimeters to individual transactions. The Jericho Forum in 2004 publicized the idea of such "deperimeterization."[18] ZTA eliminates any assumptions of implicit trust and adopts a principle of "never trust, always verify." It requires continuous validation of every stage of a digital interaction. It is not a single architecture but a set of guiding principles for workflow, system design, and operations that can be used to improve the security posture by protecting networks and enabling digital communication through strong authentication methods, leveraging network segmentation, and restricting lateral movement.[19]

ZTA prioritizes and relies on strong authentication of user identities, "least access" policies, and verification of user device integrity. Since no application is implicitly trusted, it requires continuous, real-time monitoring to validate every online behavior. Routers, switches, the cloud, the IoT, and digital supply chains are subject to the same security and validation principles. On January 26, 2022, OMB issued a memorandum establishing a federal ZTA standard for all departments and agencies by the end of fiscal year 2024.[20] They were directed to consolidate identity systems, authenticate and encrypt all traffic, employ multifactor authentication, authorize users, tighten access controls, collect and analyze metadata about users, and require users to log into applications rather than networks. You would have thought that all of that should have been happening long before 2022. And there is no guarantee that it will happen by 2024 as the memo directs.

Some believe that a better-engineered worldwide wireless internet makes more sense from a security perspective than spending trillions on upgrading costly physical infrastructures.[21] For example, one next-generation internet proposal—an alternative global network (AGN)—would be an IP rather than the traditional TCP/IP stack–based network that would be capable of updating the entire network within seconds of the first attack to prevent a problem from recurring or spreading.[22] However, proposals that allow everyone to code open architectures raise other vulnerabilities that impact the security of software.[23]

GOVERNANCE

Various forms of internal and external governance could create a safer internet. No major organization operates without a leadership structure and internal procedures. Who builds a world with no rules, no enforcement mechanism, no police, and no penalties for bad behavior? That is pretty much what cyberspace looks like, so it is no surprise that it lacks order. Sure, there are endless technical rules and standards established by organizations such as the Internet Corporation for Assigned Names and Numbers (ICANN), the Internet Society,[24] the International Organization for Standardization (ISO),[25] the European Union Agency for Cybersecurity (EUAC),[26] the National Institute of Standards and Technology (NIST),[27] the Information Systems Audit and Control Association's (ISACA) Control Objectives for Information Technologies (COBIT),[28] the International Electrotechnical Commission (IEC),[29] the Institute of Electrical and Electronics Engineers (IEEE),[30] and dozens of U.S. agencies like the Federal Communications Commission, the Department of Homeland Security, and the Federal Reserve. But universal application and enforcement is difficult or nonexistent.

The Internet Society was established in 1992 to facilitate open development of standards, protocols, administration, and the technical infrastructure of the internet on an international basis. It has created a code of conduct.[31] But to bring some sense of order to cyberspace, there would need to be wholesale improvements in its design, operation, and the standardization and enforcement of rules. That could be implemented by a governing country, a group of countries, a private organization, a consortium of organizations, intelligent machines, or some combination of all of them that was acceptable to the democratic nations of the world. Creating internet governance on a country-by-country basis when everyone shares the same networks would only lead to more virtual chaos, cyber-balkanization, and needless inefficiencies. And there is no doubt that any bureaucracy created by such a governance system would grow exponentially over time, as would its rules and regulations, inevitably slowing down the speed of the internet and perhaps technological innovation itself.

The task of obtaining global acceptance and harmonization of these rules of governance and security would be Herculean. Imagine every country, state, and local government being asked to agree to governance and security standards. Things such as this have happened in the past as exemplified by the Budapest Convention of 2001. It would perhaps be as challenging as the creation of cyber police forces that are able to identify, find, and punish those who violate the rules. In addition, the involvement of government entities in this aspect of internet governance would raise complex questions about the very nature of cyberspace, harking back to the early days of internet regulation when the U.S. government flirted with requiring the use of a surveillance chip. The "Clipper Chip" as it was known was a cryptographic device for which government agents held the decryption keys in escrow so that if necessary the government could access encrypted private communications. The algorithm, known as Skipjack, was developed by the NSA,[32] but because of the enormous backlash from internet companies, users, and privacy advocates, the Clipper Chip was jettisoned by 1996. Since then, the internet has proven itself to be largely ungovernable.

The hard choice that the world would eventually have to make is whether it prefers a path to virtual anarchy or consensus governance. Both paths are fraught with challenges and danger. Again, society is confronted with choosing the best worst option.

CHOKE POINTS

The choke-point method of regulation is a traditional way of implementing standards and rules of governance. In the case of the internet, ISPs, backbone providers, social media sites, financial institutions, payments system gateways, domain and internet routing services, pre-network application filters, and undersea cables present the most obvious physical choke points. They could be used to regulate internet traffic and enhance security through the authentication of identities, the imposition of rules, and restrictions on malicious activity. In 2008, damage to undersea cables cut off millions in the Middle East from U.S. internet servers.[33] China already uses junctions to oversee the flow of data into and out of the country. Government-sponsored digital blackouts have occurred in countries like Egypt, Burma, and Nepal.[34] However, these kinds of online controls raise serious political and legal issues depending on who is doing the regulating and who is being regulated. What might start as and look like the enhancement of security and privacy could morph into abusive exercises of authoritarianism. There is no doubt that it is easier not to have to address this issue, but the result of that easier route is the maintenance of a porous system that may itself morph into an authoritarianism nightmare. Ironically, both paths could lead to the same unsavory result.

If choke points aided by algorithms and artificial intelligence were used to enhance online security, it would again raise a host of complex social and political questions that would have to be asked and answered. Which choke points should be used, how much authority should they have, how ubiquitous should choke points be, and who should regulate the choker? The depth of the choke would also need to be determined. Should choke points engage in deep packet inspection that transforms routers and other network backbone devices from being digitally agnostic transmitters to data voyeurs? If one were simply focused on the effectiveness of internet security, law enforcement, and the intelligence agencies, the answer would be a resounding "Yes—we need this kind of an internet!"[35] But as we know, the analysis is far more complex, and the potential reallocation of power is quite significant. Similarly, the creation of master kill switches to disappear malicious travelers and software would raise a myriad of issues that would be difficult to resolve.

DESIGN STANDARDS

Government and private organizations such as those listed above publish global security standards.[36] There are no shortages of standards or standard setting bodies; it is the uniformity of application and enforcement of those standards that is uneven. In the United States, there are laws that require systems to be secure, but they generally do not mandate how companies must attain that goal.[37] Efforts to impose even more rigorous safety, security, and approval standards would meet with stern opposition to the extent that they might alter the nature of cyberspace. Why would companies want to slow the pace at which they reach financial Nirvana by having to subject their products to lengthy FDA-type approvals before they could introduce new software? After all, like credit card fraud, many of the costs of insecure networks are absorbed by the users, either directly or indirectly. Effective security demands reasonable levels of standardization, oversight, enforcement, and testing, but movement in that direction in a world that values virtual social contact and dismisses the likelihood of virtual venereal diseases seems unlikely.

Similarly, the world of source codes and application algorithms is equally unfettered by oversight or regulation. But if the circulation of insecure software or the sale of defective or spy-ridden hardware is dangerous to our health and economic welfare, there would seem to be public policy arguments for the application of standards and routine testing. In the same way, standards for product resiliency, adaptability, and data redundancy would add levels of security. The creation of a library of source codes would perhaps create as much risk as it sought to eliminate to the extent that it would create a most attractive target for hackers. But if we are to build a more secure internet, standardized, enforced design standards for software and devices can be voluntarily adopted or compelled by governments as the landscape continues to get more dangerous.

HARMONIZATION OF INTERESTS

What any one person, company, or country does to improve the security of the internet can only be most effective if it is under an umbrella of global harmonization. The internet will be as secure as the least secure link in the chain. The need for more detailed and enforced domestic and global treaties seems to increase each day as more and more hacks and breaches occur. The recent disagreement over updating the Budapest Convention between countries that want to better secure the internet and those that want to use it as a tool of authoritarianism is a stark reminder of the challenges that global uniformity standards create, even if it is as simple as defining what constitutes a cyber annoyance, crime, or act of war. A more secure internet would require cyber glossaries, peace treaties, and terms of operation and enforcement so that everyone would know who was aligned with whom and how jurisdictions would go about resolving cyber disagreements, conflicts, and warfare. That would require gaining the support of democracies around the world, and that will take leadership that fully appreciates the stakes.

EDUCATION

The knowledge gap is perhaps the greatest intellectual challenge to improving online security. Most of us don't know what we are doing online, approaching computers like our cars—we turn or push something, and it starts. We are the virtual equivalent of online drunk drivers, ever ready to do significant damage to ourselves or others around us as we recklessly speed through cyberspace. Digital security education would have to be generally available in schools so that users could learn the basics of cyberspace and be familiar with the risks of online technologies. Perhaps most important would be increased education of policy makers, political leaders, and regulators, many of whom are to computers as Donald Duck is to a nuclear reactor. Let them operate a nuclear reactor for one afternoon, and I'm sure they could create massive threats to anyone within a fifty-mile blast radius.

To make the current internet safe, users and policy makers must understand the critical importance of basic cyberspace hygiene and security. Identification and multifactor authentication are not options or nuisances that slow down the user's online experience. They are minimal, mandatory first steps to getting online in a safe way. When users understand how to update, patch, and lessen threat vulnerabilities in the software and networks that they control, security will naturally improve. That may seem overly simplistic, but many users and even businesses don't always update software or hardware or use the defensive tools they have in their arsenal to protect themselves and their networks. Better online education cannot be optional in a safer future.

TRANSPARENCY

Closely related to the education required to better secure the internet would be a much greater level of transparency and disclosure by government and private sector players. As new super science technologies and the IoT are deployed at an ever-increasing pace, better forms of disclosure would have to be required to accompany them. The government, for example, could start by more clearly explaining to the public how it isn't protecting them in cyberspace—in effect, what private citizens and businesses must be responsible for doing themselves. Like financial regulation, whenever there is a system of oversight set up and it either doesn't explain the limits of its job or doesn't do its job, the public will be head-faked into relying on a level of protection that simply doesn't exist.

Similarly, there would have to be more than the painfully little disclosure that exists today about the capabilities—good and bad—of new technological products. Do you know how much or who Alexa or your toaster is telling what you say and do? Appliances might be required to come with disclosures about what data they collect. To make an informed decision about the level of security they want, consumers would have to be told whether manufacturers are dissecting the data for their own marketing purposes and then reselling it to data brokers who might slice, dice, and resell it repeatedly to companies that might have either the best or the worst of intentions. A refrigerator, for example, might be required to disclose its security/vulnerability rating derived from a global rating system so that consumers could decide for themselves whether they are at risk beyond the occasional lapse the refrigerator may have in keeping food cold.

Increasingly, systems and users are connected to a nearly endless list of other systems and users who may not be strongly authenticated and can climb over or through someone or some electronic barrier without authorization to gain access to a network. It is a virtual game of leapfrog. Moreover, security must also confront a sort of "quantum mechanics syndrome" challenge. As we dig through the levels of the software, hardware, applications, and coding that make up networks, we find subcomponents and sub-subcomponents whose security may be suspect, built by countries that may have less than altruistic agendas. Every component is separately designed, implemented, and built, but unlike foods where the manufacturer is required to share the many different ingredients from which they are made, software and hardware do not come with a list of ingredients. There is no software bill of materials (SBOM) or device bill of materials (DBOM).[38] The DHS Information and Communications Technology Supply Chain Task Force has established working groups focused on such hardware and software bills of materials, and standards are emerging, but they are voluntary.

Network security is a game of roulette, and the rules are akin to requiring strict identification of everyone living on an island that makes its livelihood from encouraging anonymous tourism. The stakes will only get more challenging as the IoT becomes the norm. Everything that is connected will be vulnerable, and soon everything will be connected. Leaving consumers in the dark is the best way to prevent and delay the improvements and enhancements the internet so badly requires.

ENCRYPTION

Encryption is often viewed as an impenetrable curtain that keeps out malicious applications and nefarious actors. We tend to want to believe that when we send a message or traverse the internet, everything is encrypted and only someone who has the key can decipher the data. But if an encryption algorithm is not strong enough, it will eventually succumb to a well-armed persistent attacker. Encryption is a mathematical puzzle that hides the answer from someone with less mathematical or computing power. There is always a solution behind the curtain. Moreover, even when data is encrypted, it may not remain encrypted throughout its journey as it moves from one network, server, router, or system to the next. And hackers are increasingly creating fake sites with their own digital certificates and icon padlocks next to the site name in the hope of attracting unsuspecting visitors. We may think the padlock guarantees security. It does not. It only means that the site is secured with a digital certificate and that information sent between a browser and the website *should* be secure.

To strengthen the internet, strong encryption protocols would be necessary at every juncture in the millisecond trip that data takes from one point to another. This may mean that the encryption, digital certificates, and algorithms used would have to be regulated and constantly tested. Again, as a society, we would have to balance between security, privacy, and freedom and then decide who would establish those protocols and enforce them. Moreover, encryption can be too much of a good thing. If it were used to encrypt metadata, for example, efforts to decipher the digital breadcrumbs that could lead to hackers could be hindered.

OPERATING STANDARDS

Generations of Americans grew up relying on the seal of approval on a product from Underwriters Laboratories or Good Housekeeping, which they would check for before they purchased a toaster, vacuum cleaner, or chain saw. Those imprimaturs told a less than knowledgeable public that they could rely on the safety and quality of the product. In those cases, the government or the private sector was performing a societal role by conducting the diligence on a product that consumers could never be expected to do themselves in a cost-effective manner.

No such universal system of approvals exists in cyberspace. There are numerous security and ethical standard setting organizations that have sprouted up encouraging member companies to voluntarily agree to adhere to various security principles, but voluntary adherence to a vague and largely unenforceable set of guidelines will not impact bad actors in the least. There simply are not sufficient global protocols for high levels of internet security; tech product labeling; vulnerability testing; or the use, elimination, or suspension of algorithms and applications. Structures and rules could be created to oversee technical, coding, operational, and other relevant standards geared at increasing security, but the bureaucracy would be significant and the challenge of reaching global consensus imposing.

One effort that Congress has initiated under the Secure and Trusted Communications Networks Act of 2019[39] prohibits the use of any federal subsidies to companies that the FCC lists as producing communications equipment or services that pose an unacceptable risk to the national security of the United States. In effect, it creates a negative certification standard. Five Chinese companies are currently on that list: Huawei Technologies Co., ZTE Corp., Hytera Communications Corp., Hangzhou Hikvision Digital Technology Co., and Zhejiang Dahua Technology Co.[40] But the ability to trace and list prohibited products and their components throughout the world is an extraordinarily cumbersome and difficult way to achieve a higher degree of security. It may satisfy the requirements for political cover, but consider how the quality of every product, component, or widget manufactured in or passing through China or Chinese hands could be monitored from a cybersecurity perspective.

CYBER INSURANCE

Insurance could be used to fill some of the holes created by digital insecurity. Reimbursement of consumers damaged by breaches in security or losses incurred because of cyberattacks would address some of their concerns. The imposition of mandatory costs on companies in the form of premiums could impact the levels of security that their products and networks attain by imposing a form of economic carrot and stick into the process. Insurance could also put the underwriting of security in the hands of insurance companies instead of governments or individual companies to encourage improvements, with less concern for the creation of intrusive government bureaucracies.

While insurance could reimburse someone for the financial harm incurred from a cyberattack or data breach, it could not replace or repair personal data, relationships, political alliances, or lives that might be lost due to the insecurity of a computer, network, or system. And no insurer would likely be willing to assume responsibility for the collapse of the Federal Reserve, Treasury, or a major bank at a premium deemed acceptable without government reinsurance behind it. Moreover, insurance might actually increase cyberattacks such as ransomware by creating a known pool of funds for attackers to go after. Recall that Italy all but eradicated the kidnapping of corporate executives in the 1990s by freezing the assets of families that became victims, eliminating their ability to access funds to pay the kidnappers. Each payment of ransomware finances the future of ransomware, so making payment assured through insurance could solidify that business model.

DATA STORAGE

Enhanced internet security could begin with better security of data rather than networks. Such a change would have to deal with the end-to-end collection, storage, and use of personal, business, and other data by a host of tech companies, data brokers, social media companies, merchandisers, and nation-states. Security standards that deal with the protection of stored data as opposed to its collection and use would need to be greatly improved. Diverse regional, national, state, and global requirements that seek to control the collection and therefore the use of data would need to be harmonized and strengthened to prevent them from simply being ignored. In the *Age of Surveillance Capitalism*, Shoshana Zuboff, and in *The Ugly Truth*, Sheera Frenkel and Cecilia Kang pointedly make the case that tech and social media companies have a documented history of not being completely honest with consumers about their data collection and use practices. Surely, if the largest and most high-profile of such companies are not to be believed as they suggest, lesser companies should be assumed to be even greater risks to our personal and economic safety.

While internet security would greatly benefit from uniform global data collection, storage, deletion, and use protocols, even the thought of such new standards would of course raise the issue of governmental intervention and oversight that could chill innovation and slow down the creation and assimilation of new technologies. The interest of businesses that make billions of dollars harvesting and marketing date would no doubt have to be considered, complicating the issues even more. Decentralization of data from networks and the cloud back to users is an alternative that we will consider a bit later.

THREAT INFORMATION AND EVENT SHARING

Since most of the internet is controlled and used by the private sector, the task of securing it requires a seamless effort between the public and private sectors around the globe—quite a daunting challenge indeed. No government can solve the problem of network insecurity by itself. As a result, the sharing of threat information and the details of ongoing attacks between and among governments and companies has become a fundamental part of internet security.

As we have discussed, the financial services sector has created organizations to perform those tasks. But while sharing is an obvious solution, improving the current system requires universal private and public sector collection and sharing on a real-time basis.[41] It would require a centralized authority; data sharing mandates that go beyond mere volunteerism and industry boundaries; and enforcement mechanisms overseen by government bureaucracies, collective business organizations, or both. Moving to that next plateau in threat information sharing would require agreement among legislators and businesses around the globe about greater security standards, notwithstanding the added expense and loss of speed and efficiency. Hopes for such a consensus even in the United States were not helped when Congress left Washington, DC, in December 2021 and could not even agree on meager cyber incident reporting requirements.[42]

When it comes to threat information sharing, the financial sector has some unique hurdles. First, data sets may represent sensitive information that companies would not be eager to share with their competitors. After all, in an increasingly technologically enabled business environment, companies will increasingly compete based on the security and levels of trust they can offer customers relative to their competitors. That is especially true in banking and finance where confidence and trust are part of what consumers look for. Second, heavily regulated financial services companies naturally distrust their government regulators and Congress. They would have little incentive to share information with government entities and regulators that might use it against them in the right (or wrong) political or business circumstances, possibly even to prosecute them and impose fines. Third, since financial institutions operate based in part on the confidence and trust that customers have in them safeguarding their money, their inclination is to be cautious about reporting every threat, attack, or incursion no matter the size or significance since it might be unnecessarily setting them up for a fall.

Trust is an indispensable ingredient in the financial services universe. Financial institutions that report significant internal operating deficiencies and inappropriate or illegal behavior may ultimately not survive. For example, most of the nearly two dozen financial institutions that were the subject of some sort of discriminatory lending practices suits brought by the Clinton administration in the 1990s[43] were subsequently sold and acquired by other banks. The charges brought against them may not have been the sole reason for their disappearance, but it is hard to flourish in the financial services industry if the government hangs a disparaging label on the company and the public loses trust in it. As a result, disclosure responsibilities are usually handled with the greatest of expertise and discretion.

Threat information sharing is an important component of any enhanced internet security program, but if the resulting database is going to be reliable, rules will be required. They would include the creation of mandatory sharing requirements, conditions for the sealing of certain information among a credentialed group of officials, immunity from prosecution under antitrust and privacy laws, and a precise glossary setting forth the cyber incursions that would have to be reported. The complexities baked into these solutions because of the insecurity of the internet and the frequency of cyberattacks suggest that these goals will be difficult to attain. It would be far better to have a less penetrable internet with less to report.

REGULATION

Regulation of online security has been proposed, but it flies directly in the face of the very nature of the unregulated internet. It would require government oversight, or at least government supervision of private sector watchers. Once governments were involved, the internet would rapidly change as bureaucracies took hold of the virtual Wild West, creating an ever-increasing blizzard of rules and record-keeping requirements.

Consider the Biden administration's new rules to *prevent* ransomware. Rules such as these that require notices, sharing, and record keeping by legitimate companies will not eliminate hacking and ransomware attacks that have their genesis in the fundamental defects of the internet, such as poorly coded software and hardware manufactured by a host of unregulated players around the globe. The new rules may help, but they will increase the burden and costs on the legitimate, law-abiding businesses that use the internet to make money. As an example, over the last forty years, the regulatory costs of a vast array of money laundering laws and rules have been enormous for financial institutions. The rules are difficult to comply with and require detailed vertical integration and record keeping throughout the organization. Interestingly, the enforcement actions and penalties imposed by regulators have largely been based on process and procedure violations—the failure to comply with the record-keeping requirements. Few have been based on actual money laundering. And that is in a world where money laundering is ubiquitous and the quintessential bridge that every criminal must cross.

The question of who writes the rules for greater security is a controversial gating issue. While the market is always a good first choice, market-driven rules might not prove to be enough. We have noted the other choices: governments, private organizations, intelligent machines, or some combination of all of them. The use of intelligent machines obviously would have to be initiated by humans somewhere, endowing them with algorithmic intelligence and instructions to undertake the task of identifying violators. The accuracy of that process would clearly be a work in progress. And each new form of government surveillance and policing of the internet would reduce its speed and efficiency and perhaps discourage the adoption of future innovations, things that will be viewed as technological heresy by many. These are tricky tradeoffs that are unlikely to be seriously considered as long as the rosy side of technology is emphasized.

ENFORCEMENT

Enforcement is a key to the safe and sound operation of any system of rules and protocols, in effect, putting cops on the beat. Ultimately, however, it is hard to catch cyber thieves and saboteurs sitting in sanctuary cities or nations around the globe. Similarly, there are relatively modest punishments for selling insecure software or hardware or incurring a breach that exposes sensitive data to theft and misuse. Somehow, the nature of the internet has encouraged us to assume (as may be the case) that there was nothing that the company could have done to prevent the hack. The defeatism built into such a system is frightening. But enforcement is equally horrifying to many who see the internet as the representation of a culture that is free of government interference.

Building effective enforcement mechanisms would require, among other things, that we decide who decides who does the enforcing, on what authority they could act, what their jurisdiction would be, and what investigative tools they could use to find and prosecute violators. That would be a huge task on a global basis. Then a uniform glossary of terms would have to be established, including what constitutes annoyances, civil violations, crimes, or cyber wars. Finally, any system of order would have to define the kinds of civil and criminal penalties that will be available to the enforcers, how due process would be provided, how nation-state perpetrators would be penalized, and whether and when kill switches may be used to immobilize illegal software, hardware, and networks.

If countries, companies, and corporate executives were forced to bear responsibility for a range of actions that impact internet security and compliance programs became mandatory, immediate benefits from a security point of view would likely follow. An ability to extricate hackers and servers from foreign countries would be a game changer, but the prospects of such an agreement with adversarial nation-states are close to zero. The great hurdle with a kill switch or any degree of enforcement would be who or what should have access to it and can trigger it in a world where not all the players have the same incentives or goals. In addition, such immediate and potentially costly actions would require prompt resolution of the due process issues by an appropriate tribunal. As Professor David Clark has pointed out, "management of the trust relationship, the expression and manifestation of that relationship, and the balance between centralized and localized decision-making about trust . . . becomes the defining feature of a successful scheme."[44]

Some enforcement could also be delegated to users. Laws could be adopted to effectively give consumers of tech products and users of networks the authority and standing to bring lawsuits for violations of various standards, protocols, and rules. The prospect of rampaging strike suit lawyers and consumers coming after companies or nation-states for security breaches and other defects that could result in compensatory and punitive treble damages would provide an immediate economic incentive to improve performance. But it would also create inefficiencies and costs to the extent that it encouraged cyber vigilantism and frivolous lawsuits that further reduce the speed and implementation of new technologies.

THE ROLE OF PEOPLE

People are still a large part of the solution to an insecure internet. There is an increasingly acute need, for example, for more cybersecurity experts. Not people who identify as cybersecurity experts, but people who really are. Rigorous, uniform, globally accepted certification standards that authenticate experts who have the knowledge, skills, experience, and ability to build secure networks are few and far between. A similar set of standards would in the best case also be applicable to those who enforce the law in cyberspace. Some have called for a federally funded research and development center to focus on creating an elite cyber workforce for the federal government.[45] There is an equal need for the private sector to do the same.

In addition, more effective cybersecurity would have to focus on incentivizing people's online behavior both economically and through sanctions. Users, providers, and system participants would have to know that they would bear a financial and social cost for negligent or intentional bad online behavior. To the extent that methods of identifying the people and behavior that cause havoc were developed, cyberattacks and online ghoulishness could be reduced. Unfortunately, the online activity of too many people and nation-states has proven that they cannot handle the privilege of anonymity.

FORWARD PREPARATION

To maximize long-term security, as each new technology is adopted, there would have to be processes in place that would evaluate and prepare for the next levels of technological advancement. Hopefully that might become a part of the prudent judgment that businesses naturally exercise, if for no other reason than to avoid being forced to do so by expansive new laws. In the current online environment impacted by the euphoria of new technological products, such forward preparation is not a common occurrence. But the implementation of cloud computing and artificial intelligence, for example, should not be expected to distract from the consideration of how the next generation of super sciences might obsolete those technologies and the security of systems being built today. The introduction of quantum computing would be accompanied by plans to deal with the insecurities it may lead to when the next new form of technology emerges. That process is never ending, but if we really valued internet security, we would have a "one-step-ahead" viewpoint to manage it. The alternative is that technology will manage and control us.

So, what are the chances that all or most of these changes might become reality and transform the internet into something more secure? I think we all know it is highly unlikely given that there is no critical mass of government officials, businesses, and experts in agreement about whether and how it should happen. There is little to no effective leadership on these issues. Even if the United States could get its act together and make many of these improvements, it represents but a segment of the internet. Given current economic realities, it is impractical to pretend that this internet or the critical infrastructures it houses can be made anything close to safe in a reasonable amount of time. That leaves another alternative to consider—creating a new internet or internets that operate alongside the current internet and handle data and transactions that require heightened security. Every day that goes by that we don't undertake the construction of new more secure internets, the further away we get from solving what may become the greatest threat to the future of humanity.

Part V

ALTERNATIVE INTERNETS

18

IF YOU BUILD IT,
THEY WILL CLICK

The fundamental problem with internet security is that it is largely constructed from a post-attack point of reference. The inevitability of incursions is assumed, so most resources are focused on mitigating the damage and getting the company or government system back on its feet quickly. But we seem to be ever so slowly coming to the conclusion that this glide pattern needs to be adjusted. The war in Ukraine and the ever-increasing threats of cyberattacks have forced some of the United States' top cybersecurity officials to conclude that it is time for a change. But the change that they seem to default to is a stronger government role and more pervasive regulation of cybersecurity.[1] That may be part of it, but history teaches us that more regulation, particularly if it is only proffered by some countries and is not uniform, is rarely the most effective or efficient way to solve a problem. Policy makers need to be thinking outside the box of the same old internet. A successful cybersecurity strategy would be able to prevent attacks, but that will require a new internet. It is possible to build a secure new network on top of or alongside insecure ones.

This is not a new idea, but it is one that has withered each time any serious effort has emerged in the face of the weight of the status quo and the attendant costs, whether they be monetary or emotional. The difference now is that we are running out of time to reconfigure the virtual chaotic amusement park that cyberspace has become. There must be a strategy to safeguard the most sensitive data inside secure networks, deploy better technologies for walling them off, and introduce far more reliable encryption technologies. A commission appointed by President Obama after the Snowden revelations composed of intelligence officials, industry executives, and academics called for the United States to do just that.[2] It is insane to keep piling everything of value on networks that are vulnerable when the stakes are so high.

Several significant macro-challenges confront the construction of a new internet. First, the current internet is already morphing into its next iteration, something most users will not appreciate until it has overtaken them. Second, to transmit and store sensitive data or value securely, networks must be highly secure from end to end, something that takes coordination and consistency. Some of the enhancements we have discussed are incorporated in the private cloud networks that financial institutions already use. But since they do not own those networks, having the same level of security provided by tech providers cannot be assured. Third, as far down the path as we are with the current internet, it will be hard to achieve a consensus among global business and legislative leaders to abandon or replace it unless they see that the private and public business of the country cannot realistically be done any longer without information security. Unfortunately, just about every word of every report warns of the future risks of an insecure internet and offers hosts of defensive strategies. That is a path that is unlikely to ever lead to an acceptable level of security.

One solution is a clean-start reboot that transitions everything to an entirely new and more secure internet, but that is highly unlikely to occur. Legislators, regulators, businesses, and individual users would have to create something that in the end would displease almost everyone. There have been efforts toward that goal, but they have all petered out. Alternatively, the focus could be altered to secure data rather than networks, in some cases by moving the data away from central repositories and servers to reside only on a user's hard drive. That will be unpopular with data-hoarding companies and difficult to enforce. Building supplemental new sidecar or layered private networks to improve the capacities and security of those that are already in use by critical infrastructures makes more sense and is more likely to happen. But it is far from a slam dunk. It is simply the best worst option.

This idea has been proposed and ignored before. Michael Hayden, President George W. Bush's CIA director, promoted a new internet infrastructure to reduce the threat of cyberattacks in 2011. For the same reasons, U.S. Cyber Command chief General Keith Alexander suggested a separate ".secure" network for critical services such as banking that could be walled off from the public and require users to use certified credentials. This idea was criticized for raising Fourth Amendment search and seizure issues to the extent that network operators were authorized to do deep packet inspections of traffic content as is done in countries like China.[3] But faced with the increasing prospect of online economic anarchy, we should be smart enough to know how to build out enhanced security networks that are constitutional. As we will discuss below, a .bank top-level domain name regime with enhanced security already exists, but it is not by any means universally used.

When evaluating the levels of security required for different types of online interfaces, it may help to think in terms of the action, data, and value transferred from the perspective of both the sender and the receiver. For example, if the action is playing a video game with limited data and value being transmitted, lesser security protection is required. If the action is reviewing and paying a bill through an online banking account, the data transmitted will include sensitive personal information, accompanied by the transference of money. The level of security required for that should be much higher. The problem that arises however is that a hacker may gain access to a computer or network via an interface such as a video game that has less security and then once inside cross over to the online banking network the next time it is accessed. So the gateways that communicate with new secure private networks (SPNs) must be highly protected.

THE POLITICS OF A NEW INTERNET

There is no doubt that a new internet that has rules and police would represent to many a gross heretical transformation of cyberspace. After all, the internet is the virtual equivalent of the Woodstock music festival—a place where anonymity and unfettered freedom of movement and actions are cherished hallmarks and allow anyone to do whatever they want. But, as in the analog world, there is no free lunch in cyberspace, and there are many evil and malicious actors there. So something will have to give to avert the reckoning that is coming. There will be hard choices and difficult trade-offs to negotiate, including who regulates and polices this new internet. The establishment of self-regulatory standards and organizations is one way of heading off direct government regulation.

In 2017, several tech companies agreed to exchange information with the Global Internet Forum to Combat Terrorism (GIFCT) to self-regulate online conduct.[4] Founded by Facebook, Microsoft, Twitter, and YouTube, GIFCT was designed to prevent terrorists and violent extremists from exploiting digital platforms and to counter the spread of terrorist and violent extremist content online. In January 2022, Tiffany Xingyu Wang launched the nonprofit Oasis Consortium to encourage a more ethical internet. Toward that end, Oasis published user safety standards[5] that companies could voluntarily implement. They included hiring a trust and safety officer, employing content moderation, integrating the latest research in fighting toxicity, and creating a grading system for the public to evaluate.[6] Oasis also proposed using machine learning to detect harassment and hate speech. These efforts are commendable, but general and idealistic principles focused on preventing offensive content and promoting civil discourse do not address the fundamental problems of online insecurity or hinder the efforts of malicious abusers of the internet.

There has been a massive failure of political will to address the problem. Perhaps it has not been quite threatening enough. Maybe it's not a great campaign issue or way to raise election funds. But that is changing by force more than choice. In February 2022, as Russia sat poised to invade Ukraine, cyberattacks swept across the smaller country. In what can only be described as a display of what can be done if we really apply ourselves, the United States and the UK announced within forty-eight hours that they had technical information linking the Russian Main Intelligence Directorate (GRU) with the attacks. In effect, the GRU had been caught transmitting high volumes of communication to Ukraine-based IP addresses and domains. Reporting for *MIT Technology Review*, Patrick Howell O'Neill noted that the speed at which officials were able to apportion blame reflected an "enormous change" in thinking over the last decade that has led to a shift in political will, in large measure because of the increasing realization that attacks would continue to threaten the country in unimaginable ways.[7] Welcome to the twenty-first century, Washington, DC!

WHAT A SAFER INTERNET LOOKS LIKE

It will undoubtedly take a Herculean cooperative effort by the public and private sectors around the world—something that has never occurred before—to fix the internet's problems or reconstruct it from scratch. That is why I think the creation of parallel internet networks is the better and more realistic option. The insecure one that we have can continue to function, but new SPNs should link to it only through secure, monitored gateways that would allow critical infrastructures such as finance and banking to transmit data and conduct business in a safer environment.

SPNs would include several important characteristics:

1. users and applications would be required to be licensed with internet passports and authenticating digital IDs;

2. passports and IDs could be secured by collateral that would be surrendered if the terms were violated;
3. access would be through specialized, secure browsers;
4. software and hardware security standards and review processes would be established and made mandatory;
5. zero-trust architecture security would supplement current authentication and authorization standards;
6. human and artificial intelligence safeguards and checkpoints would be installed;[8]
7. security rules and operational protocols would be enforced by

   a. a cyber police force (human, machine, or both),
   b. the activation of kill switches embedded in licensed software and hardware,
   c. regular testing requirements, and
   d. certifications by cybersecurity professionals and organizations;

8. participating countries, jurisdictions, and servers from which malicious online activity emanated would be held accountable and actors extradited for prosecution; and
9. partnerships between governments and the private sector would be mandated to develop protocols and share information to defend against attacks in real time.

To build a safe online space, no user would be able to use an SPN unless it was identifiable, licensed, and subject to enforcement proceedings, including the triggering of kill switches. Since the governing authorities of SPNs would be able to act swiftly and unilaterally when for instance a kill switch was triggered, such actions would have to be subject to immediate judicial review by appropriate tribunals to ensure due process. We will get to the question of who the governing authorities might be later.

Beyond private networks, ideas have been floated and in some cases implemented to create new internets. Some seek to limit the data available to corporations, some to create more secure networks, some to facilitate illicit activities, and others to control online traffic for authoritarian purposes. Perfect security is not achievable, nor is it necessary. From the perspective of finance and banking, it just needs to be effective enough to protect the integrity of those critical infrastructures and maintain the country's ability to make decisions to prevent dramatic national consequences.[9]

Private networks may be disconnected from the internet to maintain bandwidth capacity or to provide high-volume access points such as those used in high-speed trading platforms, which cannot afford to have high levels of latency or delays in transmissions because every microsecond may equal hundreds of millions of dollars in revenues.[10] Disconnecting also provides a form of security. Fragmentation or modularity of physical networks is a form of SPN and is exemplified by IP-based networks that do not connect with the public internet or that maintain "air gaps" between networks. But blocking interconnection with other networks is not an impenetrable defense given the impact of human actions. Professor Christopher Yoo notes that the Stuxnet virus that damaged Iranian centrifuges in 2010 was transmitted by an infected memory stick inserted by a Siemens employee to update software.[11]

If you are questioning whether the concept of SPNs can work, you need only examine the segregated internet channels that already exist or are being planned. For example, the Dark Web—a virtual universe that most people don't know about and have never experienced—flourishes as a separate internet on the other side of special browsers that can access it anonymously. China has effectively created its own version of the internet for its own authoritarian reasons, controlling access to and from it. And experts around the world have been theorizing about and working for two decades on a new internet that has better accessibility, utility, economics, security, and alternative decentralized data architectural designs.[12] But to date, a combination of economic incentives, cultural independence, and a willingness to live with the status quo have dulled support for a new, more secure concept of the internet. It essentially boils down to a failure of vision, will, and leadership.

A CLEAN SLATE INTERNET

Formal and informal efforts have been launched to describe what a different and more secure internet might look like. None have produced a serious effort to actually build one. Given the resources and time that would have to be devoted globally to do so, the increasing threats appear not yet to present a convincing enough case to replace an internet that seems to work well enough. In short, it is difficult to marshal the will and resources to build something new and better until the economics or personal risks of continued insecurity become more compelling. Instead, virtual reality expands each day, much like the universe and dark matter, with the expectation that its insecurities will be patched by someone along the way. But some smart people think that should change.

In 2003, a collaborative report involving the MIT Laboratory for Computer Science, the Air Force Research Laboratory, and the UC Berkeley ICSI Center for Internet Research concluded that the future of networking was being shaped by short-term thinking and modest changes to the internet to fix specific short-term problems. It saw no overarching vision of what a future internet should be, a comment that seems equally applicable today.[13] Nevertheless, noting that the internet had been developed for purposes other than it had come to serve, the researchers determined that increased security was the most pressing need for the internet. However, there was no clear or specific description of the "security problem" that needed to be fixed. That may in large measure be a function of the fact that everyone's views of the enhancements required will always be impacted by the many conflicting interests that need to coexist in cyberspace. "Users want to have a private conversation, while law enforcement wants the ability to carry out lawful intercept. Users want protection from spam; spammers try to circumvent their protections. Conspiring users share music; the rights-holders try to prevent this."[14]

In 2005, Stanford University, in conjunction with the National Science Foundation (NSF), launched a "Clean Slate" project to determine how an internet should be built knowing what everyone knew at that time. A summary issued in connection with its workshop in July 2005 noted that despite the internet's "critical importance," it suffered under "the weight of incessant attacks ranging from software exploits to denial-of-service," primarily because it was designed with a "benign and trustworthy environment" in mind.[15] Nick McKeown, an associate professor of electrical engineering and computer science at Stanford and leader of the Clean Slate project, was reported to have described the serious deficiencies in the internet as follows: "If air-traffic control was carried on the Internet, I, for one, wouldn't fly."[16]

The goals of the Stanford project included finding a way to filter traffic between computers that lacked the latest security patches, overhauling the wireless spectrum to allow wireless devices to locate and use pockets of unused spectrum, and adding a security layer to the protocol stack. This latter goal was ultimately determined to be an impractical way to secure the various layers and the unanticipated interaction between multiple protocols.[17] A 2007 report identified that the most significant challenge to the clean slate approach was the ability to create a methodology to determine when the newly designed architecture was "sufficiently good" to attract real users and their traffic.[18] The assumption seemed to be that an internet for all was the goal rather than SPNs that served infrastructures requiring higher levels of security.

In 2008, a researcher at the Georgia Institute of Technology speculated that if new protocols were developed, they might not be deployed in less than five to ten years, with the result that technology might have already passed the relevancy of any clean slate architecture by. He also underscored the testing challenges, indicating that no matter how expensive or fast they were, they would not capture the complex interactions of the internet, exhibit the "diversity and heterogeneity of the technologies and applications used," or reproduce the "closed-loop multiscale characteristics of Internet traffic." He concluded that the best use of resources would be to invest in the creation of multiple competing solutions, relying on the real world to select the fittest innovation given the constraints and requirements at the time.[19]

In June 2009, another effort affiliated with Boston University determined that the internet's architecture originally did not take security into consideration, but that a "clean-slate Recursive Internet Architecture" could be designed to resist most security attacks in a "more manageable, on-demand basis, in contrast to the rigid, piecemeal approach of TCP/IP."[20] In 2011, a study supported in part by a grant from Intel Corporation and the NSF concluded that the internet was facing "unprecedented challenges" requiring a "new clean-slate architecture." The most recent private study of a clean slate approach to the internet in 2015 characterized the internet as an arms race between increasingly sophisticated attacks and continuing protocol fixes.[21]

Technology, economics, and familiarity have conspired after so many years to create powerful forces of inertia that will be difficult to budge. However, the increasing concern over the control of data by internet giants like Google and Meta/Facebook is creating a new energy to change the internet. The advent of 5G wireless communications and quantum computing are now also tripping government alarms about the insecurity of the current internet. The Department of Energy (DOE) has begun to consider the construction of a quantum internet to provide a more secure virtual environment for financial, health care, and other sectors where greater security is mandatory. We will return to that in a bit.

NETWORK DUPLICATION

Is it possible to slice off or clone pieces of the internet? Apparently so. In the early days of ARPANET, the precursor to the internet established by the U.S. Defense Advanced Research Projects Agency (DARPA), the communication between computers relied on the allocation of computer names and numbers organized by one man—Jon Postel. He maintained the hosts.txt file and, in conjunction with contracts with the U.S. government and the Stanford Research Institute, ran the internet. As it grew, however, other hosts emerged, leading to the creation of the Internet Assigned Numbers Authority (IANA), which Postel then ran. In 1992, Network Solutions, which also sold domain names, won a U.S. government contract to control the root, which directs all traffic to its appropriate website. In 1998, Postel challenged the loyalty of administrators of eight of the twelve regional DNS servers, asking them to recognize his computer as the root and effectively dividing the internet in two, creating a governance dispute that continues to this day.[22] It is obviously more complicated to migrate to a new severed internet that would be more secure, but the lack of will seems to be more challenging than a lack of know-how.

RESTRICTED TOP-LEVEL DOMAINS

The need for a separate and more secure internet for critical financial infrastructures has been recognized in the banking industry in the last decade as a gated domain was launched through the deployment of the .bank top-level domain (TLD) name. fTLD Registry Services LLC was granted the right to operate the .bank TLD on September 25, 2014, and it became publicly available in June 2015. As a coalition of banks, insurance companies, and financial services trade associations from around the world, it operates what is described as a trusted, verified, more secure, and easily identifiable online location for these financial companies and their customers.[23]

fTLD's .bank TLD is a proxy for a more secure internet. It uses the pipes, cables, and providers of the internet, but it imposes heightened security for passage to and from .bank entities. To be awarded the right to use a .bank domain name, institutions are required to undergo a thorough verification process and agree to comply with registry policies ensuring ongoing compliance with fTLD's security requirements, which allows customers to know that the site or email is trusted, verified, and more secure. Users must also accept domain name system security extensions, which verify those reaching the web page of the institution hosted on .bank name servers. Banks using the .bank TLD are continuously subject to compliance with relevant security requirements published by fTLD. Email authentication must be employed to protect against phishing and spoofing emails. Registered .bank institutions must also (1) be reverified by fTLD annually, or when certain registration information changes; (2) utilize multifactor authentication for access to their registrar to change the bank's registration information; (3) agree not to make proxy/privacy registrations; and (4) implement the encryption standards of NIST Special Publication 800-57.[24]

Despite the effectiveness of such a system to reduce fake websites and phishing attacks, only about 650 of the nearly 5,000 banks in the United States as of January 1, 2022, had adopted the .bank domain name, most of them community banks.[25] As a former regulator, it is hard to imagine why the bank regulators have not required banks to use these kinds of technologies to increase security. fTLD offers a similar service to insurance companies through its .insurance TLD.

DECENTRALIZATION

The most powerful driver of a new internet today is not a sudden realization that there is an alarming absence of security. It is the growing concern over the concentration, location, and control of data in the hands of Big Tech companies. While completely justified in its own right, it has created a strange but potentially effective partnership between consumers, privacy activists, governments, and competing businesses. It is part of the impetus behind the Digital Markets Act provisionally agreed to in late March 2022 by the European Parliament and Commission to curb the market power of gatekeepers—big technology companies that provide "core platform services." It will prohibit those gatekeepers from using search engines, social media networks, e-marketplaces, and web browsers to engage in certain "unfair practices" with regard to how collected data may be used and market power may be exerted over other businesses.[26]

In the 1990s, the first generation of the internet—Web1—democratized access to information, but it offered only modest navigability. In the mid-2000s, Web2 made it easy to connect and transact business online. It also marked the next phase of how companies could amass economic and political power through the collection of enormous amounts of data about users. As Web3 emerges, one of the goals being pushed is the decentralization of the internet, and the corresponding reduction of the power Big Tech has amassed, by creating a new network that can run alongside Web2. Some Web3 models would create social networks, search engines, and IoT-connected marketplaces that have "no company overlords" and operate on personal clouds or blockchain platforms such as cryptocurrencies use.[27] Some visualize internets where no one needs permission from a central authority to post anything. While freeing users from the tyranny of data control, this formulation is unfortunately the antithesis of a secure internet.

Distributed ledger technologies (DLTs) allow for data decentralization and create a transparent and perhaps more secure environment, allowing everyone to own their data and reduce the threat of surveillance and exploitative control by data hoarders.[28] NFTs, or nonfungible digital tokens of collectibles, are an example of how DLT tokenization is being incorporated into virtual lives. NFTs are precursors of a metaverse existence where consumers can buy and sell goods without fear that their digital products can be stolen or replicated thanks to "immutable" (for the time being) DLTs. Twitter is studying ways that integrate these technologies so that users might be able to tweet from their cryptocurrency account.

A free and unfettered internet that is decentralized and reduces the all-encompassing and intrusive roles of Big Tech companies seems to be the goal for many. Technologist Mark Nadal labels his concept a "dWeb,"[29] where DLTs run applications, such as those that run Bitcoin and Ethereum, to create the foundation of a new internet that allows data to reside with users rather than tech companies. Nadal's dWeb engineers a better internet by using a graph universe node (GUN) to provide a foundation for decentralized sites and apps running on devices that have access to the internet. The arkOS project is another effort that would allow individuals to create their own personal clouds and control their data.[30] With increasing political pressure on Meta/Facebook, Google, Amazon, Microsoft, and Apple, decentralized internet proponents see an operational and political opportunity to offer an alternative in which the cloud converts to a network of personal laptops, phones, and smart appliances around the world.[31]

Frank McCourt, the former owner of the Los Angeles Dodgers, is also concerned about the centralization of data by large tech companies. He believes that the internet is broken[32] and complains that it has been transformed "from something that was designed to create value for society to something that creates value from it."[33] His Project Liberty, to which he pledged $250 million, was formed to create new social networks where users, not platforms, controlled their own data, undercutting digital data monopolies that "might use data and algorithms to sow discord, suppress innovation and compromise privacy."[34] Similarly, Dominic Williams, the founder and chief scientist at the Dfinity Foundation, developed an internet computer that runs on a blockchain network that works at "web speed."[35] Whether cryptocurrencies survive long term to change the financial world is debatable, but the DLT applications they sit on seem more likely to change the future of the internet given their ability to decentralize the storage and transmission of data and peer-to-peer authentication.

In 2018, Solid, a decentralized platform created by Tim Berners-Lee, the director of the World Wide Web Consortium, attempted to give users more power over their data by decoupling applications from the data they produce.[36] Using this concept, a bank could provide customers with a program to run economic wealth creation models, but the data input, analysis, and output would reside on the user's machine, and since the bank would not control it, no one else should be able to access it either. The theory is that centralized databases would begin to disappear, and with them, hopefully some of the security problems that exist online today.[37]

The decentralization of networks and data creates new challenges, however, including prohibiting users from creating their own malicious or malignant networks or avatars. Maximum financial security might be attainable, for example, if banks were required to own and control the pipes, routers, cables, and servers that made up their own SPNs. But that would require, among other things, impractical redundant costs. The alternative would be SPNs that continued to be owned and controlled by outside providers who, like banks, would become subject to close supervision to ensure the safety and stability of the financial system. Any way you slice it, greater security raises complex governance questions, such as who watches and regulates whom in a decentralized Web3.[38]

DATA TRUSTS AND MEDIATION

Data trusts offer a way to create better security and also encourage data interoperability by ensuring that individuals have consented to the various uses of their data, removing data bias, and "de-identifying" personal data.[39] Beginning in 2017, data trusts that would use trustees to manage and protect the information rights of their beneficiaries were first proposed in various parts of the world to, among other things, train artificial intelligence programs and give communities greater control over their data.[40] An article in *MIT Technology Review* suggested as an example that a group of Meta/Facebook users might entrust their representative trustee to determine under what conditions Meta/Facebook would be permitted to collect and use their data.[41] If Meta/Facebook violated those rules, the trust might retract access to its members' data or take other action to enforce their rights. Like health insurance providers with many customers, data trusts could theoretically negotiate more favorable terms of use for their beneficiaries, assuming that Meta/Facebook would even agree to deal with them. Like most things in life, it would be a question of leverage.

Europe is forging ahead in this area. A new European data governance strategy would make the EU an active player in facilitating the use and monetization of its citizens' personal data. Unveiled by the European Commission in February 2020, the strategy outlines policy measures and investments to be rolled out over the next five years.[42] It will create a pan-European market for personal data through a data trust that manages people's data on their behalf and has fiduciary duties toward its clients. The first initiative put forth by the new EU policies was to be implemented by 2022. Global technology companies would not be allowed to store or move Europeans' data. Instead, they would be required to access it via the trusts, which would pay "data dividends" that include monetary or nonmonetary payments from companies that use their personal data, creating the world's largest data market.[43]

In 2017, Sidewalk Labs, a Google affiliate, outfitted Toronto's Quayside waterfront, a smart neighborhood, with sensors to enhance shoppers' and diners' experience. To deal with the ensuing criticism of unauthorized surveillance, it suggested the creation of a data trust under the supervision of a community review board to ensure that the data collected would be used properly pursuant to the issuance of licenses for the placement of each sensor.[44] Sidewalk Labs abandoned the Quayside project in May 2020 given the complexities of the plan and its implementation.[45]

Data trusts substitute one form of intermediary for another, but they can also create intermediaries where there were none before to provide essential online services. The challenges lay in creating large enough groups of users that are willing to pool their data for commercial and privacy purposes, finding agreement on the data and rules involved, and then identifying new trusted intermediaries and a corresponding governing bureaucracy. Those intermediaries and systems will need to be able to foster the same sense of confidence that the financial and other traditional intermediaries that are being disintermediated today by the democratization of data and the emergence of peer-to-peer networks were able to. Ericsson, headquartered in Stockholm, Sweden, has been proposing such approaches for many years.[46]

CHINA'S OTHER INTERNET

China effectively operates its own internet, at least with regard to access and available content. This is part of its long-term plan that considers the internet as fundamental to its digital foreign policy and use of censorship tools.[47] China controls digital traffic entering the country through a software algorithm that doubles as a digital Great Wall of China to block certain URLs and any IP addresses associated with them. In his book *The Great Firewall of China*, James Griffiths also describes how China controls ISPs and web companies within the country through its Golden Shield Project, which again ensures that only *appropriate* content can be seen inside China.[48] This is all possible since global internet traffic, as it does in every country, generally connects to China through several cables and choke points. In China's case, there are only three: Beijing, Shanghai, and Guangzhou, which is near Hong Kong.[49] Physical control of internet access points has allowed China to maintain its own version of cyberspace. So, for example, while Meta/Facebook dominates social media throughout the globe, WeChat is the alternative app Chinese users "favor." It sends messages and can be used to pay bills and purchase a wide variety of services.

But China wants to expand the concept of its internet. In September 2020, Chinese scientists, including some from Huawei, endeavored to persuade delegates in Geneva from more than forty countries of the value of a new infrastructure for the internet that would transfer power from individuals and a handful of U.S. corporations back into the hands of nation-states. China's geopolitical goals were transparent. Such an internet would promote data-sharing schemes across governmental entities. China's presentation painted a picture of a "digital world in 2030 where virtual reality, holographic communication and remote surgery are ubiquitous—and for which our current network is unfit."[50] Users of this new internet would need permission from their ISP to do anything via the internet, and administrators would decide who had and did not have access to it. Users would connect to segmented national internets, each with their own rules. China intended to test such a new structure despite significant global criticism.[51] While the concept can facilitate greater online security, ubiquitous government control of these internet structures would not be acceptable in most countries.

These and other efforts by the Chinese, alongside Russian proposals to adopt a new Budapest Convention, are the pieces of a plan to wrest control of the internet from the United States and Big Tech and deliver it to governments presumably acting on behalf of the general good. At best, this model gums up the progress of technology in government red tape. At worst, it underwrites the authoritarian methods that countries like China and Russia are increasingly using to surveil and control their citizens as well as those in other countries where they are impacting the control of data, natural resources, and national economies.

Most recently, Russia's invasion of Ukraine may as a matter of necessity foster the concept of a new internet or internets. When Russia blocked certain internet transmissions and participants—Meta/Facebook being the most notable—and other Big Tech companies voluntarily withdrew from Russia, the seeds of a "splinternet" were planted. The next steps would only need to deploy different, incompatible protocols with separate governing authorities, and the foundations of new and separate internets emerge.[52] Indeed, the actions of China, Iran, and North Korea are consistent with the maintenance of separate internets. Is this the window of opportunity that the world needs to remake the internet into something more secure and reliable?

THE DARK WEB

Consider the Dark Web and what it tells us about the ability to operate parallel internets. Most users have not been on the Dark Web—they don't need to and may not know how to access it. On the Dark Web, digital traffic is anonymized and encrypted, making it private and capable of circumventing geographically imposed restrictions on content.[53] Access to it is limited to users who have the required software browsers to transport and transform them into completely anonymous travelers consistent with the rules of the Dark Web. The simple fact that it exists suggests that alternative internets are not only possible but are already flourishing, albeit some on the seedy side of cyberspace.

The Dark Web is part of the Deep Web, much of which includes criminal and illicit behavior ranging from terrorism to child pornography. Tor ("The Onion Router" project) is an open-source software browser that is freely available to access encrypted and anonymous entry points and pathways into the Dark Web. The core principle of Tor was developed in the mid-1990s by U.S. Naval Research Laboratory employees, mathematicians, and computer scientists to protect U.S. intelligence communications online.[54] It camouflages users' identities and locations by routing them through intermediate servers in a global network of thousands of relays that anonymize a user's true IP address. It also makes it very difficult to identify the websites a device has visited.

Tor blankets the sender's message in layers of encryption and then relays it through "nodes," or other computers operated by Tor users, which then remove the layers of encryption. Each node only knows the identity of the previous and forwarding node. Nodes cannot identify others in the chain.[55] Tor is the browser of choice to roam the open web anonymously, including accessing hidden services sites, because it is difficult to track messages from their origin to terminal points. Most hidden services sites contain illicit material and are accessible only by invitation from the administrators.

Tor users include those who wish to keep their internet activities private; who may be concerned about cyber spying; or who are activists, journalists, and others seeking to evade online censorship. While the role of the Dark Web, Tor, and hidden services sites can be debated under moral and legal standards, the fact remains that similar software, protected sites, and secure communications should just as easily be able to be created to support a more secure alternative internet if economic and security concerns demanded it.

THE QUANTUM SOLUTION

The fact that many smart people have considered how a more secure internet or similar networks could function suggests that staying with an insecure internet is a choice, not a necessity. The most recent party to join the chorus of voices choosing to create a new internet is the U.S. government.

As we have seen, the insecurity of the current internet is a function of a combination of factors, including software that is not hack proof, hardware that is not constructed to standard security protocols, networks that are not secure end-to-end, asymmetric key exchanges and public key digital signatures that are breakable, human error, sloppiness, and a range of people who believe that it works well enough. If quantum computing eventually emerges, it will theoretically provide the computing power to enable a generation of machines that can perform calculations with lightning speed based on the probability of an object's unmeasured quantum state in a mixed superposition. While some commercial protocols allow a preshared key option and others the combination of preshared and asymmetric keys that can reduce the threat of quantum attacks that breach current cryptographic security, these can be unusually complex processes to implement.[56] A better solution is required.

The bottom line is that quantum computing or whatever the next generation of computing power looks like may obsolete every form of security that businesses and individuals are using today and require new forms of security. The fact that the U.S. government has taken notice and is in the process of trying to create quantum-proof encryption by 2024 suggests that the threats created by quantum computing are real, particularly if more benevolent countries are not the first to attain quantum supremacy.[57]

A quantum internet will be able to transmit large volumes of data across immense distances at rates that exceed the speed of light in a way that is impervious to cyber hacking. It would require new quantum routers, repeaters, gateways, hubs, and other quantum resources. It would not displace the current internet. Communications and transactions of the financial, military, and power grid sectors, which required more secure transmissions, would be migrated to these more secure quantum internets. The unauthorized interruption of communications between a sender and receiver that are encrypted using quantum cryptography would theoretically change the key and be noticed immediately, making its security highly impenetrable—until the next generation of computing power appeared.

It is anticipated that quantum networking will connect quantum computers through a quantum network built around quantum-based encryption and be able to provide computing capacity at a scale currently impossible to achieve in a single quantum computer.[58] This is triggering yet another technological arms race. The U.S. Department of Energy announced a project to create a working prototype of an *unhackable* quantum internet within the next ten years under the National Quantum Initiative Act. A quantum internet backbone is expected to be able to connect researchers to powerful new tools through the eventual networking of distributed quantum computers interfacing with satellite links or classical fiber-optic networks.[59] This network would run parallel to the current internet as a supplementary alternative for critical infrastructures. The first fifty-two-mile "quantum loop" in the Chicago suburbs is to be connected to the DOE Fermilab in Batavia, Illinois, to establish an eighty-mile test bed, with all seventeen DOE national laboratories forming the quantum internet backbone.[60] While we should take some comfort from this effort on behalf of the government, the relatively small amounts of money being devoted to it over the next decade are not encouraging.

As you might guess, the United States is not alone in developing quantum networks. China is its most significant rival with a 1,263-mile quantum link already in place between Beijing and Shanghai. China's link is not a fully realized quantum connection yet. Perhaps even more significant, however, is that Chinese scientists have established the world's first integrated quantum communication network. It includes more than 700 optical fibers on the ground with two ground-to-satellite links to achieve quantum key distribution over 4,600 kilometers, connecting more than

150 industrial users across China, including state and local banks, municipal power grids, and government websites.[61]

Scientific American notes that "China has more total patents across the full spectrum of quantum technology, but U.S. companies have a dramatic lead in quantum computing patents. And of course, China has a more sophisticated quantum network and claims the top two quantum computers."[62] It is making bets on quantum computing, feverishly collecting encrypted data today in the hopes of decrypting it when quantum computing arrives tomorrow. It is important that repressive governments not achieve quantum supremacy first given what is at stake for the world.

I have learned to be skeptical when anyone tells me that anything derived from a mathematical or scientific formula is unbreakable. Most mathematical formulas are resolvable, and most security built on mathematics is eventually breakable. The science underlying a quantum internet is likely to be harder to hack, which is all that anyone can realistically expect.[63] But a more powerful technology will always follow, changing the dynamics of security yet again. Governments around the world should use the opportunities provided by the emergence of quantum technologies to move security rather than speed and innovation to the forefront. Every step taken in the technological evolutionary ladder will impact geopolitical issues and shift power based on who controls those emerging technologies. We need to be much smarter about dominating, directing, and channeling new technologies.

SECURE ALTERNATE INTERNET ARCHITECTURES CAN EXIST

The fact that there are so many public and private sector entities trying to construct a new concept of the internet is strong evidence that the idea has merit. Building a new more secure internet may be as straightforward as reconstructing networks to insert air gaps and offline systems. The air-gap approach can be applied to any type of communication or network, essentially disconnecting or blocking certain communications. But to understand the capabilities and limitations of alternative networks, it is important to understand the architecture of the internet.

One of the best analyses of its architecture is contained in David D. Clark's 2018 book *Designing an Internet*. Professor Clark was involved in the Clean Slate project discussed above. His review of alternative internet architectures includes a metanet proposal from two decades ago that would rely on regions as the central building blocks of the next generation of networking. Similarly, around 2003, the Plutarch project experimented with ways of creating cross-region architectures and modifying routing to link regions, as did Sirpent (Source Internetwork Routing Protocol), which relied on routing packets of data so that the destination node would have a source address.[64] Other such alternatives as TRIAD, DONA, NDN (named data networking), PURSUIT (publish/subscribe internet routing paradigm), and Netinf (network of information) have all faced the challenge of creating a routing scheme that could accommodate the global scale required.[65] These architectures may be resurrected to the extent that they are better suited to the IoT which connects everything.[66]

These experiments faded given the dominance of the current internet and the complexity of connecting architectures. Clark concluded that these factors made it essentially "impossible" to move to a better solution.[67] The impossibility was not a technical one but one driven by the weight of the status quo. But the status quo is changing today driven by concerns for data concentrations, security, and government repression.

Companies are currently working on bringing wireless internet connectivity solutions to places around the globe that do not have traditional access, using high-altitude balloons and solar-powered drones. Menny Barzilay, the CTO of the Interdisciplinary Cyber Research Center at Tel Aviv University and the CEO of FortyTwo, says, "Though daring, a worldwide wireless Internet is inevitable. It simply makes more sense than spending trillions on upgrading super-costly physical infrastructures."[68] His view is that a next-generation internet—an alternative global network (AGN)—would provide a "worldwide wireless Internet access solution" to support a new way of networking. Instead of using the traditional TCP/IP stack-based network, an AGN would be IP based but built upon a new connectivity model that is more secure. For example, if a malicious entity sought to exploit an AGN protocol to launch a denial-of-service attack, when the first attack occurred and had been analyzed, the AGN provider would update the entire network in seconds to prevent the initial attack from recurring.[69] A network could consist of any set of devices connected directly to the AGN, which would authenticate users for any application that required it.

InterComputer Corporation's InterComputer Network (ICN) attempts to solve the dilemma caused by the fact that the internet and the World Wide Web were designed to share rather than securely store information, a goal that has naturally led to enormous online vulnerabilities.[70] It offers a network authentication application and digital message transfer system that is separate and apart from the Web but can be integrated with legacy systems to protect the security of data, while at the same time allowing websites and applications to be imported and run within it. It uses three-factor authentication that essentially confirms who users are, what they know, and what they have, claiming to surpass the security of two-factor verification and the "mutability" of Blockchain. Network solutions like this are capturing the attention of insurance companies that are increasingly concerned about the cyber risks they are underwriting and are actively pushing the insured toward more secure solutions like ICNs. We can only hope that the market follows these cues.

THE MOTHER OF ALL INVENTIONS

If such alternative, more secure internets are possible, why have we stayed so long with one that is insecure? Scale has been one issue, inertia another. Just as the United States stayed with magstripe credit cards more than a decade after more highly secure chips were available, the more insecure version of the internet has prevailed because of the hurdles, including the cost, of replacing or supplementing it. The capital invested in the status quo is a natural impediment to change and often doesn't budge until there is a more pressing economic need or security disaster.

When I refer to the status quo, I mean that in the broadest of political, economic, and social contexts. It is not just big business. The status quo includes large companies that can afford to expend large amounts of money on cybersecurity, as well as small businesses that have invested in expensive hardware and software and are reticent to upgrade at considerable cost. It includes users who are familiar with one system and are inclined to continue relying on it until a crisis-level event occurs. It also includes governmental entities that create bureaucracies that employ thousands of people focused on a specific goal or task who are not eager to rock the boat in a way that impacts their jobs. Importantly, it includes the many new cybersecurity professionals who make up a business now projected to generate $350 billion in revenue by 2026.[71] There is a lot of wealth and jobs riding on the continued insecurity of the internet.

Necessity will eventually prevail—I hope. A crisis-level event is approaching if it is not here already. Perhaps the unbounded nature of the metaverse will trigger a change. Ultimately, however, we will have to choose between beating it to the punch or allowing it to swallow us in a technological and geopolitical tsunami. To do that, we will need the leaders and political will to get us there. Almost everyone sees the problem and the need for a more secure internet. Short-term thinking and real-time patching only go so far. Technology is playing the long game, and if we don't control it, it will control us. The great frustration is that we can do this if we have the will.

I had lunch a while back with a friend who had been around financial services and technology even a few more years than me. He asked me what the solution was to the obvious challenges I had laid out. I told him that I had given it much thought throughout the years and believed that my five decades of financial markets and regulation experience led me to conclude that we have the technological and engineering capacity to build a more secure internet, but technical capacity alone won't change the internet. The solution will require the will to change, and that will require the kind of extraordinary leadership that comes around perhaps once in a lifetime.

I believe that given the overarching economics created by commerce on the internet, our inability to convert to a more secure internet is largely a problem of vision and leadership. Technological reconstruction of the internet is doable. It will be expensive, uncomfortable, unpopular, and disruptive. Businesses won't spend the money to create a more secure virtual world until they come face-to-face with the reality that not doing so will cost them more than they are making, or until they are directed by law to do it. Regulators don't have the authority, resources, or know-how to do it. Legislators are generally not technology literate and are driven by the views of constituents who are making gobs of money off the status quo and are more interested in political issues that can be used to raise campaign funds. But it is doable.

Unless one of several disaster scenarios occurs, it seems unlikely that we will do anything other than muddle up to the point of no return and roll the dice, hoping that we haven't gotten too close to the precipice. Someone or some group of influential people will have to demonstrate the kind of old-world leadership that has changed the world in the past. The Cyberspace Solarium Commission made a valiant effort at it, but it wasn't enough to change the world. Alternatively, it may take a financial Armageddon or digital Pearl Harbor that will scare everyone into action. That is hardly the best route to get to a more secure internet since it is unclear what would be left to be converted at that point. Economics could ultimately drive the need to create a new internet as technology evolves and subsumes or cannibalizes parts of itself. It will take a solid financial case that demonstrates the profit potential of a more secure virtual environment.

Who must lead in these efforts? Like the events of Bretton Woods that forged a reluctant consensus about the dollar as the global reserve currency in 1944 after World War II, leadership will have to come from the United States in an alliance with other democratic countries. Why the United States? This kind of sea change in the future of cyberspace and whatever other spaces follow it will require the kind of leadership that a superpower can offer to the world. Economic, military, and technological power still carries influence in this world as they did at Bretton Woods. And the United States is still for the moment the preeminent superpower of the world with the finest democracy. But the time for the United States to step up may be limited. Like Bretton Woods, if done right, opposing countries will know that being left behind, no matter how strenuously they object, will leave them technologically and economically isolated. If China seizes global technological and economic superiority, the United States may lose its chance to forge an alliance that has enough power and resolve to create a new, more secure internet that others would ignore at their peril.

My friend also asked me what the most difficult issue was that I had considered while writing this book. I told him it was how this all gets done in the complicated, crazy world we live in today. I told him that I had done something on a much smaller scale, and it had worked. To work, things like this have to come from the ground up and be driven by business concerns, the law, and economics. That is what drove the twenty-country cyberspace jurisdiction report that the Cyberspace Law Committee issued in 2000. There was a crushing need for change, and there was a built-in economic incentive to champion that change—lawyers who wanted to seize the opportunity to get in on the ground floor of the next big thing. That is why those lawyers from twenty countries gave their time freely to work on the cyberspace jurisdiction report to change the world—they wanted to be the ones who changed the world and be there to consult and advise businesses on how to implement and live with the changes. It was all free labor, but highly motivated.

The same grassroots efforts could work to rebuild and fix the internet. Governments should empanel groups of lawyers and investment bankers and charge them with coming up with a solution they can eventually make money from, and they will do it for free, whatever it entails. And they will then work tirelessly to lobby clients, legislators, and decision makers everywhere to execute the plan. Their payoff will be clients and business for life. It has worked before; it can work again. But the clock is ticking.

19

THE NEW WATCHERS

There is no order without oversight. As much as most would like the internet to continue as a virtual free-for-all, if it is going to be used to transport and store the world's most sensitive business, social, and governmental data, we must reimagine what it is and how it can operate safely and effectively. Someone or something will have to govern and police a new internet that has a governance structure and rules. I realize that is a horrifying thought, but there is no place in society where there are not enforcers of rules. And there is no alternative but a kind of virtual anarchy where criminals, hostile nation-states, and fanatics will have more incentive to act out online. It is merely about risk distribution. Enforce the rules in the analog world and not in the virtual world, and crime will gravitate to the virtual world.

Who watches and enforces what goes on, and what governance structures and rules are applied, will ultimately be a function of what society wants the internet or its successor to be. A thousand people would likely answer the question in a thousand different ways, most of them reasonable and commercially viable. At a minimum, any new internet should reflect what it has become—the virtual alternative to analog life. Therefore, it should be at least as secure and safe. We should be able to research and educate ourselves, socialize, transact business, transfer and invest money, and live virtually without fear that any of these things can be easily compromised by MUTs or repressive governments. That is an extremely complex goal, but it is achievable if we have the will. If we don't, we remain on an inevitable track to virtual anarchy and a potentially crushing cyber collapse of the economy. These are extraordinarily complicated choices, but they must be made. Finding the best worst option is excruciating.

233

If there is going to be a new internet with segregated databases and SPNs, there will have to be centralized oversight and enforcement of licensing for users and providers, safe passage transmissions, codes of conduct, enforcement against violations, and the implementation and execution of kill switches. In her book *The Age of Surveillance Capitalism*, Shoshana Zuboff asks not only who should be the watchers, but also who should watch the watchers. This is the most difficult part of reconfiguring cyberspace and realizing a more secure environment. To authenticate users, there would need to be a heightened system of certification authorization, but no matter how many certificate authorities (CAs) are added to the vertical authentication ladder, there will be a question of trust. Why should China trust American CAs and vice versa? It is unlikely that an agreement will ever be reached on this question that would satisfy both democratic and authoritarian governments. This suggests the limited sidecar approach and a "my way or the highway" strategy implemented by democratic nations with the economic and political might to compel a second Bretton Woods–type agreement. That is not stuff for the faint of heart or leaders with weak backbones, but neither is a digital Pearl Harbor.

Effective oversight, governance, and policing of a new set of internet SPNs can be divided into several categories, with the most effective and efficient models including components of each.

MARKET INCENTIVES

Economics and competition are effective market overseers, particularly when they reflect a public-private consensus and properly incentivize behavior. Income statements and balance sheets drive companies, processes, and policies. If there is no profit, there is no capital to provide liquidity for operations. If there are no operations, there are no jobs. Once a government sets rules, establishes a taxing regime, and prohibits anticompetitive behavior, markets react and determine the most efficient way to produce the largest profit under the new rules and boundaries. If internet security is indirectly embedded into the profit model through rules that impact a company's profits, regulatory burdens, taxes, and licensing, security will naturally improve.

The more financial incentives are used to regulate markets, the less SPNs will require costly regulatory structures and bureaucracies. Virtual commerce must move toward a model where competition and profits are directly impacted by the security of a company's products and delivery channels. If they had been already, some very large companies might not be nearly as profitable or large as they are.

INSURANCE AND QUALITY CERTIFICATIONS

I spent five years as a commercial insurance underwriter learning how I could impact a company's risk profile by how I underwrote, engineered, and priced its insurance policies. When I wrote workers' compensation and excess liability insurance for the construction of the Alaska pipeline, the company knew that each time a light plane crashed inspecting the pipeline, its premiums went up. The cost of scrupulously maintaining those planes was not considered so much a burden in comparison.

As we have discussed, cybercrime insurance could cover the financial damages that occur from an attack or breach. It would not directly stop cybercrime, but it would reduce it to the extent that greater security measures were imposed on companies by their insurers. Requiring companies to have cybersecurity insurance would change the current model in three ways. First, it would impose a cost on companies based on the insecurity of their products and networks. That alone would be a step in the direction of incentivizing the development of a more secure virtual world. Second, it would create a new layer of oversight in the form of insurance underwriters and engineers who could conduct risk analyses for the purposes of deciding whether to accept the coverage, how the security of the business should be modified, and what the appropriate premium levels should be. Retaining insurance at a reasonable premium naturally motivates companies to satisfy the risk concerns of the insurer. Third, if either insurance companies or other outside experts were authorized by governing bodies to award security certifications to companies, they could display such Good Housekeeping–type credentials to allow them to compete based on the greater security of their products, including qualifying for bidding on government and private sector contracts.

Requiring that the existence and results of insurance coverage examinations be disclosed would also create another natural incentive toward better security. Governments could create and underwrite cybersecurity insurance requirements by providing reinsurance for catastrophic losses. That said, as I have mentioned, insurance can also be a means of financing cyberattacks. When cybercriminals engaging in ransomware know there is a pot of gold at the end of the rainbow, they will be all the more willing to take the risk and shoot for it. They run a business, and ransom is the payday that finances it.

PUBLIC-PRIVATE OVERSIGHT

Government regulation is a well-known commodity, particularly in the financial services sector. In the case of the internet, there will have to be private sector buy-in of a shared regulatory structure that is less redundant and more universal. This model would be unique, but it becomes more preferable each day as the prospects for regulation by the government increase.

As we have discussed, a human-governed internet is the paramount goal in the construction of any new internet. Machines would assist but be subservient to humans in the development of rules and the oversight of networks, the validity of licenses, and the identification of those who break the rules. Laura DeNardis does a wonderful job of laying out schemes of internet governance in her book *The Global War for Internet Governance* as she describes the current layers of governance on the internet despite its ungoverned appearance. Her analysis demonstrates just how fragmented, disorganized, and decentralized internet governance is, to the delight of everyone. There is a price to pay for such a chaotic structure.

Governance and order in cyberspace should look much like it does in the analog world, but without the warts, redundancies, and defects that currently exist. It should not be controlled by either the private or public sectors. A government-controlled entity will become a bloated bureaucracy that ultimately slows down innovation. If the internet is controlled by the private sector, it will focus too much on innovation and profit and not enough on enforcement and security. If the choices are left to technologists, they will be myopically slanted toward technological solutions that may not work in the real world.[1] Banking and finance SPNs should be governed by government and private sector people who have deep banking, finance, securities, financial utilities, and payments systems experience, along with technologists who have worked in or advised the financial sector and know how money is created and moved. SPN police forces would be similarly staffed by professionals who know what they are doing and can do it in real time.

MACHINES TO THE RESCUE

A new supervised virtual world will operate according to a new set of standards, licenses, and disclosure responsibilities. Every license could come with a kill switch that eliminated offending technologies or travelers. Machine learning, artificial intelligence, and advanced super sciences could assist in this oversight to efficiently identify and deal with breaches of codes; uses of malicious technology; and attacks on businesses, networks, payments systems, and internet infrastructures in real time in ways that humans could not without unduly slowing traffic. The triggering of kill switches, enforcement mechanisms, and remedial actions might be taken by a consortium of human overseers enabled by their machine subordinates working according to consensus protocols. Because action would usually have to be taken in real time, it would require immediate judicial review to identify and remediate any constitutional or other legal insufficiencies in the process. Understanding these rules, legitimate businesses would have backup plans in place to deal with enforcement disruptions. Illegitimate entities would simply slip away and hope to reemerge another day.

All of this sounds extremely difficult, but after all, we have created a new universe—or discovered a new dimension. Building something new takes time, resources, infrastructure, and governance. We have been foolish to think we could skip those steps in building the internet. Whatever the next virtual world that is created as the super sciences begin to sweep over us, humans must never become subordinated to machine intelligence. That may seem obvious, but human subjugation is inevitable if we continue to deploy new forms of the super sciences, fly through cyberspace on automatic pilot, and fail to develop standardized rules. In the current default position that is driven by a euphoric tidal wave of new technologies with few if any global standards, society will naturally delegate more and more to machines, which can handle tasks more quickly and efficiently than humans. But through that process, machines will increasingly become more intelligent, and as they do, if they are unguided, they will eventually make decisions that benefit themselves, no matter what the outcome is for humans. If we are not proactive, we will never know that we have crossed the line to nonbiological control until it is too late, with no way to return.

Professor Frank Pasquale suggests as much when he says that "the stakes of technological advance rise daily. Combine facial recognition databases with ever-cheapening micro-drones, and you have an anonymous global assassination force of unprecedented precision and lethality."[2] He warns that we now have the means to channel technologies of automation rather than being captured or transformed by them and must act before those means waste away. That is a message that MIT's Professor Max Tegmark also espouses in *Life 3.0*, his intriguing analysis of how artificial intelligence will impact humanity.[3] Both men seem convinced that "the horrific outcomes predicted by futurists are all the more likely if we do not aggressively intervene now."[4] Ray Kurzweil provides a full-throated endorsement of this view in his book *The Singularity Is Near*. Look around. We are already living in a "scored society" that ranks us socially, financially, and politically, and we are only at the beginning of what artificial intelligence and the super sciences will be able to do.[5]

OVERSIGHT AND ENFORCEMENT FUNDING

The funding for these new oversight and enforcement bureaucracies could come from governments and private organizations that participate in internet governance. It might be disfavored, but would not be unusual to impose assessments on ISPs, other backbone providers, carriers, licensors, and even users of the internet. Another source of funding would be enforcement penalties and the sale or liquidation of seized internet licenses, passports, and collateral. Funding responsibilities should be allocated among private, public sector, and enforcement sources so as not to skew governance or enforcement in any particular direction. We already have enough government authorities that use zealous enforcement to fund their operations. There should be no free ride because no ride is free. Someone gets something out of every ride we take into cyberspace.

CONTROL OF THE MEANS AND
LOCATION OF PRODUCTION

One of the most regulated aspects of banking is the ownership and control of a bank. It is not a business where just anyone can walk up and buy one, largely because entities that take deposits are insured by the FDIC. The safety and soundness of the banking system is critical to the economy, so the law effectively restricts hoodlums, adversaries, and inexperienced people from controlling more than 10 percent of the stock of a bank in the United States without government approval.

As the need for greater security becomes appreciated and the internet becomes more supervised, *control* of it and the companies that run it should become a predominant concern. The means of producing cyberspace—from ISPs to undersea cables—need to be under the control of trusted entities in democratic nations if security is to mean anything. The United States has taken small steps in that regard through the empowerment of the FCC and the Department of Commerce, for example, to publish lists of countries whose products and technologies are suspect.

But more comprehensive filtering and oversight of the creators and owners of virtual backbones will have to accompany any attempt to enhance security and ensure democracy. Control of technological infrastructures becomes more relevant with each passing day as superiority in artificial intelligence, 5G, quantum computing, the IoT, and the super sciences becomes the measure of geopolitical power. Control of the internet and superiority in these and other technological developments will essentially decide the fate of the world, so entities that cannot pass muster should not be permitted to own or control internet infrastructure.

A UNIQUE FINANCIAL OPPORTUNITY

The specter of a financial meltdown from the malicious use of the internet creates an incredible opportunity to modernize the way we visualize regulation and enhance the safety and soundness of virtual communications. It is an opportunity that must not be squandered. Toward that end, financial regulation in the analog world will also have to be reformatted. While U.S. federal banking regulators are the cream of the crop when it comes to assessing loan quality and risk management of financial credits and investments, they do not have the resources to nurture enough experts to remain current in the regulation of financial technology and cybersecurity. In February 2022, the chief innovation officer at the FDIC resigned, saying as much in a *Bloomberg* editorial, concluding that the lack of technological expertise in the agency was contributing to a national security risk.[6] Delegating responsibilities to regulators that they are not able to perform creates systemic risk in and of itself.

Even bankers have reluctantly reached this conclusion. A report by the Bank Policy Institute (BPI) representing the two dozen largest banks concluded in 2017 that cyber threats continue to grow in frequency and sophistication but, "unfortunately, U.S. and global banking regulators—and only banking regulators—are doing more harm than good in this area and need to stop."[7] Between 2014 and 2017, U.S. bank regulators proposed or issued at least forty-three cybersecurity-related regulations, standards, examination expectations, and other requirements, causing large multinational banks to spend as much as 40 percent of their cybersecurity efforts demonstrating compliance with regulatory standards. One agency pronouncement included "84 proposed standards addressing eight risk categories, all without any analysis of why the current regulatory framework is inadequate or how the proposed additional standards mitigate existing deficiencies."[8] The BPI report underscored that the December 2016 "Report on Securing and Growing the Digital Economy" by the Presidential Commission on Enhancing National Cyberspace Security singled out financial regulators for criticism, recommending that they conform their regulation and guidance to NIST standards rather than duplicating or conflicting with those standards.[9]

As regulation and enforcement take hold in the Wild West of cyberspace, policy makers will have to be vigilant to the inclination of authoritarian governments to attempt to use any remodeling of the oversight of the internet to serve their own parochial purposes. The possibility of ubiquitous government surveillance and control is perhaps equally frightening as the threat of cyberattacks that can collapse the financial free world. But trying to determine who the policy makers will be who can take on this task is daunting. In the United States, there are so many legislators and committees involved in cyberspace issues and so little political agreement on anything that it is hard to imagine any constructive solutions being produced in the near term. If we have any hope of insightful legislation laying out the framework for a new internet, there needs to be a select joint committee in Congress that supersedes everyone else.

Internet enthusiasts will be horrified by the possibility of less online freedom and greater surveillance, and they are right to be concerned about that. But I am sure they will be less horrified than if their money were to vanish from their bank accounts tomorrow morning. It is all about trade-offs and choosing the best worst option. This is a unique opportunity for humanity to ensure that its next chapter is safe, secure, and prosperous.

20

THIS SHOULDN'T BE THE
WAY THE WORLD ENDS

When my wife Karen read the manuscript for this book, she said it was a bit dark and spooky. I thought about that. I told her that I had tried to portray the reality I had seen and lived with throughout my career. And after all, I suggested that the book has a happy ending. There are market solutions driven by reasonable doses of oversight and common sense. Solutions to the problems that technology is creating are attainable. She didn't seem convinced. That told me something.

Evolving technologies and the changes that they create can be dark and spooky to many. But we can't stop progress; we can only channel it. It is easy to go with the flow and enjoy the ride while it seems beneficial, profitable, and fun. Everyone enjoys the ride to the top. But someone eventually must step forward and lead to ensure that new technologies are used to enhance the general good. Someone has to say, "Hey, this isn't working—what are we doing?"

In the epilogue of Nicole Perlroth's book *This Is How They Tell Me the World Ends*, she admits that she has painted a dark picture, noting that in the two decades since 9/11, it has become easier to sabotage the software of a Boeing 737 than to hijack it and crash it into buildings. She goes on to answer the question of why she wrote her book, saying that ignorance of the issues had become our greatest vulnerability that governments count on as they avoid doing what needs to be done to keep citizens safe: "There are no silver bullets; it is going to take people to hack our way out of this mess."[1]

We can build a better virtual existence, we can prosper from it, and we do not need to trade our privacy, data, or security to reach it. Machines are there to do what humans tell them to do. Code can be written to be more secure, encryption can be stronger and more efficient, and data can be collected and stored in safer ways. Software can be produced without so many secret side and back doors, and networks can be constructed that communicate data securely. People, computers, software, and applications can be authenticated and monitored. Malicious characters can be punished and eradicated. All of this can be done if we want to and if we are willing to lead.

I have worked for five decades with businesspeople, consultants, technologists, lawyers, and politicians earning my technology stripes on the job. I am not a computer scientist. But in my time, I have helped financial companies at the highest levels build and understand internet platforms and the insecurities they create. I have come to learn that too many people feign excellence and have advanced degrees in techno-jargon. They can spit out a glossary of terms meant to impress, but they are the digital-age equivalents of snake oil salesmen. Too many are simply selling something to perpetuate the problem rather than solve it. There is a whole lot of cybersecurity services dollars to be made these days. Unfortunately, as I have learned, very few people thoroughly understand banking, finance, and payments systems as well as technology. That tends to skew the advice received and conclusions reached so that the real problems never actually get solved.

When a bank CEO calls a consultant and asks if she can help build a virtual platform in conjunction with a new tech startup to deliver an interactive wealth analysis product to the bank's customers, the answer will always be yes. But the first question the CEO should have asked is not "Can we build it?" It should have been "Should we build it?" Is it safe for our customers and the bank? Will this product expose the bank or its customers to additional risks, and what are they? Once the "can we" question is asked, everyone is focused on getting it done and being paid. That is a train that is rarely derailed, no matter how defective the rails may be. That explains in some ways how the current internet was built and why it is insecure.

I challenge critics to dispute that the internet is not insecure or that we don't have the technological know-how to build a more secure virtual ecosystem. If I am wrong, then to the extent that the questions and solutions raised in this book have triggered a debate, we all will have benefited. If I am even partially correct, this debate may just result in making the future more manageable for humanity.

Stop and consider what we would wish we had done today if we were looking back from a worst-case cyberattack in March 2030. The president might slowly walk to the podium that day in the White House and deliver a short statement:

> My fellow Americans, as you know, the country has been under unprecedented cyber-attacks over these last two days. The power grid is

only sporadically operational, and the nation's financial services infrastructure has been temporarily destroyed due to thousands of pieces of malicious software that have been unleashed into every part of the system. I realize that most of you cannot access your money, and you may not have electricity, but I promise you that we are on top of the problem and will have it fixed shortly. In the meantime, please stay in your homes as much as you can so as not to overtax our residual highway systems, limited energy supplies, police resources, and food supply chain. The secretaries of the treasury, homeland security, and commerce are here with me today to answer your questions. God bless America.

As the president departs and the secretaries move to the podium, I can hear the back-and-forth with reporters mimicking the dialogue at the very end of *Raiders of the Lost Ark* as the ark is rolled into a warehouse the size of Rhode Island and Indiana Jones asks two government officials what is going to happen to it.

"We have top men working on it right now," the official says.

"Who?" Indy asks.

"Top men!" the official spits back with great indignation as we see the ark being placed in a cavernous warehouse never to be touched again.

I think there is a better than 50 percent chance that a president will have to deliver that message in the future if action is not taken that results in real progress toward a safer virtual existence. What would we say to ourselves in 2030 about what we should have done in 2023 to avoid a financial Armageddon? Every week that we don't do something to fix the insecurity of the internet and the increasing encroachment of technology on people, the odds increase that a president will have to step to the podium and deliver this statement. So, to whoever is planning on being president in 2030, be prepared, understand that your predecessors have enjoyed the luxury of hurling words and reports at the public for three decades, but you won't be able to. Standing there and wishing that someone had done something besides talking for all those years won't cut it with the American people at that point.

Appendix A

TRADITIONAL MONEY
AND PAYMENTS SYSTEMS

To better understand and compare the risks that the financial sector faces as new forms of money and value are created, consider the current forms of money and payments systems that exist, with particular focus on the number of participants, intersections, and junction points that can create cyber vulnerabilities as we transition to cyber payments.

THE ROOT OF ALL MONEY

The $50 bill in your pocket, let's call it "Don," is the analog version of Bitcoin, except it is effectively backed by the standing of the U.S. government. It has not been backed by gold since the 1970s. Don has likely journeyed through a web of people, businesses, and financial networks after he was minted by the Bureau of Engraving at the Treasury and was put into circulation by the Federal Reserve Bank of New York. Don was sent to a commercial bank member of the Federal Reserve and then transferred to a local pharmacy to stock its cash register. Don wandered into your pocket as change when you used a $100 bill (Debbie) to purchase a prescription medication costing $45. Later that day, you deposit Don into your checking account at First National Bank. There he rejoins Debbie, whom the pharmacy deposited earlier that afternoon in the same bank. They both spend the night together in the vault.

At home, knowing that Don is resting safely in your checking account, you write a check on your account to pay your monthly credit card bill intending to mail it the next day. You also wire some money from your account to your online broker to purchase shares of a mutual fund for your 401(k). Don never moves from the vault; he has digital representatives that do the running around for him. He stays put as the bank sends paper or electronic representations of him to your credit card company and your broker. When you deposited Don into your account, you lost your right to control Don himself and gained an interest in $50 of the Dons sitting on the bank's balance sheet. So, while Don lounges in the vault sipping piña coladas with Debbie, the value he represents gets transferred through book entry processes and other electronic means that are cleared and settled through the Federal Reserve or other private clearinghouse networks where the bank's daily transactions are netted out against accounts it has there with other banks that also have accounts.

Don begins this circle of life all over again when Eric cashes his paycheck and gets $250 in cash from the bank. Don spends a few days in Eric's pocket. He is likely sentenced to a life of circulating in a rather localized circle of people and small businesses for the rest of his natural life, unless he somehow strays into the drug trade or other cash-driven businesses where he might need to be gathered up with many other friends and laundered through a legitimate business to return home to the bank and not be hauled away as evidence.

As of December 2020, there was $2.02 trillion of Federal Reserve notes in circulation in the United States.[1] Coins accounted for about another $50 billion.[2] Cash, denominated as legal tender such as U.S. dollars, is fiat government-issued currency that is no longer backed by a physical commodity such as gold or silver. Neither is it backed by the full faith and credit of the United States as a matter of law, but it is treated as if it is. The strength and stature of the U.S. economy made it the world's reserve currency at the Bretton Woods Agreement in 1944. Over the last two centuries, money in the United States has ping-ponged between specie— gold and silver—coins, banknotes, greenbacks, Federal Reserve notes, and a variety of tokens and representative types of cash. Today it is morphing into various forms of digital cash and cryptocurrencies.

For our purposes of evaluating security, we need only focus on a few important attributes of cash. It is highly insecure, but there is a limit to how much that insecurity can cost anyone. It is anonymous, simple to use, and final once transferred. Since it is difficult to move large amounts, its use is often restricted geographically. When you spend a dollar bill and it moves to a friend, the government, or a merchant, there is no formal record that links you to that dollar bill once it leaves your hand. Your privacy is infallibly protected, unless of course you spend $10,000 in cash. In that case, financial disclosures and filings will be triggered under various federal money laundering laws with the Financial Crimes Enforcement Network

(FinCEN) at the Department of Treasury so that their supercomputers can track those funds. It is the anonymity of cash that makes it the darling of criminals and why the prevention and detection of money laundering is so difficult. The finality of cash is equally applicable if it is lost, mutilated, stolen, or counterfeited. If you lose it, there is no way to trace it.

The ultimate value of fiat money is a function of the stability of the issuing government. Unstable countries can issue as much fiat currency as they want, and often do to finance their operation. When fiat money is used, whether in the form of cash, coin, check, book entry, or wire, the government is an indirect party in the transaction. It is effectively guaranteeing the legitimacy and reliability of the money used. Parties do not worry about the reliability of the value of the dollar, whereas in Venezuela, parties worry about the inflating value of bolivares, which have become nearly worthless. As a result, Venezuela has become one of the largest users of cryptocurrencies since individuals and businesses there have limited confidence in the government or its money.

For similar reasons, El Salvador has adopted Bitcoin as a government currency. Its experience has been a bumpy one, as the treasury lost $200 million due to falling Bitcoin prices in February 2022, and consumer adaption proved to be anemic.[3]

This suggests an important metaphysical characteristic of money that digital technology has yet to entirely achieve. *The Summit* by Ed Conway analyzes world dynamics surrounding the events at Bretton Woods when the dollar was anointed as the world's reserve currency despite the objections of Britain's John Maynard Keynes and the Russians. Because the United States emerged from World War II as the only superpower and the country that had the most gold in its vaults, it could essentially dictate the terms of the Bretton Woods Agreement. It was the political and economic elephant in the room, and try as they did, other countries could not muster the arguments nor the economic power to do anything to derail the dollar from becoming the preeminent form of money in the world. In the early 1970s, President Nixon terminated the dollar's linkage to gold, leaving it backed solely by confidence in the U.S. economy.

That was a critical and seminal moment that altered the view of U.S. money. The delinkage of the dollar from gold was ironically the precursor of cryptocurrencies like Bitcoin. If money does not need to be anchored to any metal or commodity, then why can't it exist only in cyberspace and have the value that markets give it? Putting aside the creditworthiness of the United States, which has continued to erode each year over the last thirty years, the dollar bill represents nothing that you can cash in for anything else. It is the genesis point of its own value; it cannot be transferred to the Federal Reserve for $1 worth of the Grand Canyon.

This gets us to the fundamental essence—the prime matter, if you will—of money. It is a representation of a system that garners enough confidence so that it can be used to buy and sell goods and services. Governments provide the confidence that fiat money needs. Most people do not hesitate to hitch their economic wagons to the dollar and continue to be leery of crypto assets that may come and go with no government or central bank ever blinking an eye.

CHECKS

Since money is inconvenient to transport in bulk, alternative forms of transmitting it were created throughout time. Carrying large bags of coins, gold, silver, or banknotes was at best inconvenient and at worse an irresistible temptation for thieves. Trusted paper representations of money in a bank that could flow freely in commerce became an obvious solution. A check is a private negotiable instrument drawn on a depository institution such as a bank, savings institution, or credit union in the form of payment from a payor to a payee. It can be drawn for any amount up to the balance of the account on which it is drawn and made payable to anyone. The roots of checks are traceable to medieval times, and perhaps earlier.[4]

Checks need payments systems to be processed. The payments system in the United States has evolved in the analog world over decades, stitched together one product and one system at a time. Typically the individual communicating components were proprietary networks to which access was scrupulously guarded. Those networks were not impenetrable to the outside world but were certainly more secure that the current spiderweb of networks connecting thousands of parties who communicate, do business, and transmit money through and over the internet.

Advancements in the movement of money had to develop as people in the United States moved ever westward and the country became more industrialized. Checks written on bank accounts to pay for goods in a city other than where the payor's bank was located were deposited in a bank in a different city or region, requiring checks to be physically transported to the banks whose accounts they were drawn on for acceptance, clearing, and settlement. It was a clumsy but workable process later made somewhat more effective by local clearinghouses. They were associations of banks that agreed to clear each other's checks and in distressed times lend to and financially support each other. The first check clearinghouse—the New York Clearinghouse—was established in 1853 and would prove to be vital in times of economic distress, financially supporting members by providing liquidity.[5]

Once the Federal Reserve entered the picture in 1913, this process took on a more centralized and efficient look. The twelve Federal Reserve banks acted as central repositories, crediting and debiting the netted amounts of checks processed each day to the accounts of their member banks. The Federal Reserve effectively functions as a government-sponsored clearinghouse, running the largest check and check-imaging clearinghouse in the country. This was the beginning of what would become the digital transfer of money long before most people ever thought about it.

Checks found their raison d'être in the nineteenth century. Congress enacted the National Bank Act of 1865, which among other things imposed up to a 10 percent tax on the use of banknotes issued by state banks as currency. The purpose of the tax was to force those notes out of circulation in favor of national banknotes, which were being issued by the newly organized national banks in the country beginning in 1863. It was presumed or hoped that this tax would also eliminate state banks, which at the time were meagerly capitalized, loosely regulated, and poorly operated.

State banks took a dim view of this tax and found a work-around, which turbocharged the issuance and use of checks. They permitted depositors and borrowers to issue checks that could be used to draw upon customers' money in the bank. Since checks were not banknotes, they were not taxable under the act. This began an interconnected system of payments that has lasted to the present time and resulted in the creation of a massive infrastructure built around banks, the Federal Reserve, and clearinghouse organizations that transmit, process, clear, and settle checks written on bank accounts.

In the late 1950s and early 1960s, a new technology, magnetic ink character recognition (MICR), included printed numbers on checks that could be read by machines to enable high-speed check processing. Imaging technologies reduced some of the burden of moving checks in the 1990s and reduced the related costs. When Congress enacted Check 21—the Check Clearing for the 21st Century Act—in October 2004 allowing electronic check images to substitute for a physical check, it solved the problem of having to transport large numbers of checks each day for settlement.[6] The Federal Reserve banks play several roles in the check clearing and settling process. They regulate the payments systems and write the rules that govern them; operate the largest clearinghouse in the country; and as the manager of the National Settlement Service, provide settlement services to clearinghouses.[7]

When Julie writes a check to pay her monthly mortgage and mails it to her mortgage company, it probably goes to its "lockbox." That is a bank-operated mailing address that allows a third-party vendor to facilitate the process of crediting those funds to the mortgage company's account. If Julie had used the online banking feature offered by her bank, it likely would have sent a physical check to a lockbox or transmitted a digital image of a check to the mortgage company. The mortgage company's bank would post the physical or scanned image of the check and apply payment against Julie's outstanding mortgage obligation. The mortgage company's bank would then proceed to clear and settle the check, in effect making sure that the funds exist in and are debited from Julie's bank and moved into the mortgage company's bank account. If the mortgage company's bank is also Julie's bank—called an "on-us" transaction—the settlement process is internal to that bank and occurs more quickly.

Instead of shipping checks through the mail or by courier as used to be done, the mortgage company's bank may receive and process an image of the front and back of Julie's check transmitted electronically through a local clearinghouse exchange, a larger correspondent banking institution, or the local Federal Reserve bank. If the electronic presentment file is through a Federal Reserve bank, the account of Julie's bank at the Federal Reserve would be credited or debited based on the daily flow of checks. Like the Federal Reserve, clearinghouses provide a vehicle for their members to exchange checks and payments in bulk, instead of on a check-by-check basis. They net out the balances at the end of the day and settle up among the members through the Fedwire, an electronic funds transfer (EFT) system that handles large-scale check settlement between U.S. banks. This process is not completed on a real-time basis, requiring holds on the use of the funds until complete. This is one feature that digital money and new real-time payments systems seek to improve, since time is money.

This process, even using electronic copies of checks, has many parties, junction points, and places where a malicious user of technology could insert itself to spy, alter, mutilate, steal, replicate, or destroy the various instruments passing through them. There's the mortgage company, its bank, the clearing entities, and Julie's bank, all using a combination of proprietary and open networks to transmit the information. Each of those participants in the system will have outside vendors who will have their own outside vendors and providers that they will use and rely upon, multiplying the number of players and vulnerabilities in the system that could be attacked beyond the analog-world risks of forgery, destruction, and system failure. Each party will have its own systems, technologies, and security standards and protections, which may or may not be completely effective or compatible with each other. And each party and network will have different forms of encryption that may or may not cover the entire journey of the check.

Although once the principal alternative to cash, the number and amount of checks has been steadily decreasing as the digitization of payments systems increases. Between 2015 and 2018, the number of checks processed in the United States decreased by 3.6 billion to 14.5 billion annually, as the value decreased by $3.4 trillion to $25.80 trillion.[8] As the number of checks has decreased, check fraud, by value, declined from $1.11 billion in 2012 to $710 million in 2015. The fraud rate, by value, of checks also decreased from 0.41 basis points to 0.25 basis points over the same period.[9] There have been premature predictions of the disappearance of checks over the last thirty years, but checks have proved to be unexpectedly resilient given their reliability. Checks are a great example of necessity being the mother of invention. And that invention is still used today.

CREDIT CARDS

The total number of credit card payments grew to 131.2 billion with a value of $7.08 trillion in 2018, up 29.7 billion and $1.56 trillion since 2015.[10] Credit cards have been the alternative money of choice for many years, albeit they are not money but debit transactions in which the consumer accesses a preapproved, unsecured line of credit. But as ubiquitous as usage is today, the history of credit cards is marked by periods when their future was very much in doubt.

The tortured birth of credit cards began when Bank of America tried to introduce them in select markets in the 1950s and 1960s following the introduction of installment charge cards by Sears, American Express, and Diners Club. Credit cards ran into many human and psychological barriers trying to win over a generation of post-Depression Americans taught to consider debt as something to be avoided at all costs. The initial failures and convoluted path to success that credit cards took underscores the important role that confidence and customer habits play in financial services. This story is artfully told by Joe Nocera in his 1994 book *A Piece of the Action: How the Middle Class Joined the Money Class.*

When Gideon stops by his local McDonald's to purchase a Big Mac and fries and offers his credit card as payment, he probably has no idea what happens behind the scenes. McDonald's obtains approval over the internet or a phone line from the bank that issued Gideon's Mastercard when he presents it, usually by swiping the magstripe on the back, inserting the chip on the front into the McDonald's point-of-sale (POS) terminal, or tapping a contactless card on that terminal. This all used to be done over a closed, proprietary network, and much of it still is. But customers like Gideon are now able to tap into the network of their credit card online to check balances and make payments. McDonald's bank or card processor forwards Gideon's information to the credit card network, which begins the process of clearing payment from Gideon's bank. It begins by verifying his account, a process that usually involves an independent sales organization (ISO) that provides card acceptance functionality to the merchant, a specialized processor known as a gateway, and front-end and back-end processors beyond the card-issuing bank and the merchant acquiring bank.

Once the merchant receives the authorization, usually in seconds, Gideon's bank places a hold on Gideon's account in the amount of the purchase. McDonald's provides the customer a receipt to complete the sale, and the transaction is posted to Gideon's monthly credit card billing statement. The POS terminal collects all approved authorizations to be processed in a "batch" at the end of the business day, and they are routed to the credit card network for settlement. The credit card network forwards each approved transaction and then transfers the funds over the next day or two, minus an interchange fee that it shares with parties in the credit card network. McDonald's bank credits McDonald's account for Gideon's purchase less a merchant discount rate. Gideon's bank posts the transaction information to his account so that he can pay the bill in the next month, or pay only part of it and allow the remaining balance to roll and pay interest on it.[11]

Talk about a crowded playing field with many touch points and junctions that could be hacked or digitally sabotaged. Apart from the risks of loss of the card and everyday fraud that proliferates, credit card systems are subject to risks ranging from the creation of false identities and transactions to operating system failures. The new risks of hacking users and systems that arise from marrying credit card payments systems with the open architecture of the internet suggests that enormous resources must constantly be devoted to system security. Many people, providers, and processes must function in a secure fashion each day to avoid these threats.

You can judge for yourself how successful that effort has been. The credit card business is already drenched in fraudulent transactions. There were 271,823 cases of credit card identity fraud in 2019, an increase of 72.4 percent from 2018.[12] Losses from fraud reached $27.85 billion in 2018 and are projected to rise to $35.67 billion in five years and $40.63 billion in ten years. Credit card fraud rose by 44.7 percent between 2019 and 2021, rising to 393,207 reports.[13] The losses derived from all this fraud is money that credit card users effectively cover in the form of high interest rates and other costs, answering the question why credit card annual percentage rates can be 25 percent or more. Users pay for the insecurity in the system, which today equates to somewhere around $30 billion per year, if not more. And it is only getting worse, as 2019 saw more than 100 million online customer accounts compromised by data breaches.

The number of providers, users, advisors, and third-party vendors in addition to the merchants and banks involved create an endless number of junction points, interfaces, handoffs, and people-related vulnerabilities that are impossible to stop. The system can either retool, which would be very expensive, or incur the risks and related costs and pass them on to the users, merchants, and banks. The latter course, the one chosen by default, is incredibly dangerous since the risks are growing faster than the introduction of their solutions. But as long as users cover the cost of fraud, it reduces the incentive to spend large amounts of money to build systems that are fraud proof.

PAYMENTS SYSTEMS AND NETWORKS

Money is limited in what it can do unless there are systems that can move it from buyer to seller, investor to investment, financier to counterparty, and back to a financial institution or central bank. Money in its various forms enjoys a vagabond lifestyle, cramming every last minute of travel into its life. That requires many parties to operate the channels that transport money or representations of it. It is a complicated and crowded ecosystem that has been transitioning from the analog to the digital world over the last three decades. The pressure to convert to real-time digital formats has been magnified by the increasing speeds that companies and consumers demand for transacting business, and by the sheer enormity of the money to be made from doing so. The insecurities created by connecting these thousands of users, providers, servicers, regulators, and processors have been important but subservient priorities to the interests of innovation and profit. The bill for that quarter century of myopia is coming due.

Now that we have summarized the basic forms of money in use, appreciating how they are transmitted and how they are being impacted by technology is critical to understanding the potential vulnerabilities in the system. A comprehensive description of U.S. payments systems and clearing networks is included in a white paper by the Bank for International Settlements.[14]

ATMs

Everyone loves new technology toys. In August 1977, one hit the market in the form of RadioShack's TRS-80 Micro Computer System, which had a keyboard, a sixty-four-column video monitor, and could be programmed by tapes that could play on a home cassette player. At the Office of the Comptroller of the Currency, we were working on what we thought were groundbreaking bank technology issues. That effort produced an opinion that authorized national banks to deploy automated teller machine (ATM) networks that would not be considered branch banks. This would open up a new world for financial institutions, and the closest thing yet to a video game for financial institution customers.

ATMs do not make or represent money. They are electronic banking outlets that accept, transmit, and dispense it. Customers may complete basic banking transactions through an ATM using a credit or debit card, including making a deposit, obtaining a cash withdrawal, initiating bill payments, and transferring funds between accounts. ATMs are often embedded in the walls or drive-through facilities of financial institutions but can be free-standing machines anywhere. They may be leased or owned and operated by banks or third-party intermediaries and located anywhere from airports to 7-Elevens. Fees are commonly charged for cash withdrawals by the banks involved if the ATM is not operated by the customer's bank. There are 400,000 ATMs in the United States handling about 5 billion transactions annually valued at about $800 billion.[15]

While they operate on proprietary networks, the involvement of customers added a new security risk that we had never focused on. When customers insert their ATM, debit, or credit cards into the ATM and enter a personal identification number (PIN), they enter this network. The mainboard that houses the CPU and memory connects to the system so that it can approve that PIN issued by the customer's bank. Each card is required to be able to access multiple networks so that there is a backup if the initial network is unable to process the transaction. The ATM network sends the information to the card issuer (usually a bank) for approval of the transaction. If the transaction is a cash withdrawal, once processed, the bank debits the customer's account for the amount, and the mainboard then initiates the dispensing of cash by the machine.[16]

ATMs hold substantial amounts of cash and, as featured in the hilarious movie *Barbershop*, are targets of thieves who may physically remove them or attach (skimming) or insert (shimming) card readers that steal the information when cards are inserted. The value of all card payments and ATM withdrawals rose 21.2 percent in 2015, but the value of fraudulent card payments and ATM withdrawals grew 63.8 percent to $6.46 billion, making fraud substantially greater for every dollar spent by card than by ACH or check.[17]

The insertion of a card that carries other embedded digital information and has been used on various other systems and networks creates a host of security issues that were never contemplated when the original approval was issued. That action in effect connects them indirectly to every network, open or closed, when it comes to the deployment of malicious technology. While it was not obvious at the beginning, this was the beginning of the end of financial proprietary networks, and the beginning of a new age of insecurity for banks, savings institutions, and credit unions.

FEDWIRE SYSTEM

The Fedwire system is an electronic system maintained by the Federal Reserve banks that debits and credits bank members' accounts on as close to a real-time basis as one could expect. No end-of-day netting of gross settlements among various members' accounts is necessary since each transfer is an individual transaction from a bank sender to a Federal Reserve bank receiver. There are extensive rules, regulations, and civil and criminal laws that apply to the transfer of money through the Fedwire. The system is as good as the money in the sending account and the Federal Reserve's security, which relies on encrypted transmissions and token-based authentication devices and software.

There are security issues that range from unauthorized access to Fedwire machinery to capturing and using real or fraudulent account numbers. And these security issues are expanding as new scams and networks connect the Fedwire with numerous users and open-architecture networks that bring their security issues with them. Total wire fraud in the country from all systems is running at about $1.5 billion annually.[18] In June 2021, the Federal Reserve Bank of New York warned about scammers claiming to be with the Federal Open Market Committee (FOMC) and offering to assist in the recovery of "lost" cryptocurrency for a fabricated fee for the alleged recovery of the funds plus attorney's fees.[19]

The Federal Reserve is currently modernizing the Fedwire system, having announced in January 2021 that over 110 organizations were participating in the FedNow Service pilot program to develop a real-time payments processing system with greater cross-border interoperability to be launched in 2024.[20] This effort trails that of the Clearing House owned by large banks, which was launched in 2017. The private sector, which is seeking to capitalize on the benefits and profitability of deploying a modernized wire system, argues that there is neither authority nor a business need for the Federal Reserve to occupy an area in the payments world that the private market can adequately address. This debate will continue, as the Federal Reserve has demonstrated a keen desire to keep all the jurisdiction that it ever had or could get, no matter how much modern technology obviates the need for its continued role.

AUTOMATED CLEARING HOUSE TRANSFERS

The Automated Clearing House (ACH), a bank-owned proprietary payments system begun in 1970 is operated under the trusteeship of NACHA, the National Automated Clearing House Association, which manages and regulates it through published operating rules. I represented NACHA for a time with regard to developing digital currencies. The ACH is wired to every demand deposit account in the country, making it convenient for every type of customer to make transfers of money through the communication of the MICR number and account number on a check, rather than the check itself. The ACH handles both "push" transactions initiated by the payor and "pull" transactions initiated by the receiver of the funds through a third-party operator that performs the switching from the originating to the receiving bank. ACH operators calculate net settlements daily and submit them to the Federal Reserve for settlement, resulting in a "zero float" among the participating banks and their customers.[21] The first-quarter 2021 volume of 7.1 billion payments was an increase of 11.2 percent from the same period in 2020. That represented a value for those payments of $17.3 trillion, a nearly 19 percent increase from a year earlier.

ACH fraud is comparatively rare these days but easy to execute in the analog world since all that is needed is an account number and a bank routing number, data that can be taken in a data breach or spear-phishing scam where the individual is redirected to a website infected with malware that records keystrokes, including passwords. ACH fraud typically targets a specific bank account and takes advantage of the time delay that occurs in ACH processing. A 2018 Federal Reserve payments study found that payment fraud represented only 0.08 basis points, or eight cents for every $10,000 in payments.[22] ACH fraud is mitigated by the ACH network, which includes the originating depository financial institution (ODFI) and receiving depository financial institution (RDFI), the Clearing House,

NACHA, and the Federal Reserve. Based on the NACHA operating rules, each ACH entity must adhere to a set of guidelines to ensure that the ACH file that is being processed is checked for compliance with the guidelines. With more customers accessing the system through their bank, which they communicate with through the internet and their online banking application, these bridges between the ACH proprietary network and the open and more insecure architecture of the internet are likely to be a target of increasingly sophisticated hackers.

CLEARING HOUSE INTERBANK PAYMENTS SYSTEM

The Clearing House Interbank Payments System (CHIPS) is a private automated network that is a real-time alternative to the Fedwire. It is owned by the Clearing House, an association established in 1853, which is today owned by twenty-four large banks.[23] CHIPS clears and settles 240,000 transactions totaling $1.8 trillion in domestic and international payments per day using a patented algorithm that matches and nets payments, resulting in an extremely efficient clearing process. CHIPS is not as accessible as Fedwire, having only fifty direct participants.[24]

CHIPS enables banks to transfer and settle international payments on the same day by replacing official bank checks with electronic bookkeeping entries processed with intraday or end-of-day payment finality through a real-time system that is determined by the balances held by participating banks. Historically, CHIPS specialized in settling the dollar portion of foreign exchange transactions. Its services now include electronic data interchange (EDI) capability allowing participants to transmit business information (such as the purpose of a payment) along with their electronic funds transfers.[25]

PEER-TO-PEER PAYMENT SYSTEMS

The PayPal, Zelle, and Venmo payment networks allow people to directly transfer funds between and among themselves using email addresses or mobile phone numbers. These peer-to-peer transactions move money directly from a linked bank account or card to a recipient's linked account or card, typically taking only minutes to process. Unlike Zelle, which only moves funds from a bank account, a balance can be maintained in a separate Venmo electronic wallet. So, if an account doesn't have enough funds to complete a transaction, money can be drawn from the Venmo wallet to complete it. Venmo includes a social aspect also. It allows users to follow friends on its timeline and see what they are doing and paying for, a social feature to some and a scary feature to others.

SWIFT

The Society for Worldwide Interbank Financial Telecommunication messaging service was founded in the 1970s to create a global financial network and a common language for international financial messaging. SWIFT went live with its messaging services in 1977, the year that ATMs were authorized, replacing telex technology with standards to allow for data to be transmitted seamlessly across linguistic and system boundaries. SWIFT, though technically not a payments system, is a global financial infrastructure in more than 200 countries and territories supporting payments-related transactions by more than 11,000 institutions.[26]

After the $81 million fraud attack on Bangladesh Bank in 2016, SWIFT installed new safeguards. However, a survey of 200 banks worldwide found that four in five of these banks had experienced at least one SWIFT fraud attempt since 2016, with cyber fraud on the rise since 2016. And a "significant portion" of the banks surveyed did not have prevention policies in place to address SWIFT fraud. One troubling trend is the increase of "insider risk," where hackers are combining forces with employees at banks in charge of sending or receiving SWIFT payments. A SWIFT fraud report states that one in seven banks has experienced at least one SWIFT fraud attempt involving an employee, with the problem appearing to be particularly acute in the Asia-Pacific region with 83 percent of money stolen from the SWIFT network forwarded to accounts in Asia, and 10 percent to Europe. In October 2018, SWIFT announced new "payments control" features and has offered "stop and recall" that allows banks to stop a SWIFT transaction and reverse it once they have flagged it as suspicious.[27] The organization came into the spotlight when Russia invaded Ukraine and cutting Russia off from SWIFT became a controversial sanctions issue given how hard it could impact the German people who had become dependent on Russian energy.

These payments systems and exchanges have thousands of users, participating brokers, regulators, service providers, advisors, and consultants, all linked to the internet and to each other through computer networks of wildly differing security aptitudes. There are as many vulnerabilities as there are participants, junctions, and interchange points. It has been nearly impossible to seal all the front, back, and side doors through each participant and system in the current internet configuration.

FINANCIAL UTILITIES AND EXCHANGES

Financial market utilities (FMUs) are multilateral systems that provide the back-office infrastructure for transferring, clearing, and settling payments, securities, and other financial instrument sales and purchases. Securities and related financial instruments are traded through approximately thirty SEC-registered exchanges such as the New York Stock Exchange. There are more than 100 million broker-to-broker transactions per day with a value of nearly $1.2 trillion.[28] The Federal Reserve supervises FMUs that have been designated as systemically important by the Financial Stability Oversight Council (FSOC). Professor Paolo Saguato has spent considerable time analyzing the regulation and systemic risks posed by these companies.[29] They include the following:

- the Clearing House Payments Company LLC;
- CLS Bank International;
- Chicago Mercantile Exchange, Inc.;
- the Depository Trust Company;
- the Fixed Income Clearing Corporation;
- ICE Clear Credit LLC;
- the National Securities Clearing Corporation; and
- the Options Clearing Corporation.[30]

The number of daily transactions, participants, and users of these systems speak to the complexity of maintaining secure networks that intersect with the open architecture of the internet and thousands of junction points.

Appendix B

SELECT FINANCIAL SERVICES CYBERSECURITY REPORTS AND EVENTS: 1996–2022

(As of July 2022)

	DATE	DOCUMENTS
1	1996	Critical Infrastructure Protection Executive Order 13010 July 15, 1996 https://www.govinfo.gov/content/pkg/FR-1996-07-17/html/96-18351.htm
2	1997	Report by President's Commission on Critical Infrastructure Protection: Overview Briefing (Marsh Commission) June 1997 https://www.hsdl.org/?view&did=487492
3		Critical Foundations Protecting America's Infrastructures: The Report of the President's Commission on Critical Infrastructure Protection October 1997 https://fas.org/sgp/library/pccip.pdf
4	1999	Protecting America's Critical Infrastructure Presidential Decision Directive (PDD) 63 Updated February 1999 https://www.hsdl.org/?abstract&did=3544
5		Financial Services Information Sharing and Analysis Center (FS-ISAC) Established in response to a White House request to coordinate sector responses to incidents such as major financial services attacks. https://www.fsisac.com/who-we-are

	DATE	DOCUMENTS
6		Defending America's Cyberspace: National Plan for Information Systems Protection Version 1.0: An Invitation to a Dialogue January 2000 https://fas.org/irp/offdocs/pdd/CIP-plan.pdf
7	2000	GAO: Comments on the National Plan for Information Systems Protection February 2000 https://www.hsdl.org/?view&did=341
8	2001	Convention on Cybercrime, Budapest European Treaty Series, No. 185, Council of Europe November 23, 2001 https://rm.coe.int/1680081561
9		Critical Infrastructure Protection in the Information Age Executive Order 13231 October 16, 2001 https://www.dhs.gov/xlibrary/assets/executive-order-13231-dated-2001-10-16-initial.pdf
10	2002	Financial Services Sector Coordinating Council (FSSCC) Seventy of the largest financial institutions and their industry associations representing banking, insurance, credit card networks, credit unions, exchanges, financial utilities in payments, clearing, and settlement. https://fsscc.org
11	2003	The National Strategy to Secure Cyberspace February 2003 https://us-cert.cisa.gov/sites/default/files/publications/cyberspace_strategy.pdf
12		Interagency Paper on Sound Practices to Strengthen the Resilience of the U.S. Financial System Federal Reserve Board, Office of the Comptroller of the Currency, Securities Exchange Commission April 8, 2003 https://www.occ.treas.gov/news-issuances/bulletins/2003/bulletin-2003-14.html

	DATE	DOCUMENTS
13		The National Strategy for the Physical Protection of Critical Infrastructures and Key Assets White House February 2003 https://www.dhs.gov/xlibrary/assets/Physical_Strategy.pdf
14	2004	National Strategy to Secure Cyberspace National Security Presidential Directive 38 (NSPD 38) July 7, 2004 (never officially made public)
15	2008	Comprehensive National Cybersecurity Initiative (CNCI) launched by President George W. Bush in National Security Presidential Directive 54/Homeland Security Presidential Directive 23 (NSPD-54/HSPD-23) January 2008 https://fas.org/irp/offdocs/nspd/nspd-54.pdf
16	2010	Department of Homeland Security Computer Network Security and Privacy Protection February 19, 2010 https://www.dhs.gov/xlibrary/assets/privacy/privacy_cybersecurity_white_paper.pdf
17–27	2012	Financial Stability Oversight Council (FSOC) Annual Reports, 2012–2021 https://search.usa.gov/search?affiliate=ofr&query=ANNUAL+REPORT&commit=
28	2013	Improving Critical Infrastructure Cybersecurity Executive Order 13636, 3 CFR 13636 February 12, 2013 https://obamawhitehouse.archives.gov/the-press-office/2013/02/12/executive-order-improving-critical-infrastructure-cybersecurity
29	2015	FS-ISAC Sheltered Harbor established www.fsisac.com/sites/default/files/news/SH_FACT_SHEET_2016_11_22_FINAL3.pdf https://shelteredharbor.org

	DATE	DOCUMENTS
30		Blocking the Property of Certain Persons Engaging in Significant Malicious Cyber-Enabled Activities Executive Order 13694 April 1, 2015 https://obamawhitehouse.archives.gov/the-press-office/2015/04/01/executive-order-blocking-property-certain-persons-engaging-significant-m
31	2016	Financial Systemic Analysis and Resilience Center (FSARC) (now a subsidiary of FS-ISAC) https://www.prnewswire.com/news-releases/fs-isac-announces-the-formation-of-the-financial-systemic-analysis--resilience-center-fsarc-300349678.html
32		2016 Financial Stability Report Office of Financial Research, U.S. Department of Treasury https://www.financialresearch.gov/financial-stability-reports/files/OFR_2016_Financial-Stability-Report.pdf
33		Financial Services and Cybersecurity: The Federal Role Congressional Review Service March 23, 2016 https://crsreports.congress.gov/product/pdf/R/R44429
34		The Committee on Payments and Market Infrastructures and the International Organization of Securities Commissions (CPMI-IOSCO) Guidance on Cyber Resilience for Financial Market Infrastructures June 2016 https://www.bis.org/cpmi/publ/d146.pdf
35		Report on Securing and Growing the Digital Economy Commission on Enhancing National Cyberspace Security December 11, 2016 https://www.nist.gov/system/files/documents/2016/12/02/cybersecurity-commission-report-final-post.pdf
36	2017	Cybersecurity and Financial Stability: Risks and Resilience Office of Financial Research, U.S. Department of Treasury February 15, 2017 https://www.financialresearch.gov/viewpoint-papers/files/OFRvp_17-01_Cybersecurity.pdf

	DATE	DOCUMENTS
37		G7 (FSB): The Fundamental Elements for Effective Assessment of Cybersecurity in the Financial Sector Stocktake of Publicly Released Cybersecurity Regulations, Guidance, and Supervisory Practices October 13, 2017 https://www.fsb.org/wp-content/uploads/P131017-2.pdf
38		Building a Defensible Cyberspace New York Cyber Task Force Columbia School of International and Public Affairs 2017 https://www.sipa.columbia.edu/sites/default/files/3668_SIPA%20Defensible%20Cyberspace-WEB.PDF
39		Regulatory Approaches to Enhance Banks' Cyber-Security Frameworks Bank for International Settlements (BIS) Juan Carlos Crisanto and Jermy Prenio August 2, 2017 https://www.bis.org/fsi/publ/insights2.htm
40		Cyber Risk, Market Failures, and Financial Stability International Monetary Fund (IMF), Working Paper WP/17/185 Emanuel Kopp, Lincoln Kaffenberger, and Christopher Wilson August 7, 2017 https://www.imf.org/~/media/Files/Publications/WP/2017/wp17185.ashx https://www.imf.org/en/Publications/WP/Issues/2017/08/07/Cyber-Risk-Market-Failures-and-Financial-Stability-45104
41		Cyber Security and Financial Stability: How Cyber-Attacks Could Materially Impact the Global Financial System Institute of International Finance Martin Boer and Jaime Vazquez September 2017 https://www.iif.com/Portals/0/Files/IIF%20Cyber%20Financial%20Stability%20Paper%20Final%2009%2007%202017.pdf?ver=2019-02-19-150125-767

	DATE	DOCUMENTS
42		Forces Shaping the Cyber Threat Landscape for Financial Institutions SWIFT Institute Working Paper, No. 2016-004 William Carter October 2, 2017 https://www.swiftinstitute.org/wp-content/uploads/2017/10/SIWP-2016-004-Cyber-Threat-Landscape-Carter-Final.pdf
43	2018	Rise of the Machines: Artificial Intelligence and Its Growing Impact on U.S. Policy Subcommittee on Information Technology, Committee on Oversight and Government Reform, U.S. House of Representatives September 2018 https://www.hsdl.org/?view&did=816362
44		Protecting Financial Institutions from Cyber Threats Cyber Policy Initiative Working Paper Series, Cybersecurity and the Financial System, No. 2 Erica D. Borhard September 2018 https://carnegieendowment.org/files/WP_Borghard_Financial_Cyber_formatted_complete.pdf
45		The Future of Financial Stability and Cyber Risk Brookings Cybersecurity Project Jason Healey, Patricia Mosser, Katheryn Rosen, and Adriana Tache October 2018 https://www.brookings.edu/wp-content/uploads/2018/10/Healey-et-al_Financial-Stability-and-Cyber-Risk.pdf
46		FSB Cyber Incident Response and Recovery—Survey of Industry Practices October 2018 https://www.fsb.org/wp-content/uploads/P110719.pdf
47	2019	The Cyber Threat Landscape: Confronting Challenges to the Financial System Adrian Nish and Saher Naumaan March 2019 https://carnegieendowment.org/2019/03/25/cyber-threat-landscape-confronting-challenges-to-financial-system-pub-78506

	DATE	DOCUMENTS
48		Financial Sector's Cybersecurity: A Regulatory Digest World Bank May 2019 http://pubdocs.worldbank.org/en/208271558450284768 /CybersecDigest-3rd-Edition-May2019.pdf
49		Securing the Information and Communications Technology and Services Supply Chain Executive Order 13873 May 15, 2019 https://www.federalregister.gov/documents/2019/05 /17/2019-10538/securing-the-information-and-com munications-technology-and-services-supply-chain
50		Decentralised Financial Technologies: Report on Financial Stability, Regulatory and Governance Implications Financial Stability Board (FSB) June 6, 2019 https://www.fsb.org/wp-content/uploads/P060619.pdf
51		Capacity-Building Tool Box for Cybersecurity and Financial Organizations Carnegie Endowment for International Peace Tim Maurer and Kathryn Taylor July 2019 https://ceipfiles.s3.amazonaws.com/pdf/FinCyber/Cyber cecurity+Toolkit_Full.pdf
52		Cyber Risk Scenarios, the Financial System, and Systemic Risk Assessment Carnegie Endowment for International Peace Lincoln Kaffenberger and Emanuel Kopp September 30, 2019 https://carnegieendowment.org/2019/09/30/cyber-risk -scenarios-financial-system-and-systemic-risk-assessment -pub-79911
53		Lessons Learned and Evolving Practices of the TIBER Framework for Resilience Testing in the Netherlands Carnegie Endowment for International Peace Petra Hielkema and Raymond Kleijmeer October 2019 https://carnegieendowment.org/2019/10/07/lessons -learned-and-evolving-practices-of-tiber-framework -for-resilience-testing-in-netherlands-pub-80007

	DATE	DOCUMENTS
54	2020	Cyber Risk and the U.S. Financial System: A Pre-Mortem Analysis Federal Reserve Bank of New York Staff Report No. 909 Thomas M. Eisenbach, Anna Kovner, and Michael Junho Lee January 2020 https://www.finextra.com/finextra-downloads/newsdocs/fedcyberattack.pdf
55		Operational and Cyber Risks in the Financial Sector BIS Working Papers No. 840 Iñaki Aldasoro, Leonardo Gambacorta, Paolo Giudici, and Thomas Leach February 2020 https://www.bis.org/publ/work840.pdf
56		From Long-Distance Entanglement to Building a Nation-wide Quantum Internet Report of the DOE Quantum Internet Blueprint Workshop SUNY Global Center, New York, NY February 5–6, 2020 https://www.energy.gov/sites/prod/files/2020/07/f76/QuantumWkshpRpt20FINAL_Nav_0.pdf
57		Secure and Trusted Communications Networks Act of 2019 Public Law 116-124 116th Congress March 12, 2020 https://www.congress.gov/116/plaws/publ124/PLAW-116publ124.pdf
58		Cyberspace Solarium Commission Report March 2020 https://drive.google.com/file/d/1ryMCIL_dZ30QyjFqFkkf10MxIXJGT4yv/view
59		Cyber Mapping the Financial System Carnegie Endowment for International Peace Jan-Philipp Brauchle, Matthias Göbel, Jens Seiler, and Christoph von Busekist April 7, 2020 https://carnegieendowment.org/2020/04/07/cyber-mapping-financial-system-pub-81414

	DATE	DOCUMENTS
60		Effective Practices for Cyber Incident Response and Recovery Financial Stability Board April 20, 2020 https://www.fsb.org/wp-content/uploads/P200420-1.pdf
61		The Drivers of Cyber Risk BIS Working Papers No. 865 Iñaki Aldasoro, Leonardo Gambacorta, Paolo Giudici, and Thomas Leach May 2020 https://www.bis.org/publ/work865.pdf
62		Addressing the Threat Posed by TikTok, and Taking Additional Steps to Address the National Emergency with Respect to the Information and Communications Technology and Services Supply Chain Executive Order 13942 August 6, 2020 https://www.federalregister.gov/documents/2020/08/11/2020-17699/addressing-the-threat-posed-by-tiktok-and-taking-additional-steps-to-address-the-national-emergency
63		Addressing the Threat Posed by WeChat, and Taking Additional Steps to Address the National Emergency with Respect to the Information and Communications Technology and Services Supply Chain Executive Order 13943 August 11, 2020 https://www.federalregister.gov/documents/2020/08/11/2020-17700/addressing-the-threat-posed-by-wechat-and-taking-additional-steps-to-address-the-national-emergency
64		GAO, Critical Infrastructure Protection: Treasury Needs to Improve Tracking of Financial Sector Cybersecurity Risk Mitigation Efforts September 2020 https://www.gao.gov/assets/710/709455.pdf

	DATE	DOCUMENTS
65		Proposal for a Regulation of the European Parliament and of the Council on Digital Operational Resilience for the Financial Sector and Amending Regulations (EC) No. 1060/2009, (EU) No. 648/2012, (EU) No. 600/2014, and (EU) No. 909/2014, Articles 28–29 European Commission September 24, 2020 https://eur-lex.europa.eu/legal-content/EN/TXT/PDF/?uri=CELEX:52020PC0595&from=EN
66		International Strategy to Better Protect the Financial System against Cyber Threats Carnegie Endowment for International Peace in collaboration with the World Economic Forum Tim Maurer and Arthur Nelson November 18, 2020 https://carnegieendowment.org/files/Maurer_Nelson_FinCyber_final1.pdf
67		Enduring Cyber Threats and Emerging Challenges to the Financial Sector Cyber Policy Initiative Working Paper Series, Cybersecurity and the Financial System, No. 8 Adrian Nish, Saher Naumaan, and James Muir November 2020 https://carnegieendowment.org/files/NishNaumaan_FincyberThreatsChallenges_v3.pdf
68		G-7 Fundamental Elements of Cyber Exercises Programmes November 2020 https://home.treasury.gov/system/files/216/G7-Fundamental-Elements-of-Cyber-Exercise-Programs-November-2020-Published.pdf
69		Capacity-Building Tool Box for Cybersecurity and Financial Organizations Carnegie Endowment for International Peace Tim Maurer, Kathryn Taylor, and Taylor Grossman December 2020 https://swiftinstitute.org/wp-content/uploads/2021/01/CyberToolkit_EXECUTIVE-SUMMARY_11_25.pdf

	DATE	DOCUMENTS
70	2021	Transition Book for the Incoming Biden Administration CSC White Paper No. 5 U.S. Cyberspace Solarium Commission January 2021 https://www.solarium.gov/public-communications/transition-book
71		Addressing the Threat Posed by Applications and Other Software Developed or Controlled by Chinese Companies Executive Order 13971 January 8, 2021 https://www.federalregister.gov/documents/2021/01/08/2021-00305/addressing-the-threat-posed-by-applications-and-other-software-developed-or-controlled-by-chinese
72		National Security Commission on Artificial Intelligence Final Report March 2021 https://www.nscai.gov/wp-content/uploads/2021/03/Full-Report-Digital-1.pdf
73		Insider Threat Mitigation for U.S. Critical Infrastructure Entities: Guidelines from an Intelligence Perspective National Counterintelligence and Security Center (NCSC) March 2021 https://www.dni.gov/files/NCSC/documents/news/20210319-Insider-Threat-Mitigation-for-US-Critical-Infrastru-March-2021.pdf
74		Executive Order on Improving the Nation's Cybersecurity Executive Order 14028 May 12, 2021 https://www.whitehouse.gov/briefing-room/presidential-actions/2021/05/12/executive-order-on-improving-the-nations-cybersecurity
75		Protecting Americans' Sensitive Data from Foreign Adversaries Executive Order 14034 June 9, 2021 https://www.federalregister.gov/documents/2021/06/11/2021-12506/protecting-americans-sensitive-data-from-foreign-adversaries

	DATE	DOCUMENTS
76		EBA Analysis of RegTech in the EU Financial Sector June 2021 https://www.eba.europa.eu/sites/default/documents /files/document_library/Publications/Reports/2021 /1015484/EBA%20analysis%20of%20RegTech%20in %20the%20EU%20financial%20sector.pdf
77		Chinese State-Sponsored Cyber Operations: Observed TTPs Cybersecurity Alert National Security Agency, Cybersecurity and Infrastructure Security Agency, Federal Bureau of Investigation July 19, 2021 https://media.defense.gov/2021/Jul/19/2002805003/-1 /-1/1/CSA_CHINESE_STATE-SPONSORED _CYBER_TTPS.PDF
78		National Security Memorandum on Improving Cybersecu- rity for Critical Infrastructure Control Systems White House Executive Memo; President Joe Biden July 28, 2021 https://www.whitehouse.gov/briefing-room/statements -releases/2021/07/28/national-security-memorandum -on-improving-cybersecurity-for-critical-infrastructure -control-systems
79		Conducting Due Diligence on Financial Technology Companies: A Guide for Community Banks OCC, FDIC, FRB August 2021 https://www.occ.gov/news-issuances/news-releases/2021 /nr-ia-2021-85a.pdf
80		Federal Cybersecurity: America's Data Still at Risk Committee on Homeland Security and Governmental Af- fairs, United States Senate, Staff Report August 2021 https://www.hsgac.senate.gov/imo/media/doc/Federal %20Cybersecurity%20-%20America's%20Data%20Still %20at%20Risk%20(FINAL).pdf

	DATE	DOCUMENTS
81		Updated Advisory on Potential Sanctions Risks for Facilitating Ransomware Payments U.S. Department of Treasury September 21, 2021 https://home.treasury.gov/system/files/126/ofac_ransom ware_advisory.pdf
82		Public Policy Principles for Retail Central Bank Digital Currencies G7 UK 2021 October 14, 2021 https://assets.publishing.service.gov.uk/government/up loads/system/uploads/attachment_data/file/1025235 /G7_Public_Policy_Principles_for_Retail_CBDC _FINAL.pdf
83		Computer-Security Incident Notification Requirements for Banking Organizations and Their Bank Service Providers OCC, 12 CFR Part 53 (Docket ID OCC-2020-0038), RIN 1557-AF02; Federal Reserve System, 12 CFR Part 225 (Docket No. R-1736), RIN 7100-AG06; FDIC, 12 CFR Part 304, RIN 3064-AF59 November 17, 2021 https://www.federalreserve.gov/newsevents/pressreleases /files/bcreg20211118a1.pdf
84	2022	Memorandum for the Heads of Executive Departments and Agencies, M-22-0 Executive Office of the President, Office of Management and Budget January 26, 2022 https://www.whitehouse.gov/wp-content/uploads/2022 /01/M-22-09.pdf
85		Ensuring Responsible Development of Digital Assets Executive Order 14067 March 9, 2022 https://www.whitehouse.gov/briefing-room/presidential -actions/2022/03/09/executive-order-on-ensuring -responsible-development-of-digital-assets

	DATE	DOCUMENTS
86		S.3600—Strengthening American Cybersecurity Act of 2022 (part of the Omnibus Funding Act of 2020) 117th Congress (2021–2022) March 2020 https://www.congress.gov/bill/117th-congress/senate-bill/3600/text
87		Cyberattacks and Financial Stability: Evidence from a Natural Experiment Finance and Economics Discussion Series 2022-025 Board of Governors of the Federal Reserve System Antonis, Kotidis, and Stacey L. Schreft March 23, 2022 https://doi.org/10.17016/FEDS.2022.025
88		Enhancing the National Quantum Initiative Advisory Committee Executive Order May 4, 2022 https://www.whitehouse.gov/briefing-room/presidential-actions/2022/05/04/executive-order-on-enhancing-the-national-quantum-initiative-advisory-committee
89		National Security Memorandum on Promoting United States Leadership in Quantum Computing While Mitigating Risks to Vulnerable Cryptographic Systems May 4, 2022 https://www.whitehouse.gov/briefing-room/statements-releases/2022/05/04/national-security-memorandum-on-promoting-united-states-leadership-in-quantum-computing-while-mitigating-risks-to-vulnerable-cryptographic-systems
90		Timeline of Cyber Incidents Involving Financial Institutions Carnegie Endowment for International Peace Ongoing updates https://carnegieendowment.org/specialprojects/protectingfinancialstability/timeline

Source: Thomas P. Vartanian, 2022

NOTES

INTRODUCTION: CANDLESTICKS ARE NICE

1. Aaron S. Evans and Lee Armus, "Galaxy Collisions Preview Milky Way's Fate," *Scientific American*, December 2012, https://www.scientificamerican.com/article/galaxy-collisions-preview-milky-ways-fate.

2. Michelle Starr, "Scientists Figured Out How and When Our Sun Will Die, and It's Going to Be Epic," ScienceAlert, September 5, 2021, https://www.sciencealert.com/scientists-figured-out-how-and-when-our-sun-will-die-and-it-s-going-to-be-epic.

3. William Carter, "Forces Shaping the Cyber Threat Landscape for Financial Institutions," SWIFT Institute Working Paper No. 2016-004, October 2, 2017, 11–12, https://www.swiftinstitute.org/wp-content/uploads/2017/10/SIWP-2016-004-Cyber-Threat-Landscape-Carter-Final.pdf.

4. Patrick Howell O'Neil, "How a Russian Cyberwar in Ukraine Could Ripple Out Globally," *MIT Technology Review*, January 21, 2022, https://www.technologyreview.com/2022/01/21/1043980/how-a-russian-cyberwar-in-ukraine-could-ripple-out-globally.

5. Jillian K. Melchior, "The Cyberspace Front in the Attacks on Ukraine," *Wall Street Journal*, February 18, 2022, https://www.wsj.com/articles/virtual-front-attacks-invasion-ukraine-cybersecurity-attack-threat-hackers-russia-ukraine-dos-denial-of-service-energy-notpetya-11645221274.

6. David Uberti, "Ukrainian Defense Ministry, Banks Hit by Suspected Cyberattacks, Officials Say," *Wall Street Journal*, February 15, 2022, https://www.wsj.com/articles/ukrainian-defense-ministry-banks-hit-by-suspected-cyberattacks-officials-say-11644960352.

7. "Implement Cybersecurity Measures Now to Protect against Potential Critical Threats," CISA Insights, January 18, 2022, https://www.cisa.gov/sites/default/files/publications/CISA_Insights-Implement_Cybersecurity_Measures_Now_to_Protect_Against_Critical_Threats_508C.pdf.

8. David Uberti and Dustin Volz, "Destructive Malware Is Seen in Ukraine as the Risk of Cyber Spillover Looms," *Wall Street Journal*, February 24, 2022, https://www.wsj.com/livecoverage/russia-ukraine-latest-news/card/tdxMzFCXFSxCx BSRPIpX; Juan Andrés Guerrero-Saade, "HermeticWiper: New Destructive Malware Used in Cyber Attacks on Ukraine," Sentinel Labs, February 23, 2022, https://www.sentinelone.com/labs/hermetic-wiper-ukraine-under-attack.

9. Patrick Howell, "Russia Hacked an American Satellite Company One Hour before the Ukraine Invasion," *MIT Technology Review*, May 10, 2022, https://www.technologyreview.com/2022/05/10/1051973/russia-hack-viasat-satellite -ukraine invasion.

10. Igor Bonifacic, "Anonymous Claims Responsibility for Russian Government Website Outages," *MSN*, February 26, 2022, https://www.msn.com/en-us /news/technology/anonymous-claims-responsibility-for-russian-government-web site-outages/ar-AAUlMnQ.

11. Robert McMillan, Kevin Poulsen, and Dustin Volz, "Secret World of Pro-Russia Hacking Group Exposed in Leak," *Wall Street Journal*, March 28, 2022, https://www.wsj.com/articles/trickbot-pro-russia-hacking-gang-documents -ukrainian-leaker-conti-11648480564.

12. Steven Zeitchik, "Hackers Hit Popular Video Game, Stealing More than $600 Million in Cryptocurrency," *Washington Post*, March 29, 2022, https://www.msn.com/en-us/news/technology/hackers-hit-popular-video-game-steal ing-more-than-dollar600-million-in-cryptocurrency/ar-AAVDCQv.

13. Patrick Howell O'Neill, "These Hackers Showed Just How Easy It Is to Target Critical Infrastructure," *MIT Technology Review*, April 21, 2022, https://www.technologyreview.com/2022/04/21/1050815/hackers-target-critical-infra structure-pwn2own.

14. Paul Vigna, "Crypto Thieves Get Bolder by the Heist, Stealing Record Amounts," *Wall Street Journal*, April 22, 2022, https://www.wsj.com/articles /crypto-thieves-get-bolder-by-the-heist-stealing-record-amounts-11650582598.

15. Patrick Howell O'Neill, "The Hacker for Hire Industry Is Now Too Big to Fail," *MIT Technology Review*, December 28, 2021, https://www.technologyreview .com/2021/12/28/1043029/the-hacker-for-hire-industry-is-now-too-big-to-fail.

16. Patrick Howell O'Neill, "Inside the Plan to Fix America's Never-Ending Cybersecurity Failures," *MIT Technology Review*, March 18, 2022, https://www .technologyreview.com/2022/03/18/1047395/inside-the-plan-to-fix-americas -never-ending-cybersecurity-failures.

17. Greg Baer and Rob Hunter, "A Tower of Babel: Cyber Regulation for Financial Services," Banking Perspectives, June 3, 2017, https://www.bankingper spectives.com/a-tower-of-babel-cyber-regulation-for-financial-services.

18. Eric Geller, "'TSA Has Screwed This Up': Pipeline Cyber Rules Hitting Major Hurdles," Politico, March 17, 2022, https://www.politico.com/news /2022/03/17/tsa-has-screwed-this-up-pipeline-cyber-rules-hitting-major-hur dles-00017893.

19. J. David Grossman, "The Path to a More Resilient and Robust GPS," April 22, 2021, C4ISRNET, https://www.c4isrnet.com/opinion/2021/04/22/the -path-to-a-more-resilient-and-robust-gps.

20. National Defense Authorization Act for Fiscal Year 2018, Pub. L. 115-91, 131 Stat. 1283, December 12, 2017, https://www.congress.gov/115/plaws /publ91/PLAW-115publ91.pdf; Frank LoBiondo Coast Guard Authorization Act of 2018, S-140, Pub. L. 115-282, 132 Stat. 4192, December 4, 2018, https://www .congress.gov/115/plaws/publ282/PLAW-115publ282.pdf.

21. James Ball, "Russia Is Risking the Creation of a "Splinternet"—and It Could Be Irreversible," *MIT Technology Review*, March 17, 2022, https://www .technologyreview.com/2022/03/17/1047352/russia-splinternet-risk.

CHAPTER 1. THE NEXT FINANCIAL BLACK HOLE

1. Charles W. Calomiris and Stephen H. Haber, *Fragile by Design: The Political Origins of Banking Crises and Scarce Credit* (Princeton, NJ: Princeton University Press, 2014), 3–6.

2. Jeb Henserling, "Debt and Inflation Threaten U.S. Security," *Wall Street Journal*, February 23, 2022, https://www.wsj.com/articles/debt-and-inflation-threaten-us-security-jerome-powell-supply-chain-federal-reserve-russia-china-de fense-spending-11645565624.

3. Thomas P. Vartanian and William M. Isaac, "The Next Financial Crisis Is Edging Closer. There's Time to Stop It," *American Banker*, July 14, 2021, https:// www.americanbanker.com/opinion/the-next-financial-crisis-is-edging-closer -theres-time-to-stop-it; Steve Goldstein, "The U.S. Is the Country Most Vulnerable to a New Financial Crisis, Nomura Warns," MarketWatch, June 25, 2021, https://www.marketwatch.com/story/this-new-model-flags-vulnerability-to-a -financial-crisis-and-it-finds-the-u-s-is-most-at-risk-11624617751.

4. Graham Allison and Eric Schmidt, "China Will Soon Lead the U.S. in Tech," *Wall Street Journal*, December 7, 2021, https://www.wsj.com/articles/china -will-soon-lead-the-us-in-tech-global-leader-semiconductors-5g-wireless-green -energy-11638915759.

5. See Graham Allison, Kevin Klyman, Karen Barbesino, and Hugo Yen, "The Great Tech Rivalry: China vs. the U.S.," Belfer Center for Science and International Affairs, Harvard Kennedy School, 2021, https://www.belfercenter.org/sites /default/files/GreatTechRivalry_ChinavsUS_211207.pdf.

6. Kai-Fu Lee, *AI Superpowers: China, Silicon Valley, and the New World Order* (Boston: Mariner, 2018).

7. Silvia Elaluf-Calderwood, "Huawei Data Centres and Clouds Already Cover Latin America. Chinese Tech Influence Is a Gift to Countries and Politicians That Don't Respect Human Rights," Strand Consult, February 7, 2022, https://strand consult.dk/blog/huawei-data-centres-and-clouds-already-cover-latin-america

-chinese-tech-influence-is-a-gift-to-countries-and-politicians-that-dont-respect
-human-rights.

8. David Sanger, "Managing the Fifth Generation: America, China, and the Struggle for Technological Dominance," Aspen Institute, January 20, 2020, https://www.aspeninstitute.org/blog-posts/managing-the-fifth-generation-amer ica-china-and-the-struggle-for-technological-dominance.

9. 2021 Report to Congress of the US-China Economic and Security Review Commission, 117th Cong., 1st Sess., November 2021, p. 26, https://www.po litico.com/f/?id=0000017d-26b4-d8e1-a57d-effd6d3a0000.

10. Patrick Howell O'Neill, "How China Built a One-of-a-Kind Cyber-Espionage Behemoth to Last," *MIT Technology Review*, February 28, 2022, https:// www.technologyreview.com/2022/02/28/1046575/how-china-built-a-one-of-a -kind-cyber-espionage-behemoth-to-last.

11. Michael Schuman, "China Wants to Rule the World by Controlling the Rules," *Atlantic*, December 9, 2021, https://www.theatlantic.com/international /archive/2021/12/china-wants-rule-world-controlling-rules/620890.

12. "AI-Augmented Attacks and the Battle of the Algorithms," DarkTrace, 2020, https://newtech.mt/wp-content/uploads/2020/09/AI-Augmented-Attacks -and-the-Battle-of-the-Algorithms.pdf.

13. Thomas M. Eisenbach, Anna Kovner, and Michael Junho Lee, "Federal Reserve Bank of New York, Staff Reports—Cyber Risk and the U.S. Financial System: A Pre-Mortem Analysis," Federal Reserve Bank of New York, Staff Report No. 909, January 2020, revised June 2020, 2–3, https://www.newyorkfed .org/medialibrary/media/research/staff_reports/sr909.pdf.

14. Antonis Kotidis and Stacey L. Schreft, "Cyberattacks and Financial Stability: Evidence from a Natural Experiment," Finance and Economics Discussion Series 2022-025, May 2022, Board of Governors of the Federal Reserve System, https:// doi.org/10.17016/FEDS.2022.025.

15. Richard J. Danzig, "Surviving on a Diet of Poisoned Fruit: Reducing the National Security Risks of America's Cyber Dependencies," Center for a New American Society, July 2014, 13, https://www.cnas.org/publications/reports /surviving-on-a-diet-of-poisoned-fruit-reducing-the-national-security-risks-of -americas-cyber-dependencies.

CHAPTER 2. THE ILLUSION OF ONLINE SECURITY

1. "The Extortion Economy Podcast: Exploring the Secret World of Ransomware," *Pro Publica* and *MIT Technology Review* podcast, episode 1, December 20, 2021, https://www.propublica.org/article/the-extortion-economy-podcast -exploring-the-secret-world-of-ransomware.

2. Samantha Murphy Kelly, "The Bizarre Story of the Inventor of Ransomware," CNN Business, May 16, 2021, https://www.cnn.com/2021/05/16/tech /ransomware-joseph-popp/index.html.

3. The Editors, "Hacking the Ransomware Problem," *Scientific American*, January 2022, https://www.scientificamerican.com/article/hacking-the-ransomware-problem.

4. FBI Alert Number MC-000150-MW, August 25, 2021, https://www.ic3.gov/Media/News/2021/210825.pdf.

5. Julian E. Barnes, "U.S. Military Has Acted against Ransomware Groups, General Acknowledges," *New York Times*, December 5, 2021, https://www.nytimes.com/2021/12/05/us/politics/us-military-ransomware-cyber-command.html.

6. Richard J. Danzig, "Surviving on a Diet of Poisoned Fruit: Reducing the National Security Risks of America's Cyber Dependencies," Center for a New American Society, July 2014, 11, https://www.jstor.org/stable/pdf/resrep06383.pdf.

7. Danzig, "Surviving on a Diet of Poisoned Fruit," 11.

8. Danzig, "Surviving on a Diet of Poisoned Fruit," 12.

9. Patrick Howell O'Neill, "The U.S. Military Wants to Understand the Most Important Software on Earth," *MIT Technology Review*, July 14, 2022, http://www.technologyreview.com/2022/09/14/1055894/us-military-software-linux-kernel-open-source/.

10. Danzig, "Surviving on a Diet of Poisoned Fruit," 10.

11. David D. Clark, *Designing an Internet* (Cambridge: Massachusetts Institute of Technology, 2018), 197.

12. Clark, *Designing an Internet*, 209.

13. Clark, *Designing an Internet*, 189–92.

14. Patrick Howell O'Neill, "The U.S. Is Worried That Hackers Are Stealing Data Today So Quantum Computers Can Crack It in a Decade," *MIT Technology Review*, November 3, 2021, https://www.technologyreview.com/2021/11/03/1039171/hackers-quantum-computers-us-homeland-security-cryptography.

15. Thomas M. Eisenbach, Anna Kovner, and Michael Junho Lee, "Federal Reserve Bank of New York, Staff Reports—Cyber Risk and the U.S. Financial System: A Pre-Mortem Analysis," Federal Reserve Bank of New York, Staff Report No. 909, January 2020, Revised June 2020, 2–3, https://www.newyorkfed.org/medialibrary/media/research/staff_reports/sr909.pdf.

16. Ray Kurzweil, *The Singularity Is Near: When Humans Transcend Biology* (New York: Viking, 2005).

CHAPTER 3. IS VIRTUAL ORDER POSSIBLE?

1. Patrick Howell O'Neill, "2021 Has Broken the Record for Zero-Day Hacking Attacks," *MIT Technology Review*, September 23, 2021, https://www.technologyreview.com/2021/09/23/1036140/2021-record-zero-day-hacks-reasons.

2. Bill Bostock, "A Pentagon Official Said He Resigned Because US Cybersecurity Is No Match for China, Calling It 'Kindergarten Level,'" Business Insider, October 11, 2021, https://www.businessinsider.com/pentagon-official-quit-saying-us-cybersecurity-no-match-china-2021-10.

3. Sultan Meghji, "Why I Quit as FDIC Innovation Chief: Technophobia," Bloomberg, February 22, 2022, https://www.bloomberg.com/opinion/articles/2022-02-22/i-quit-as-fdic-innovation-chief-because-of-regulators-technophobia.

4. National Security Commission on Artificial Intelligence, "Final Report," March 2021, 45, https://www.nscai.gov/wp-content/uploads/2021/03/Full-Report-Digital-1.pdf.

5. Bailee Hill, "Pentagon Official Resigns, Sounds Alarm about US Losing Technological Edge," Fox News, April 19, 2022, https://www.msn.com/en-us/news/world/pentagon-official-resigns-sounds-alarm-about-us-losing-technological-edge/ar-AAWmEuk.

6. Winnona Desombre, Lars Gjesvki, and Johann Ole Willers, "Surveillance Technology at the Fair: Proliferation of Cyber Capabilities in International Arms Markets," Atlantic Council, November 8, 2021, 1–3, https://www.atlanticcouncil.org/wp-content/uploads/2021/11/Surveillance-Technology-at-the-Fair.pdf; Patrick Howell O'Neill, "A Grim Outlook: How Cyber Surveillance Is Booming on a Global Scale," *MIT Technology Review*, November 8, 2021, https://www.technologyreview.com/2021/11/08/1039395/grim-outlook-cyber-boom-atlantic-council-report.

7. O'Neill, "A Grim Outlook."

8. Summer Said and Stephen Kalin, "Saudi Arabia Considers Accepting Yuan Instead of Dollars for Chinese Oil Sales," *Wall Street Journal*, March 15, 2022, https://www.wsj.com/articles/saudi-arabia-considers-accepting-yuan-instead-of-dollars-for-chinese-oil-sales-11647351541.

9. Richard J. Danzig, "Surviving on a Diet of Poisoned Fruit: Reducing the National Security Risks of America's Cyber Dependencies," Center for a New American Society, July 2014, 6, https://www.jstor.org/stable/pdf/resrep06383.pdf.

10. Christopher Yoo, "The Non-monolithic Internet," International Institute of Communications, *Inter Media* 41, no. 2 (May 2013): 2–3, https://www.iicom.org/wp-content/uploads/COMPRESSED-143093_Intermedia-41-2-min.pdf.

11. *Tallinn Manual 2.0 on the International Law Applicable to Cyber Operations* (Cambridge: Cambridge University Press, 2017), 2.

CHAPTER 4. FOR OUR NEXT ACT!

1. Michael Tabb, Jeffery DelViscio, and Andrea Gawrylewski, "What Is 5G? Here Is a Short Video Primer," *Scientific American*, October 13, 2021, https://www.scientificamerican.com/video/what-is-5g-here-is-short-video-primer.

2. Sascha Segan, "What Is 5G?," *PC Magazine*, August 26, 2021, https://www.pcmag.com/news/what-is-5g.

3. Tabb, DelViscio, and Gawrylewski, "What Is 5G? Here Is a Short Video Primer."

4. Milo Medin and Gilman Louie, "The 5G Ecosystem: Risks & Opportunities for DoD Defense Innovation Board," April 3, 2019, 2–3, https://www.readkong .com/page/the-5g-ecosystem-risks-opportunities-for-dod-8459529.

5. Medin and Louie, "The 5G Ecosystem," 2–3.

6. "Department of Defense (DoD) 5G Strategy (U)," unclassified document, approved by secretary of defense May 2, 2020, 1, https://www.cto.mil/wp-con tent/uploads/2020/05/DoD_5G_Strategy_May_2020.pdf.

7. "Department of Defense (DoD) 5G Strategy (U)."

8. "Department of Defense 5G Strategy Implementation Plan."

9. "Department of Defense 5G Strategy Implementation Plan."

10. Medin and Louie, "The 5G Ecosystem."

11. Medin and Louie, "The 5G Ecosystem."

12. David Sanger, "Managing the Fifth Generation: America, China, and the Struggle for Technological Dominance," Aspen Institute, January 20, 2020, https://www.aspeninstitute.org/blog-posts/managing-the-fifth-generation-amer ica-china-and-the-struggle-for-technological-dominance.

13. Sanger, "Managing the Fifth Generation."

14. Sanger, "Managing the Fifth Generation."

15. "What Is IoT?," Oracle, https://www.oracle.com/internet-of-things/what -is-iot.

16. "What Is IoT?"

17. "How Do Quantum Computers Work?," Science Alert, https://www .sciencealert.com/quantum-computers.

18. Scott Aronson, "What Makes Quantum Computing So Hard to Explain?," *Quanta Magazine*, June 8, 2021, https://www.quantamagazine.org/why-is-quan tum-computing-so-hard-to-explain-20210608.

19. Siobhan Roberts, "This New Startup Has Built a Record-Breaking 256 Qu-bit Quantum Computer," *MIT Technology Review*, November 17, 2021, https:// www.technologyreview.com/2021/11/17/1040243/quantum-computer-256-bit -startup.

20. Roberts, "This New Startup."

21. Scott Aronson, "What Makes Quantum Computing So Hard to Explain?," *Quanta Magazine*, June 8, 2021, https://www.quantamagazine.org/why-is-quan tum-computing-so-hard-to-explain-20210608.

22. Mordechai Rorvig, "Qubits Can Be as Safe as Bits, Researchers Show," *Quanta Magazine*, January 6, 2022, https://www.quantamagazine.org/qubits-can -be-as-safe-as-bits-researchers-show-20220106.

23. Ahmed Banafa, "Quantum Internet Explained," OpenMind BBVA, December 14, 2020, https://www.bbvaopenmind.com/en/technology/digital-world/quantum-internet-explained; Anil Ananthaswamy, "The Quantum Internet Is Emerging, One Experiment at a Time," *Scientific American*, July 19, 2019, https:// www.scientificamerican.com/article/the-quantum-internet-is-emerging-one -experiment-at-a-time.

24. Rorvig, "Qubits Can Be as Safe as Bits."

25. "Quantum Key Distribution (QKD) and Quantum Cryptography (QC)," National Security Agency Central Security Service, https://www.nsa.gov/what-we -do/cybersecurity/quantum-key-distribution-qkd-and-quantum-cryptography-qc.

26. Arthur Herman, "The Quantum Computing Threat to American Society," *Wall Street Journal*, November 10, 2019, https://www.wsj.com/articles/the -quantum-computing-threat-to-american-security-11573411715.

27. "Executive Order on Enhancing the National Quantum Initiative Advisory Committee," May 4, 2022, https://www.whitehouse.gov/briefing-room/pres idential-actions/2022/05/04/executive-order-on-enhancing-the-national-quan tum-initiative-advisory-committee; "National Security Memorandum on Promoting United States Leadership in Quantum Computing While Mitigating Risks to Vulnerable Cryptographic Systems," May 4, 2022, https://www.whitehouse.gov /briefing-room/statements-releases/2022/05/04/national-security-memorandum -on-promoting-united-states-leadership-in-quantum-computing-while-mitigating -risks-to-vulnerable-cryptographic-systems.

28. Portions of this discussion are derived from an article written by Mr. Vartanian and first published at *The Hill*; see "To Avoid Virtual Anarchy, We Must Move Cautiously and Fix Things," November 5, 2021, *The Hill*, https://thehill .com/opinion/cybersecurity/580202-its-time-to-move-cautiously-and-fix-things.

29. Leslie Jamison, "The Digital Ruins of a Forgotten Future," *Atlantic*, December 2017, https://www.theatlantic.com/magazine/archive/2017/12/second -life-leslie-jamison/544149.

30. Herrman and Browning, "Are We in the Metaverse Yet?"

31. Herrman and Browning, "Are We in the Metaverse Yet?"

32. "NFTs and Play-to-Earn Games Shed Light on Future of Metaverse Economy," *Wall Street Journal* (video), March 16, 2022, https://www.wsj.com /video/series/in-depth-features/nfts-and-play-to-earn-games-shed-light-on-future -of-metaverse-economy/A5540926-5121-48CC-9248-78478BEAF38D.

33. "NFTs and Play-to-Earn Games."

CHAPTER 5. HOW THE CYBER GOT INTO SPACE

1. Christopher Yoo, "The Non-monolithic Internet," International Institute of Communications, *Inter Media*, 1.

2. HP provides a short illustrated history of the Internet titled "How Does the Internet Work: A Step-by-Step Pictorial," HP Tech Takes, https://www.hp.com /us-en/shop/tech-takes/how-does-the-internet-work.

3. Laura DeNardis, *The Global War for Internet Governance* (New Haven, CT: Yale University Press, 2014).

4. "Domain Name Registration Process," ICANN, https://whois.icann.org /en/domain-name-registration-process.

5. DeNardis, *The Global War for Internet Governance*, 44–45, 48.

6. "What Is DNS/How Does DNS Work?," Cloudflare, https://www.cloud flare.com/learning/dns/what-is-dns.

7. Rus Shuler, "How Does the Internet Work?," 2002, https://web.stanford .edu/class/msande91si/www-spr04/readings/week1/InternetWhitepaper.htm.

8. DeNardis, *The Global War for Internet Governance*, 48; "Root Servers," Internet Assigned Numbers Authority (INNA), https://www.iana.org/domains/root /servers.

9. DeNardis, *The Global War for Internet Governance*, 39–41.

10. Shuler, "How Does the Internet Work?"

11. DeNardis, *The Global War for Internet Governance*, 107–9.

12. David D. Clark, *Designing an Internet* (Cambridge: Massachusetts Institute of Technology, 2018), 304–27.

13. James Griffiths, "The Global Internet Is Powered by Vast Undersea Cables. But They're Vulnerable," CNN, July 26, 2019, https://www.cnn .com/2019/07/25/asia/internet-undersea-cables-intl-hnk/index.html.

14. Griffiths, "The Global Internet."

15. Chris Stokel-Walker, "Tonga's Volcano Blast Cut It Off from the World. Here's What It Will Take to Get It Reconnected," *MIT Technology Review*, January 18, 2022, https://www.technologyreview.com/2022/01/18/1043790/tongas -volcano-internet-reconnected.

16. Griffiths, "The Global Internet."

17. Griffiths, "The Global Internet."

18. Christopher Mims, "Google, Amazon, Meta and Microsoft Weave a Fiber-Optic Web of Power," *Wall Street Journal*, January 15, 2022, https://www.wsj .com/articles/google-amazon-meta-and-microsoft-weave-a-fiber-optic-web-of -power-11642222824.

19. Helene Fouquet, "China's 7,500-Mile Undersea Cable to Europe Fuels Internet Feud," *Bloomberg Businessweek*, March 5, 2021, https://www.bloomberg .com/businessweek.

CHAPTER 6. THE NEXT GREAT TRAIN ROBBERY

1. *The Great Train Robbery* (Dino de Laurentis Company, Startling Films, United Artists, 1978), opening narration by Sean Connery, https://www.youtube.com /watch?v=ZNCnHYxkoGo.

2. Cliff Stoll, *The Cuckoo's Egg: Tracking a Spy through the Maze of Computer Espionage* (New York: Gallery Books, 1989).

3. Douglas Waller, "Onward Cyber Soldiers, *Time*, August 21, 1995, accessed February 15, 2021, http://content.time.com/time/subscriber/article/0,33009,983318,00 .html.

4. Waller, "Onward Cyber Soldiers."

5. Waller, "Onward Cyber Soldiers."

6. Elizabeth Dilts Marshall and David Henry, "JPMorgan's Dimon Blasts Bitcoin as 'Worthless,' Due for Regulation," Reuters, October 10, 2021, https://www.reuters.com/business/jpmorgans-dimon-blasts-bitcoin-worthless-due-regulation-2021-10-11.

7. MacKenzie Sigalos, "Crypto Scammers Took a Record $14 Billion in 2021," CNBC, January 7, 2022, https://www.cnbc.com/2022/01/06/crypto-scammers-took-a-record-14-billion-in-2021-chainalysis.html.

8. Steven Zeitchik, "Hackers Hit Popular Video Game, Stealing More than $600 Million in Cryptocurrency," *Washington Post*, March 29, 2022, https://www.msn.com/en-us/news/technology/hackers-hit-popular-video-game-stealing-more-than-dollar600-million-in-cryptocurrency/ar-AAVDCQv.

9. Paul Vigna, "Crypto Thieves Get Bolder by the Heist, Stealing Record Amounts," *Wall Street Journal*, April 22, 2022, https://www.wsj.com/articles/crypto-thieves-get-bolder-by-the-heist-stealing-record-amounts-11650582598.

10. Christopher Beam, "The Math Prodigy Whose Hack Upended DeFi Won't Give Back His Millions," *Bloomberg*, May 19, 2022, https://www.bloomberg.com/news/features/2022-05-19/crypto-platform-hack-rocks-blockchain-community.

11. Caitlin Ostroff and Alexander Osipovich, "Cryptocurrency TerraUSD Falls below Fixed Value, Triggering Selloff," *Wall Street Journal*, May 9, 2022, https://www-wsj-com.cdn.ampproject.org/c/s/www.wsj.com/amp/articles/cryptocurrency-terrausd-falls-below-fixed-value-triggering-selloff-11652122461.

12. They include *Cyber War* by Richard A. Clarke and Robert K. Knake, *Future Crimes* by Marc Goodman, *This Is How They Tell Me the World Ends* by Nicole Perlroth, *Click Here to Kill Everybody* by Bruce Schneier, *Sandworm* by Andy Greenberg, *The Perfect Weapon* by David E. Sanger, and *The Singularity Is Near* by Ray Kurzweil.

CHAPTER 7. WILL THE REAL
SATOSHI PLEASE STAND UP?

1. Cem Dilmegani, "What Are Crypto Derivatives? Types, Features & Top Exchanges," *AI Multiple*, October 11, 2021, https://research.aimultiple.com/crypto-derivatives. This analysis first appeared in somewhat different form in an op-ed I wrote in 2021; see Thomas P. Vartanian, "To Avoid Virtual Anarchy, We Must Move Cautiously and Fix Things," *The Hill*, November 5, 2021, https://thehill.com/opinion/cybersecurity/580202-its-time-to-move-cautiously-and-fix-things.

2. Carla Tardi, "Genesis Block," Investopedia, September 11, 2019, https://www.investopedia.com/terms/g/genesis-block.asp.

3. Coryanne Hicks, "The History of Bitcoin," *US News & World Report*, September 1, 2020, https://money.usnews.com/investing/articles/the-history-of-bitcoin.

4. Paul Vigna, "Bitcoin Creator Satoshi Nakamoto Could Be Unmasked at Florida Trial," *Wall Street Journal*, November 13, 2021, https://www.wsj.com/articles/bitcoin-creator-satoshi-nakamoto-could-be-unmasked-at-florida-trial-11636808401.

5. Paul Vigna, "Jury in Satoshi Nakamoto Suit Finds No Evidence of Partnership to Create Bitcoin," *Wall Street Journal*, December 6, 2021, https://www.wsj.com/articles/jury-in-satoshi-nakamoto-suit-finds-no-evidence-of-partnership-to-create-bitcoin-11638823714.

6. William Magnuson, *Blockchain Democracy: Technology, Law and the Rule of the Crowd* (Cambridge: Cambridge University Press, 2020), 41–60.

7. Kevin Collier, "Crypto Exchanges Keep Getting Hacked, and There's Little Anyone Can Do," NBC News, December 17, 2021, https://www.nbcnews.com/tech/security/bitcoin-crypto-exchange-hacks-little-anyone-can-do-rcna7870.

8. Andrew Perrin, "16% of Americans Say They Have Ever Invested in, Traded or Used Cryptocurrency," Pew Research Center, November 11, 2021, https://www.pewresearch.org/fact-tank/2021/11/11/16-of-americans-say-they-have-ever-invested-in-traded-or-used-cryptocurrency. Some sources report that there are 7,800 cryptocurrenices in circulation in December 2020; see Connor Freitas, "How Many Cryptocurrencies Are There?," Currency.com, December 7, 2020, https://currency.com/how-many-cryptocurrencies-are-there.

9. https://www.statista.com/statistics/730876/cryptocurrency-maket-value.

10. Bruce Schneier, "Understanding Crypto 6: Security, Trust, and Blockchain," *The Rational Reminder Podcast*, July 8, 2022, https://rationalreminder.libsyn.com/understanding-crypto-6-bruce-schneier-security-trust-and-blockchain.

11. Paul Vigna, "Crypto Stocks Perform Worse than Cryptocurrencies," *Wall Street Journal*, April 19, 2022, https://www.wsj.com/articles/crypto-stocks-perform-worse-than-cryptocurrencies-11650338283.

12. "Tulip Mania," Britannica, https://www.britannica.com/event/Tulip-Mania.

13. Bjarke Smith-Meyer, "El Salvador Looks to Crypto Bros to Stabilize Bitcoin," Politico, November 10, 2021, https://www.politico.eu/article/el-salvador-cryptocurrency-bitcoin-economy-price-stability-nayib-bukele.

14. Andres Oppenheimer, "El Salvador Was the First Country to Adopt Bitcoin. It Might Be the First to Go Broke, Too," *Miami Herald*, January 26, 2022, https://www.miamiherald.com/opinion/editorials/article257746438.html.

15. Paul Kiernan, "SEC's Gensler Does Not See Cryptocurrencies Lasting Long," *Wall Street Journal*, September 21, 2021, https://www.wsj.com/articles/secs-gensler-doesnt-see-cryptocurrencies-lasting-long-11632246355.

16. Zeke Faux, "Anyone Seen Tether's Billions?," *Bloomberg Businessweek*, October 7, 2021, https://www.bloomberg.com/news/features/2021-10-07/crypto-mystery-where-s-the-69-billion-backing-the-stablecoin-tether.

17. Faux, "Anyone Seen Tether's Billions?"

18. "CFTC Orders Tether and Bitfinex to Pay Fines Totalling $42.5 Million," Commodity Futures Trading Commission, Release No. 8450-21, October 15, 2021, https://www.cftc.gov/PressRoom/PressReleases/8450-21.

19. "Consultative Report Application of the Principles for Financial Market Infrastructures to Stablecoin Arrangements," Committee on Payments and Market Infrastructures Board of the International Organization of Securities Commissions, October 2021, 5–6, https://www.iosco.org/library/pubdocs/pdf /IOSCOPD685.pdf.

20. "Regulation, Supervision and Oversight of 'Global Stablecoin' Arrangements: Progress Report on the Implementation of the FSB High-Level Recommendations," Financial Stability Board, October 7, 2021, 3, https://www.fsb.org /wp-content/uploads/P071021.pdf.

21. "Report on Stablecoins," President's Working Group on Financial Markets, FDIC and OCC, November 2021, 15–18, https://home.treasury.gov/system /files/136/StableCoinReport_Nov1_508.pdf.

22. Financial Stability Oversight Council, *2021 Annual Report*, December 17, 2021, 172, https://home.treasury.gov/system/files/261/FSOC2021AnnualReport .pdf.

23. Greg Baer, "Making Stablecoins Stable: Is the Cure Worse than the Disease?," Bank Policy Institute, September 27, 2021, https://bpi.com/making-stable coins-stable-is-the-cure-worse-than-the-disease.

24. "What Is a Central Bank Digital Currency? Is the Federal Reserve Moving toward Adopting a Digital Dollar?," Federal Reserve, https://www.federalreserve .gov/faqs/what-is-a-central-bank-digital-currency.htm.

25. This discussion first appeared in my article, "No. 1 Rule for Central Bank Digital Currencies: Do No Harm," *American Banker*, April 21, 2021, https://www .americanbanker.com/opinion/no-1-rule-for-central-bank-digital-currencies-do -no-harm.

26. Greg Baer, "Central Bank Digital Currencies: Costs, Benefits and Major Implications for the U.S. Economic System," Banking Policy Institute, BPI Staff Working Paper, April 7, 2021, 11–15, https://bpi.com/wp-content /uploads/2021/04/Central-Bank-Digital-Currencies-Costs-Benefits-and-Major -Implications-for-the-U.S.-Economic-System.pdf.

27. "BIS Annual Economic Report," June 2022, https://www.bis.org/publ /arpdf/ar2022e.pdf.

28. "Executive Order on Enhancing the National Quantum Initiative Advisory Committee," May 4, 2022, https://www.whitehouse.gov/briefing-room/pres idential-actions/2022/05/04/executive-order-on-enhancing-the-national-quan tum-initiative-advisory-committee; "National Security Memorandum on Promoting United States Leadership in Quantum Computing While Mitigating Risks to Vulnerable Cryptographic Systems," May 4, 2022, https://www.whitehouse.gov /briefing-room/statements-releases/2022/05/04/national-security-memorandum

-on-promoting-united-states-leadership-in-quantum-computing-while-mitigating -risks-to-vulnerable-cryptographic-systems.

29. Andreas Baumhoff, "Breaking RSA Encryption—an Update on the State-of-the-Art," Quintessence Labs, June 13, 2019, https://www.quintessencelabs .com/blog/breaking-rsa-encryption-update-state-art.

30. Vartanian, "No. 1 Rule."

31. Vice Chair for Supervision Randal K. Quarles, "Parachute Pants and Central Bank Money" (speech given at the 113th Annual Utah Bankers Association Convention, Sun Valley, Idaho, June 28, 2021), https://www.federalreserve.gov /newsevents/speech/quarles20210628a.htm.

32. Quarles, "Parachute Pants."

33. Governor Christopher J. Waller, "CBDC: A Solution in Search of a Problem?," American Enterprise Institute, August 5, 2021, https://www.federalreserve .gov/newsevents/speech/waller20210805a.htm.

34. Board of Governors of the Federal Reserve System "Money and Payments: The U.S. Dollar in the Age of Digital Transformation," January 2022, https:// www.federalreserve.gov/publications/files/money-and-payments-20220120.pdf.

35. Board of Governors of the Federal Reserve System, "Money and Payments," 20.

36. "Executive Order on Ensuring Responsible Development of Digital Assets," White House, March 9, 2022, https://www.whitehouse.gov/briefing-room /presidential-actions/2022/03/09/executive-order-on-ensuring-responsible-devel opment-of-digital-assets.

37. Nurjannah Ahmat and Sabrina Bashir, "Central Bank Digital Currency: A Monetary Policy Perspective," Monetary Policy Department, Bank Negara Malaysia, Central Bank of Malaysia, September 2017, https://www.bnm.gov.my /documents/20124/826874/CB_Digital+Currency_Print.pdf; Central Bank Digital Currency Tracker, Atlantic Council, https://www.atlanticcouncil.org/cbdctracker.

38. "Report on Stablecoins."

39. Elaine Wu and Joe Wallace, "China Declares Cryptocurrency Transactions Illegal: Bitcoin Price Falls," *Wall Street Journal*, September 24, 2021, https://www .wsj.com/articles/china-declares-bitcoin-and-other-cryptocurrency-transactions -illegal-11632479288.

40. See the testimony of Brian P. Brooks, CEO, Bitfury Group, before the Committee on Financial Services, U.S. House of Representatives, December 8, 2021, 55:17–58:16, https://financialservices.house.gov/events/eventsingle.aspx ?EventID=408705.

41. Statement of Brian P. Brooks, CEO, Bitfury Group, before the Committee on Financial Services, U.S. House of Representatives, December 8, 2021, 5–6, https://financialservices.house.gov/uploadedfiles/hhrg-117-ba00-wstate -brooksb-20211208.pdf.

CHAPTER 8. DATA LOLLAPALOOZAS

1. Cliff Stoll, *The Cuckoo's Egg: Tracing a Spy through the Maze of Computer Espionage* (New York: Gallery Books, 1989), 371.

2. Federal Reserve 2019 Payments Study, January 6, 2020, https://www.feder alreserve.gov/paymentsystems/2019-December-The-Federal-Reserve-Payments -Study.htm; Erica Sandberg, "The Average Number of Credit Card Transactions per Day & Year," CardRates.com, November 9, 2020, https://www.cardrates .com/advice/number-of-credit-card-transactions-per-day-year.

3. Thomas M. Eisenbach, Anna Kovner, and Michael Junho Lee, "Federal Reserve Bank of New York, Staff Reports—Cyber Risk and the U.S. Financial System: A Pre-Mortem Analysis," Federal Reserve Bank of New York, Staff Report No. 909, January 2020, Revised June 2020, 2–3, https://www.newyorkfed .org/medialibrary/media/research/staff_reports/sr909.pdf.

4. Antonis Kotidis and Stacey L. Schreft, "Cyberattacks and Financial Stability: Evidence from a Natural Experiment," Finance and Economics Discussion Series 2022-025, Board of Governors of the Federal Reserve System, March 23, 2022, https://doi.org/10.17016/FEDS.2022.025.

5. Eisenbach, Kovner, and Lee, "Federal Reserve Bank of New York, Staff Reports," 1, 3.

6. David Heun, "Banks Combat Rising Threat of Fake Websites," *American Banker*, November 23, 2021, https://www.americanbanker.com/news/banks -combat-rising-threat-of-fake-websites.

7. Maggie Miller, "Financial Firms Facing Serious Hacking Threat in COVID-19 Era," *The Hill*, June 16, 2020, https://thehill.com/policy/cybersecurity/503051 -financial-firms-facing-serious-hacking-threat-in-covid-19-era.

8. Maria Korolov, "Banks Get Attacked Four Times More than Other Industries," CSO Online, June 23, 2015, https://www.csoonline.com/article/2938767 /report-banks-get-attacked-four-times-more-than-other-industries.html; "For Wealth Managers, Off Year Sparks Opportunity to Reignite Growth," Boston Consulting Group, June 20, 2019, https://www.bcg.com/press/20june2019 -global-wealth-report.

9. Shoshana Zuboff, *The Age of Surveillance Capitalism: The Fight for a Human Future at the New Frontier of Power* (New York: PublicAffairs, 2019).

10. Stephen Sorace, "Amazon to Opt-In Customers with Echo, Ring Devices to New 'Sidewalk' WiFi-Sharing Feature," Yahoo! Finance, December 6, 2020, https://finance.yahoo.com/news/amazon-opt-customers-echo-ring-190137438 .html.

11. Some material included in the next several pages is adapted and drawn from a published article coauthored by Mr. Vartanian that previously appeared in the *American Banker*: Thomas P. Vartanian and William M. Isaac, "The Dark Side of Financial Data Sharing," *American Banker (BankThink)*, June 1, 2022, https://www .americanbanker.com/opinion/the-dark-side-of-financial-data-sharing.

12. "Update on Open Banking," UK press release, October 1, 2021, www.gov .uk; Frank Saavedra-Lim, "Look Out for Risks in Open Banking!" Teradata, June 21, 2021, https://www.teradata.com/Blogs/Look-Out-for-Risks-in-Open-Banking.

13. Jeffrey Voas, Phil Laplante, Steve Lu, Rafail Ostrovsky, Mohamad Kassab, and Nir Kshetri, "Cybersecurity Considerations for Open Banking Technology and Emerging Standards," National Institute of Standards and Technology, Draft NISTIR 8389, January 2022, https://nvlpubs.nist.gov/nistpubs/ir/2022/NIST .IR.8389-draft.pdf.

14. Letter to National Institute of Standards and Technology from Paige Pidano Paridon, Melissa MacGregor, and Rob Morgan, March 3, 2022, https://bpi .com/wp-content/uploads/2022/03/BPI-Comment-Letter-NIST-Open-Banking -Report.pdf.

15. Jill Lepore, *If Then: How the Simulmatics Corporation Invented the Future* (New York: Liveright, 2020).

16. Andreas Baumhof, "Breaking RSA Encryption—an Update on the State-of-the-Art," Quintessence Labs, June 13, 2019, https://www.quintessencelabs.com /blog/breaking-rsa-encryption-update-state-art.

17. Baumhof, "Breaking RSA Encryption."

18. Baumhof, "Breaking RSA Encryption."

19. Drew Harwell, "IRS Wants to Scan Your Face," *Washington Post*, January 27, 2022, https://www.washingtonpost.com/technology/2022/01/27/irs-face -scans; Richard Rubin and Laura Saunders, "IRS Retreats from Facial Recognition to Verify Taxpayers' Identities," *Wall Street Journal*, February 7, 2022, https:// www.wsj.com/articles/irs-backs-away-from-facial-recognition-to-verify-taxpay ers-identities-11644264843.

20. U.S. Cyber Command, https://www.cybercom.mil/About/Mission-and -Vision.

21. U.S. Cybersecurity and Infrastructure Security Agency, "Infrastructure Security," https://www.cisa.gov/infrastructure-security.

22. U.S. Cybersecurity and Infrastructure Security Agency, "Financial Services Specific-Sector Plan 2015," 3–4, https://www.cisa.gov/sites/default/files/publica tions/nipp-ssp-financial-services-2015-508.pdf.

23. U.S. Government Accountability Office, "Cybersecurity—Clarity of Leadership Urgently Needed to Fully Implement the National Strategy," September 2020, https://www.gao.gov/assets/710/709986.pdf.

24. Josh Chin, "China's New Tool for Social Control: A Credit Rating for Everything," *Wall Street Journal*, November 28, 2016, https://www.wsj.com/articles /chinas-new-tool-for-social-control-a-credit-rating-for-everything-1480351590.

CHAPTER 9. THE EVE OF DESTRUCTION

1. "The Eve of Destruction," written by P. F. Sloan, performed by Barry Mc-Guire (Dunhill Records, 1965).

2. Dustin Volz, "U.S. Agencies Hacked in Foreign Cyber Espionage Campaign Linked to Russia," *Wall Street Journal*, December 13, 2020, https://www.wsj.com/articles/agencies-hacked-in-foreign-cyber-espionage-campaign-11607897866.

3. Dustin Volz and Robert McMillan, "Hack Suggests New Scope, Sophistication for Cyberattacks," *Wall Street Journal*, December 18, 2020, https://www.wsj.com/articles/hack-suggests-new-scope-sophistication-for-cyber attacks-11608251360.

4. SolarWinds website, https://www.solarwinds.com/federal-government/it -management-solutions-for-government.

5. SolarWinds Security Advisory, December 16, 2020, https://www.solar winds.com/securityadvisory; see also "Highly Evasive Attacker Leverages Solar-Winds Supply Chain to Compromise Multiple Global Victims with SUNBURST Backdoor," December 13, 2020, https://www.fireeye.com/blog/threat -research/2020/12/evasive-attacker-leverages-solarwinds-supply-chain-compro mises-with-sunburst-backdoor.html; "Mitigate SolarWinds Orion Code Compromise," Emergency Directive 21-01, December 13, 2020, https://www.cisa.gov /emergency-directive-21-01.

6. Christina Zhaoa, "SolarWinds, Probably Hacked by Russia, Serves White House, Pentagon, NASA," *Newsweek*, December 13, 2020, https://www.news week.com/solar-winds-probably-hacked-russia-serves-white-house-pentagon -nasa-1554447.

7. Volz and McMillan, "Hack Suggests New Scope."

8. David E. Sanger, "Russian Hackers Broke into Federal Agencies, U.S. Officials Suspect," *New York Times*, December 14, 2020, https://www.nytimes .com/2020/12/13/us/politics/russian-hackers-us-government-treasury-com merce.html.

9. Sanger, "Russian Hackers Broke into Federal Agencies."

10. Volz and McMillan, "Hack Suggests New Scope."

11. Patrick Howell O'Neill, "Recovering from the SolarWinds Hack Could Take 18 Months," *MIT Technology Review*, March 2, 2021, https://www.technolo gyreview.com/2021/03/02/1020166/solarwinds-brandon-wales-hack-recovery -18-months.

12. William Carter, "Forces Shaping the Cyber Threat Landscape for Financial Institutions," SWIFT Institute Working Paper No. 2016-004, October 2, 2017, 11–12, https://www.swiftinstitute.org/wp-content/uploads/2017/10/SIWP -2016-004-Cyber-Threat-Landscape-Carter-Final.pdf.

13. Carter, "Forces Shaping the Cyber Threat Landscape."

14. Carter, "Forces Shaping the Cyber Threat Landscape."

15. Flagstar Bank Release, https://www.flagstar.com/customer-support/accel lion-information-center.html.

16. Penny Crosman, "Flagstar's Data Breach and What Lenders Can Learn from It," *National Mortgage News*, March 15, 2021, https://www.nationalmortgagenews .com/news/flagstars-data-breach-and-what-lenders-can-learn-from-it.

17. Jordan Venet, "Microsoft's Big Email Hack: What Happened, Who Did It, and Why It Batters," CNBC, March 9, 2021, https://www.cnbc.com /2021/03/09/microsoft-exchange-hack-explained.html.

18. https://fiscaldata.treasury.gov/datasets/daily-treasury-statement/operating -cash-balance; https://fred.stlouisfed.org/series/WTREGEN.

19. R. Christopher Whalen, "Wag the Fed: Will the TGA Force Rates Negative?," *Institutional Risk Analyst*, December 14, 2020, https://www.theinstitutional riskanalyst.com/post/outlook-2020-was-bad-but-2021-looms.

20. Richard A. Clarke and Robert K. Knake, *Cyber War: The Next Threat to Cyber Security and What to Do about It* (New York: Ecco, 2010), 114–15.

21. Clarke and Knake, *Cyber War*, 114–15.

22. "The Comprehensive National Cybersecurity Initiative," White House, https://obamawhitehouse.archives.gov/issues/foreign-policy/cybersecurity/na tional-initiative.

23. Damien McGuinness, "How a Cyber Attack Transformed Estonia," BBC News, April 27, 2017, https://www.bbc.com/news/39655415.

24. "Seven Iranians Working for Islamic Revolutionary Guard Corps–Affiliated Entities Charged for Conducting Coordinated Campaign of Cyber Attacks against U.S. Financial Sector," Department of Justice, March 24, 2016, https://www .justice.gov/opa/pr/seven-iranians-working-islamic-revolutionary-guard-corps -affiliated-entities-charged.

25. Criminal Complaint against Park Jin Hyo, Department of Justice, June 8, 2018, https://www.justice.gov/opa/press-release/file/1092091/download.

26. "Chinese Military Personnel Charged with Computer Fraud, Economic Espionage and Wire Fraud for Hacking into Credit Reporting Agency Equifax," Department of Justice, February 10, 2020, https://www.justice.gov/opa/pr /chinese-military-personnel-charged-computer-fraud-economic-espionage-and -wire-fraud-hacking.

27. "Seattle Tech Worker Arrested for Data Theft Involving Large Financial Services Company," Department of Justice, July 29, 2019, https://www.justice .gov/usao-wdwa/pr/seattle-tech-worker-arrested-data-theft-involving-large-finan cial-services-company.

28. Tom Kellermann, "'Modern Bank Heists' Threat Report Finds Dramatic Increase in Cyberattacks against Financial Institutions amid COVID-19," *VMware Carbon Black* (blog), May 15, 2020, https://www.carbonblack.com/blog /modern-bank-heists-threat-report-finds-dramatic-increase-in-cyberattacks-against -financial-institutions-amid-covid-19.

29. "The Evolving Cyber Threat to the Banking Community," SWIFT Institute, November 2017, https://www.swift.com/swift-resource/176276/download.

30. Letter to shareholders from Jamie Dimon, chairman and CEO of JPMorgan Chase, April 4, 2019, 35, https://www.jpmorganchase.com/content /dam/jpmc/jpmorgan-chase-and-co/investor-relations/documents/ceo-letter-to -shareholders-2018.pdf; Hugh Son, "Jamie Dimon Says Risk of Cyberattacks

'May Be Biggest Threat to the US Financial System,'" CNBC, April 4, 2019, https://www.cnbc.com/2019/04/04/jp-morgan-ceo-jamie-dimon-warns-cyber -attacks-biggest-threat-to-us.html.

31. "U.S. Condemnation of Russia Cyber-Attack on Georgia," U.S. Mission to the OSCE, February 27, 2017, https://osce.usmission.gov/u-s-condemnation-of -russian-cyber-attack-on-georgia.

32. Gordon Correa, "UK Says That Russia's GRU behind Massive Georgia Cyber-Attack," BBC News, February 20, 2020, https://www.bbc.com/news /technology-51576445.

33. Council of Economic Advisors (CEA), "The Cost of Malicious Cyber Activity to the U.S. Economy," February 2018, 38, https://www.whitehouse .gov/wp-content/uploads/2018/02/The-Cost-of-Malicious-Cyber-Activity-to -the U.S.-Economy.pdf.

34. CEA, "Cost of Malicious Cyber Activity," 39.

35. "More than 100 Financial Services Firms Hit with DDoS Extortion Attacks," February 2021, press release, Financial Services–Information Sharing and Analysis Center, https://www.fsisac.com/newsroom/globalleaders.

36. Samantha Ravich, "Artificial Intelligence and the Adversary," *Wall Street Journal*, December 3, 2019, https://www.wsj.com/articles/artificial-intelligence and-the-adversary-11575417680.

CHAPTER 10. THE GENESIS POINT
OF VIRTUAL FINANCE: 1994–2000

1. "Critical Infrastructure Protection," Executive Order 13010, White House, July 15, 1996, https://www.govinfo.gov/content/pkg/FR-1996-07-17/html/96 -18351.htm.

2. Pub. L. 104-294, Title I, § 101(a), 110 Stat. 3488, October 11, 1996.

3. Report by President's Commission on Critical Infrastructure Protection: Overview Briefing, June 1997, https://www.hsdl.org/?view&did=487492.

4. Report by President's Commission on Critical Infrastructure Protection.

5. "Critical Foundations Protecting America's Infrastructures: The Report of the President's Commission on Critical Infrastructure Protection," October 1997, https://fas.org/sgp/library/pccip.pdf.

6. "Critical Foundations Protecting America's Infrastructures," A-38.

7. "Critical Foundations Protecting America's Infrastructures," A-38.

8. "Critical Foundations Protecting America's Infrastructures," A-39.

9. "Critical Foundations Protecting America's Infrastructures," A-39-0.

10. "Critical Foundations Protecting America's Infrastructures," A-37.

11. "Critical Foundations Protecting America's Infrastructures," A-37.

12. "Critical Foundations Protecting America's Infrastructures," A-40-1.

13. "Critical Foundations Protecting America's Infrastructures," 22–23.

14. "Protecting America's Critical Infrastructure," Presidential Decision Directive (PDD) 63, May 22, 1998, updated February 8, 1999, https://fas.org/irp/offdocs/pdd/pdd-63.htm, https://www.hsdl.org/?abstract&did=3544.

15. https://www.fsisac.com/who-we-are.

16. https://www.fsisac.com/hs-search-results?term=sheltered+harbor&type=SITE_PAGE&type=BLOG_POST&type=LISTING_PAGE.

17. Gramm–Leach–Bliley Act of 1999, Pub. L. 106-102, 113, Stat. 1338 § 501, 15 U.S.C. § 6801, https://www.govinfo.gov/content/pkg/PLAW-106publ102/pdf/PLAW-106publ102.pdf.

18. "Defending America's Cyberspace: National Plan for Information Systems Protection Version 1.0: An Invitation to a Dialogue," January 2000, https://fas.org/irp/offdocs/pdd/CIP-plan.pdf.

19. "Defending America's Cyberspace," iv.

20. "Defending America's Cyberspace," v.

21. "Defending America's Cyberspace," 145.

22. Statement of Jack L. Brock, Jr., director, Governmentwide and Defense Information Systems Accounting and Information Management Division, "Critical Infrastructure Protection: Comments on the National Plan for Information Systems Protection," Government Accountability Office, February 2000, 5, https://www.hsdl.org/?view&did=341.

23. Brock, "Critical Infrastructure Protection," 8–9.

24. Margaret G. Stewart, "Achieving Legal and Business Order in Cyberspace: A Report on Global Jurisdiction Issues Created by Cyberspace," *Business Lawyer* 55, no. 4 (August 2000): 1801–946, https://scholarship.kentlaw.iit.edu/cgi/viewcontent.cgi?article=1603&context=fac_schol.

CHAPTER 11. AN INSECURE
WORLDVIEW EMERGES: 2001–2008

1. "Critical Infrastructure Protection in the Information Age," Executive Order 13231, White House, October 16, 2001, https://www.dhs.gov/xlibrary/assets/executive-order-13231-dated-2001-10-16-initial.pdf.

2. "Critical Infrastructure Protection in the Information Age," 2–3.

3. 18 U.S.C. §1030.

4. Pub. L. 107-56, 115 Stat., October 24, 2001.

5. Jeff Kosseff, "Defining Cybersecurity Law" (2018), 103 Iowa L. Rev. 985, 1017.

6. Convention on Cybercrime, Budapest, European Treaty Series No. 185, Council of Europe, November 23, 2001, https://rm.coe.int/1680081561.

7. Chart of signatures and ratifications of Treaty 185, Convention on Cybercrime, European Treaty Series No. 185, status as of June 12, 2021, https://www.coe.int/en/web/conventions/full-list?module=signatures-by-treaty&treatynum=185.

8. Financial Services Sector Coordinating Council, https://fsscc.org.

9. Pub. L. 107-347, 116 Stat. 2899, 44 U.S.C. § 3541, December 17, 2002.

10. "The National Strategy to Secure Cyberspace," Cyberspace and Infrastructure Security Agency, February 2003, https://us-cert.cisa.gov/sites/default/files/publications/cyberspace_strategy.pdf.

11. "The National Strategy to Secure Cyberspace," vii.

12. "The National Strategy to Secure Cyberspace," viii.

13. "The National Strategy to Secure Cyberspace," 2.

14. "The National Strategy to Secure Cyberspace," ix.

15. "The National Strategy to Secure Cyberspace," xi.

16. "The National Strategy for the Physical Protection of Critical Infrastructures and Key Assets," White House, February 2003, 63–64, https://www.dhs.gov/xlibrary/assets/Physical_Strategy.pdf.

17. "National Strategy for the Physical Protection," 64.

18. "National Strategy for the Physical Protection," 64.

19. "Interagency Paper on Sound Practices to Strengthen the Resilience of the U.S. Financial System," 68 Fed. Reg. 70, 17809 et seq., April 11, 2003, https://www.occ.treas.gov/news-issuances/bulletins/2003/OCC2003-14a.pdf.

20. "Interagency Paper on Sound Practices," 17812.

21. "Cybersecurity Awareness," Federal Financial Institutions Examination Council, https://www.ffiec.gov/cybersecurity.htm.

22. "Presidential Directives & Cybersecurity," https://epic.org/privacy/cybersecurity/presidential-directives/cybersecurity.html.

23. Joshua David, "Hackers Take Down the Most Wired Country in Europe," *Wired*, August 8, 2017, https://www.wired.com/2007/08/ff-estonia.

24. Damien McGuinness, "How a Cyber-Attack Transformed Estonia," BBC News, April 27, 2017, https://www.bbc.com/news/39655415.

25. John Markoff, "Georgia Takes a Beating in the Cyber War with Russia," *New York Times*, BITS, August 11, 2008, https://bits.blogs.nytimes.com/2008/08/11/georgia-takes-a-beating-in-the-cyberwar-with-russia.

26. Pub. L. No. 108-275, Tit. II, 122 Stat. 356, September 26, 2008, codified at 18 U.S.C. §1030.

27. Richard A. Clarke and Robert K. Knake, *Cyber War: The Next Threat to Cyber Security and What to Do about It* (New York: Ecco, 2010), 116.

28. "The Comprehensive National Cybersecurity Initiative," White House, President Barack Obama, https://obamawhitehouse.archives.gov/issues/foreign-policy/cybersecurity/national-initiative.

29. Clarke and Knake, *Cyber War*, 118.

30. Clarke and Knake, *Cyber War*, 118.

31. Clarke and Knake, *Cyber War*, 127.

CHAPTER 12. DISTRACTIONS, DIPLOMACY, AND CYBER THREATS: 2009–2016

1. "Timeline of Cyber Incidents Affecting Financial Institutions," Carnegie Endowment for International Peace, https://carnegieendowment.org/specialprojects/protectingfinancialstability/timeline.

2. "Computer Network Security & Privacy Protection," Department of Homeland Security, February 19, 2010, https://www.dhs.gov/xlibrary/assets/privacy/privacy_cybersecurity_white_paper.pdf.

3. "Computer Network Security & Privacy Protection," 1.

4. "Computer Network Security & Privacy Protection," 4.

5. "Timeline of Cyber Incidents."

6. "U.S. Cyber Operations Policy," Presidential Policy Directive (PDD) 21, White House, October 16, 2012, https://fas.org/irp/offdocs/ppd/ppd-20.pdf.

7. "U.S. Cyber Operations Policy," 9–10.

8. Pub. L. 111-203, July 21, 2010.

9. Financial Stability Oversight Council, *2011 Annual Report*, 16, https://www.treasury.gov/initiatives/fsoc/studies-reports/Documents/2012%20Annual%20Report.pdf.

10. Financial Stability Oversight Council, *2012 Annual Report*, 137, https://www.treasury.gov/initiatives/fsoc/studies-reports/Documents/2012%20Annual%20Report.pdf.

11. "Improving Critical Infrastructure Cybersecurity," Executive Order 13636, White House, February 12, 2013, https://www.federalregister.gov/documents/2013/02/19/2013-03915/improving-critical-infrastructure-cybersecurity.

12. "Critical Infrastructure Security and Resilience," Presidential Policy Directive (PPD) 21, White House, February 12, 2013, https://obamawhitehouse.archives.gov/the-press-office/2013/02/12/presidential-policy-directive-critical-infrastructure-security-and-resil.

13. The agencies given responsibility included the Departments of Justice, State, Interior, Commerce, Energy, Homeland Security, and Defense; the Social Security Administration; the FBI; the National Cyber Investigative Joint Task Force (NCIJTF); the Intelligence Community (IC); the Director of National Intelligence (DNI); the General Services Administration (GSA); the Nuclear Regulatory Commission (NRC); the Environmental Protection Agency (EPA); and the Federal Communications Commission (FCC).

14. They were the FDIC, the Federal Reserve, the Office of the Comptroller of the Currency, the National Credit Union Administration, the Federal Housing Finance Agency, the Securities and Exchange Commission, and the Commodities Futures Trading Commission.

15. Financial Stability Oversight Council, *2013 Annual Report*, 136, https://carnegieendowment.org/specialprojects/protectingfinancialstability/timeline.

16. Jessica Silver-Greenberg, Matthew Goldstein, and Nicole Perlroth, "JPMorgan Hacking Affects 76 Million Households," *New York Times*, October 2, 2014, https://dealbook.nytimes.com/2014/10/02/jpmorgan-discovers-further-cyber-security-issues.

17. "Improving the Security of Consumer Financial Transactions," Executive Order 13681, White House, October 17, 2014, https://obamawhitehouse.archives.gov/the-press-office/2014/10/17/executive-order-improving-security-consumer-financial-transactions.

18. "Sony Pictures Hack," Wikipedia, https://en.wikipedia.org/wiki/Sony_Pictures_hack.

19. Financial Stability Oversight Council, *2014 Annual Report*, 120, https://www.treasury.gov/initiatives/fsoc/Documents/FSOC%202014%20Annual%20Report.pdf.

20. Financial Stability Oversight Council, *2014 Annual Report*, 12.

21. "Testimony of Bill Nelson, President and CEO of the Financial Services Information Sharing and Analysis Center (FS-ISAC)," Committee on Banking, Housing, and Urban Affairs, U.S. Senate, May 24, 2018, 9, https://www.banking.senate.gov/imo/media/doc/Nelson%20Testimony%205-24-18.pdf.

22. "Promoting Private Sector Cybersecurity Information Sharing," Executive Order 13691, White House, February 13, 2015, https://obamawhitehouse.archives.gov/the-press-office/2015/02/13/executive-order-promoting-private-sector-cybersecurity-information-shari.

23. "Blocking the Property of Certain Persons Engaging in Significant Malicious Cyber-Enabled Activities," Executive Order 13694, White House, April 1, 2015, https://obamawhitehouse.archives.gov/the-press-office/2015/04/01/executive-order-blocking-property-certain-persons-engaging-significant-m.

24. "Financial Services Specific-Sector Plan 2015," Cybersecurity and Infrastructure Security Agency, 3–4, https://www.cisa.gov/sites/default/files/publications/nipp-ssp-financial-services-2015-508.pdf.

25. Cybersecurity Resource Center, "Cybersecurity Incidents," https://www.opm.gov/cybersecurity/cybersecurity-incidents.

26. Financial Stability Oversight Council, *2015 Annual Report*, 105, https://www.treasury.gov/initiatives/fsoc/studies-reports/Documents/2015%20FSOC%20Annual%20Report.pdf.

27. Financial Stability Oversight Council, *2015 Annual Report*, 9–10.

28. Financial Stability Oversight Council, *2015 Annual Report*, 105.

29. Consolidated Appropriations Act, 2016, Pub. L. No. 114-113, Div. N, Title I, 2015, https://www.congress.gov/114/plaws/publ113/PLAW-114publ113.pdf.

30. Foreign Intelligence Surveillance Act of 1978, Pub. L. 95-511, 92 Stat. 1783, October 25, 1978, https://www.govinfo.gov/content/pkg/STATUTE-92/pdf/STATUTE-92-Pg1783.pdf#page=1.

31. Electronic Communications Privacy Act of 1986, Pub. L. 99-508, 100 Stat. 1848, October 25, 1986, https://www.govinfo.gov/content/pkg/STATUTE-100/pdf/STATUTE-100-Pg1848.pdf.

32. Kim Zetter, "That Insane, $81M Bangladesh Bank Heist? Here's What We Know," *Wired*, May 17, 2016, https://www.wired.com/2016/05/insane-81m-bangladesh-bank-heist-heres-know.

33. "Timeline of Cyber Incidents Affecting Financial Institutions."

34. Those trade associations included the American Bankers Association, BITS, the Bank Policy Institute, the Credit Union National Association, the Financial Services Forum, FS-ISAC, the Independent Community Bankers of America, the National Association of Federal Credit Unions, the Securities Industry and Financial Markets Association, and the Clearing House.

35. www.fsisac.com/sites/default/files/news/SH_FACT_SHEET_2016_11_22_FINAL3.pdf, https://shelteredharbor.org; see May 14, 2019, letter to all financial institution CEOs, https://shelteredharbor.org/images/SH/Docs/Sheltered_Harbor_Trade_Assn_Exec_Letter_Genericfinal_051619.pdf.

36. May 14, 2019, letter to all financial institution CEOs.

37. https://www.businesswire.com/news/home/20201030005462/en/Announcing-the-Formation-of-the-Analysis-Resilience-Center-ARC-for-Systemic-Risk.

38. http://www.prnewswire.com/news-releases/fs-isac-announces-the-formation-ofthe-financial-systemic-analysis—resilience-center-fsarc-300349678.html.

39. "Guidance on Cyber Resilience for Financial Market Infrastructures," Bank for International Settlements' Committee on Payments and Market Infrastructures and the International Organization of Securities Commissions (CPMI-IOSCO), June 2016, https://www.bis.org/cpmi/publ/d146.pdf.

40. "Guidance on Cyber Resilience," 1.

41. "Enhanced Cyber Risk Management Standards," 81 Fed. Reg. 207, 74315 et seq., October 26, 2016, https://www.federalregister.gov/documents/2017/01/24/2017-01539/enhanced-cyber-risk-management-standards.

42. Tim Keary, "NIST Announces Four Quantum-Resistant Algorithms," *VentureBeat*, July 5, 2022, https://venturebeat.com/security/nist-post-quantum-cryptography-standard/.

43. Office of Financial Research, *2016 Financial Stability Report*, 38, https://www.financialresearch.gov/financial-stability-reports/files/OFR_2016_Financial-Stability-Report.pdf.

44. *2016 Financial Stability Report*, 38.

45. *2016 Financial Stability Report*, 44–47.

46. "G-7 Fundamental Elements for Effective Assessment of Cybersecurity in the Financial Sector," October 2016, https://www.gov.uk/government/uploads/system/uploads/attachment_data/file/559186/G7_Fundamental_Elements_Oct_2016.pdf.

47. "Report on Securing and Growing the Digital Economy," Commission on Enhancing National Cyberspace Security, December 1, 2016, https://www.nist.gov/system/files/documents/2016/12/02/cybersecurity-commission-report-final-post.pdf.

48. "Report on Securing and Growing the Digital Economy," 10.

CHAPTER 13. RUNNING IN CYBER CIRCLES: 2017–2020

1. G20 Finance Ministers and Central Bank Governors Communiqué, Baden, March 18, 2017, http://www.g20.utoronto.ca/2017/170318-finance-en.html.

2. Nicole Perlroth and David E. Sanger, "White House Eliminates Cybersecurity Coordinator Role," *New York Times*, May 15, 2018, https://www.nytimes.com/2018/05/15/technology/white-house-cybersecurity.html.

3. "Supervisory Policy and Guidance Topics: Information Technology Guidance," Federal Reserve Board, https://www.federalreserve.gov/supervisionreg/topics/information-technology-guidance.htm; "Cybersecurity Awareness," Federal Financial Institutions Examination Council, https://www.ffiec.gov/cybersecurity.htm; "Information Technology (IT) & Cybersecurity," Federal Deposit Insurance Corporation, https://www.fdic.gov/resources/bankers/information-technology; "Cybersecurity: Joint Statement on Heightened Cybersecurity Risk," Office of the Comptroller of the Currency, January 16, 2020, https://www.occ.gov/news-issuances/bulletins/2020/bulletin-2020-5.html.

4. "Cybersecurity and Financial Stability: Risks and Resilience," Office of Financial Research, U.S. Department of Treasury, February 15, 2017, 1, https://www.financialresearch.gov/viewpoint-papers/files/OFRvp_17-01_Cybersecurity.pdf.

5. G20 Finance Ministers and Central Bank Governors, "Communiqué," University of Toronto, March 18, 2017, http://www.g20.utoronto.ca/2017/170318-finance-en.html.

6. Tim Maurer, Ariel (Eli) Levite, and George Perkovich, "Toward a Global Norm against Manipulating the Integrity of Financial Data," Cyberspace Endowment for International Peace, Cyber Policy Initiative Working Paper Series, Cybersecurity and the Financial System, No. 1, 4, March 27, 2017, https://carnegieendowment.org/2017/03/27/toward-global-norm-against-manipulating-integrity-of-financial-data-pub-68403.

7. Maurer, Levite, and Perkovich, "Toward a Global Norm," 8.

8. "Strengthening the Cybersecurity of Federal Networks and Critical Infrastructure," Executive Order 13800, White House, May 11, 2017, 82 Fed. Reg. No. 93, 22391 et seq., May 16, 2017, https://www.govinfo.gov/content/pkg/FR-2017-05-16/pdf/2017-10004.pdf.

9. Emanuel Kopp, Lincoln Kaffenberger, and Christopher Wilson, "Cyber Risk, Market Failures, and Financial Stability," IMF Working Paper WP/17/185, August 7, 2017, https://www.imf.org/en/Publications/WP/Issues/2017/08/07/Cyber-Risk-Market-Failures-and-Financial-Stability-45104; https://www.imf.org/~/media/Files/Publications/WP/2017/wp17185.ashx.

10. "Cyber Risk, Market Failures, and Financial Stability," 5, 8.

11. "Cyber Risk, Market Failures, and Financial Stability," 7, 17.

12. "Cyber Risk, Market Failures, and Financial Stability," 29.

13. Juan Carlos Crisanto and Jeremy Prenio, "Regulatory Approaches to Enhance Banks' Cybersecurity Frameworks," Bank for International Settlements, August 2017, https://www.bis.org/fsi/publ/insights2.htm.

14. Crisanto and Prenio, "Regulatory Approaches," 4.

15. Kopp, Kaffenberger, and Wilson, "Cyber Risk."

16. Martin Boer and Jaime Vazquez, "Cyber Security & Financial Stability: How Cyber-Attacks Could Materially Impact the Global Financial System," Institute of International Finance, September 2017, 4, https://www.iif.com/Portals/0 /Files/IIF%20Cyber%20Financial%20Stability%20Paper%20Final%2009%2007%20 2017.pdf?ver%3D2019-02-19-150125-767.

17. Boer and Vazquez, "Cyber Security & Financial Stability," 4–5.

18. Boer and Vazquez, "Cyber Security & Financial Stability," 6–7.

19. Boer and Vazquez, "Cyber Security & Financial Stability," 7.

20. Boer and Vazquez, "Cyber Security & Financial Stability," 9.

21. Criminal indictment (under seal), U.S. v. Wu Zhiyong, et al., January 28, 2020, https://www.justice.gov/opa/press-release/file/1246891/download.

22. "Building a Defensible Cyberspace," New York Cyber Task Force, Columbia School of International and Public Affairs, 2017, https://www.sipa.columbia .edu/sites/default/files/3668_SIPA%20Defensible%20Cyberspace-WEB.PDF.

23. "Building a Defensible Cyberspace," 1.

24. "Building a Defensible Cyberspace," 6.

25. David E. Sanger, *The Perfect Weapon* (New York: Crown, 2018).

26. "Building a Defensible Cyberspace," 8.

27. "Building a Defensible Cyberspace," 9.

28. "The Fundamental Elements for Effective Assessment of Cybersecurity in the Financial Sector: Stocktake of Publicly Released Cybersecurity Regulations, Guidance and Supervisory Practices," Financial Stability Board, October 2017, https://www.fsb.org/wp-content/uploads/P131017-2.pdf.

29. William A. Carter, "Forces Shaping the Cyber Threat Landscape for Financial Institutions," SWIFT Institute Working Paper No. 2016-004, October 2017, 6–13, https://www.swiftinstitute.org/wp-content/uploads/2017/10/SIWP -2016-004-Cyber-Threat-Landscape-Carter-Final.pdf. SWIFT is a cooperative society created under Belgian law owned by its 11,000-member financial institutions that operates in more than 200 countries and territories. It enables global financial institutions to send and receive payment order information about financial transactions in a secure, standardized, and reliable environment without engaging in the clearing or settlement of transactions. As of 2018, around half of all high-value cross-border payments worldwide used the SWIFT network, but it is currently in a battle with upstart competitors such as Ripple; Martin Arnold, "Ripple and Swift Slug It Out over Cross-Border Payments," *Financial Times*, June 5, 2018, https:// www.ft.com/content/631af8cc-47cc-11e8-8c77-ff51caedcde6.

30. Carter, "Forces Shaping the Cyber Threat Landscape," 14, 24; Danny Palmer, "New Wave of Cyberattacks against Global Banks Linked to Lazarus Cybercrime Group," *ZDNet*, February 13, 2017, http://www.zdnet.com/article /string-of-cyberattacks-against-global-banks-linked-to-lazarus-cybercrime-group.

31. Financial Stability Oversight Council, *2017 Annual Report*, 7–9, https:// home.treasury.gov/system/files/261/FSOC_2017_Annual_Report.pdf.

32. Nicholas Confessore, "Cambridge Analytica and Facebook: The Scandal and the Fallout So Far," *New York Times*, April 4, 2018, https://www.nytimes.com/2018/04/04/us/politics/cambridge-analytica-scandal-fallout.html.

33. "Rise of the Machines: Artificial Intelligence and Its Growing Impact on U.S. Policy," Subcommittee on Information Technology, Committee on Oversight and Government Reform, U.S. House of Representatives, September 2018, 1, 3–4, https://www.hsdl.org/?view&did=816362.

34. "Rise of the Machines," 1, 3–4.

35. Erica D. Borghard, "Protecting Financial Institutions against Cyber Threats: A National Security Issue," Cyber Policy Initiative Working Paper Series, Cybersecurity and the Financial System," No. 2, 5, September 2018, https://www.jstor.org/stable/pdf/resrep20722.pdf.

36. Borghard, "Protecting Financial Institutions," 6–7, 15.

37. Borghard, "Protecting Financial Institutions," 6–10, 13.

38. Borghard, "Protecting Financial Institutions," 6–7 n. 13.

39. Jason Healey, Patricia Mosser, Katheryn Rosen, and Adriana Tache, "The Future of Financial Stability and Cyber Risk," Brookings Cybersecurity Project, October 2018, https://www.brookings.edu/wp-content/uploads/2018/10/Healey-et-al_Financial-Stability-and-Cyber-Risk.pdf.

40. Healey et al., "The Future of Financial Stability," 13–14.

41. Financial Stability Oversight Council, *2018 Annual Report*, 7, 107, https://home.treasury.gov/system/files/261/FSOC2018AnnualReport.pdf.

42. "Cyber Risk Institute Launches to Mitigate Cyber Risk and Support Compliance for Financial Services," Cyber Risk Institute, 2018, https://cyberriskinstitute.org/cyber-risk-institute-launches-to-mitigate-cyber-risk-and-support-compliance-for-financial-services.

43. Tim Maurer and Arthur Nel, "International Strategy to Better Protect the Financial System against Cyber Threats," Carnegie Endowment for International Peace, in collaboration with the World Economic Forum, November 18, 2020, 2, https://carnegieendowment.org/files/Maurer_Nelson_FinCyber_final1.pdf, citing Davey Winder, "$645 Billion Cyber Risk Could Trigger Liquidity Crisis, ECB's Lagarde Warns," *Forbes*, March 10, 2020, https://www.forbes.com/sites/daveywinder/2020/02/08/645-billion-cyber-risk-could-trigger-liquidity-crisis-ecbslagarde-warns.

44. Adrian Nish and Saher Naumann, "The Cyber Threat Landscape: Confronting Challenges to the Financial System," Carnegie Endowment for International Peace, March 25, 2019, https://carnegieendowment.org/files/03_19_Nish_Naumaan_Fin_Threats_final.pdf.

45. Nish and Naumaan, "The Cyber Threat Landscape," 1–2, 8–9.

46. Nish and Naumaan, "The Cyber Threat Landscape," 8.

47. Rob McLean, "A Hacker Gained Access to 100 Million Capital One Credit Card Applications and Accounts," CNN Business, July 30, 2019, https://www.cnn.com/2019/07/29/business/capital-one-data-breach/index.html; "Timeline

of Cyber Incidents Involving Financial Institutions," Carnegie Endowment for International Peace, https://carnegieendowment.org/specialprojects/protectingfi nancialstability/timeline.

48. Consent Order, In the Matter of Capital One, N.A., McLean, Virginia, AA-EC-20-51, August 5, 2020, https://www.occ.gov/static/enforcement-actions /ea2020-036.pdf; In the Matter of Capital One Financial Corp., McLean, Virginia, Docket No. 20-014-B-HC, August 4, 2020, https://www.federalreserve.gov /newsevents/pressreleases/files/enf20200806a1.pdf.

49. "Securing the Information and Communications Technology and Services Supply Chain," Executive Order 13873, White House, May 15, 2019, https:// www.federalregister.gov/documents/2019/05/17/2019-10538/securing-the-in formation-and-communications-technology-and-services-supply-chain.

50. "Addressing the Threat Posed by TikTok, and Taking Additional Steps to Address the National Emergency with Respect to the Information and Communications Technology and Services Supply Chain," Executive Order 13942, White House, August 6, 2020, https://www.federalregister.gov/docu ments/2020/08/11/2020-17699/addressing-the-threat-posed-by-tiktok-and -taking-additional-steps-to-address-the-national-emergency.

51. "Addressing the Threat Posed by WeChat, and Taking Additional Steps to Address the National Emergency with Respect to the Information and Communications Technology and Services Supply Chain," Executive Order 13943, White House, August 11, 2020, https://www.federalregister.gov/docu ments/2020/08/11/2020-17700/addressing-the-threat-posed-by-wechat-and -taking-additional-steps-to-address-the-national-emergency.

52. "Addressing the Threat Posed by Applications and Other Software De-veloped or Controlled by Chinese Companies," Executive Order 13971, White House, January 8, 2021, https://www.federalregister.gov/documents/2021/01/08 /2021-00305/addressing-the-threat-posed-by-applications-and-other-software -developed-or-controlled-by-chinese.

53. "Secure and Trusted Communications Networks Act of 2019," Pub. L. 116-124, 116th Cong., March 12, 2020, https://www.congress.gov/116/plaws /publ124/PLAW-116publ124.pdf.

54. "List of Equipment and Services Covered by Section 2 of the Secure Net-works Act," Federal Communications Commission, updated March 25, 2022, https://www.fcc.gov/supplychain/coveredlist.

55. Zachery Basu, "U.S. Blacklists World's Largest Commercial Drone Firm for Uyghur Surveillance," Axios, December 16, 2021, https://www.axios.com /dji-drones-china-surveillance-a14cc7b4-16f9-461a-8a52-c55462bc5d63.html.

56. "Protecting Americans' Sensitive Data from Foreign Adversaries," Execu-tive Order 14034, White House, June 9, 2021, https://www.federalregister.gov /documents/2021/06/11/2021-12506/protecting-americans-sensitive-data-from -foreign-adversaries.

57. "Decentralised Financial Technologies: Report on Financial Stability, Regulatory and Governance Implications," Financial Stability Board, June 6, 2019, 6–7, https://www.fsb.org/wp-content/uploads/P060619.pdf.

58. Maurer and Nel, "International Strategy," 2, citing Mark Bendeich and Leika Kihara, "Cyber Threat Could Become Banking's Most Serious Risk," Reuters, January 24, 2019, https://www.reuters.com/article/davos-meeting -cyberkuroda/davos-cyber-threat-could-become-bankings-most-serious-risk-bojid USS8N1PK01N, and Hugh Son, "Jamie Dimon Says Risk of Cyberattacks 'May Be Biggest Threat to the US Financial System,'" CNBC, April 4, 2019, https:// www.cnbc.com/2019/04/04/jp-morganceo-jamie-dimon-warns-cyber-attacks -biggest-threat-to-us.html.

59. Tim Maurer and Kathryn Taylor, "Capacity-Building Tool Box for Cybersecurity and Financial Organizations," Carnegie Endowment for International Peace, July 2019, https://ceipfiles.s3.amazonaws.com/pdf/FinCyber /Cybercecurity+Toolkit_Full.pdf.

60. Lincoln Kaffenberger and Emanuel Kopp, "Cyber Risk Scenarios, the Financial System, and Systemic Risk Assessment," Carnegie Endowment for International Peace, September 30, 2019, https://carnegieendowment.org/2019/09/30 /cyber-risk-scenarios-financial-system-and-systemic-risk-assessment-pub-799/11.

61. Kaffenberger and Kopp, "Cyber Risk Scenarios," 5–7.

62. Kaffenberger and Kopp, "Cyber Risk Scenarios," 7.

63. Petra Hielkema and Raymond Kleijmeer, "Lessons Learned and Evolving Practices of the TIBER Framework for Resilience Testing in the Netherlands," Carnegie Endowment for International Peace, October 2019, https://carn egieendowment.org/2019/10/07/lessons-learned-and-evolving-practices-of-tiber -framework-for-resilience-testing-in-netherlands-pub-80007.

64. Thomas M. Eisenbach, Anna Kovner, and Michael Junho Lee, "Cyber Risk and the U.S. Financial System: A Pre-Mortem Analysis," Federal Reserve Bank of New York, Staff Report No. 909, January 2020, https://www.finextra.com /finextra-downloads/newsdocs/fedcyberattack.pdf.

65. Iñaki Aldasoro, Leonardo Gambacorta, Paolo Giudici, and Thomas Leach, "Operational and Cyber Risks in the Financial Sector," BIS Working Papers No. 840, February 2020, https://www.bis.org/publ/work840.pdf.

66. Aldasoro et al., "Operational and Cyber Risks in the Financial Sector," 26.

67. Iñaki Aldasoro, Leonardo Gambacorta, Paolo Giudici, and Thomas Leach, "The Drivers of Cyber Risk," BIS Working Papers No. 865, May 2020, https:// www.bis.org/publ/work865.pdf.

68. Aldasoro et al., "The Drivers of Cyber Risk," 37–38.

69. Quentin Hardy, "Where Does Cloud Storage Really Reside? And Is It Secure?," *New York Times*, January 23, 2017, https://www.nytimes.com/2017/01/23 /insider/where-does-cloud-storage-really-reside-and-is-it-secure.html.

70. Michael Tabb, Jeffery DelViscio, and Andrea Gawrylewski, "What Is 'The Cloud' and How Does It Pervade Our Lives?," *Scientific American*, December

1, 2021, https://www.scientificamerican.com/video/what-is-the-cloud-and-how
-does-it-pervade-our-lives.

71. Hal S. Scott, John Gulliver, and Hillel Nadler, "Cloud Computing in
the Financial Sector: A Global Perspective," Program on International Financial Systems, 2019, 6–10, https://www.pifsinternational.org/wp-content/up
loads/2019/07/Cloud-Computing-in-the-Financial-Sector_Global-Perspective
-Final_July-2019.pdf.

72. Vic "J. R." Winkler, *Securing the Cloud: Cloud Computer Security Techniques*
(Waltham, MA: Elsevier, 2011).

73. Tabb, DelViscio, and Gawrylewski, "What Is 'The Cloud'?"

74. "Joint Statement Security in a Cloud Computing Environment," Federal
Financial Institutions Examination Council, August 30, 2020, https://www.ffiec
.gov/press/pdf/FFIEC_Cloud_Computing_Statement.pdf.

75. Scott, Gulliver, and Nadler, "Cloud Computing in the Financial Sector."

76. United States Cyberspace Solarium Commission (USCSC), March 2020
Report, "Executive Summary," ii, https://www.solarium.gov/report.

77. USCSC Report, i–iii.

78. USCSC Report, 8–13, 46–47.

79. USCSC Report, 131–36.

80. USCSC Report, 16.

81. USCSC Report, 121–22.

82. USCSC Report, 59–60, 98.

83. The 2021 National Defense Authorization Act codified the commission's
recommendations, including (1) authorizing the Cybersecurity Infrastructure and
Security Agency (CISA) to conduct unalerted threat hunting on federal networks;
(2) tasking the Department of Defense with developing a plan for the annual
assessment of cyber vulnerabilities of major weapons systems; (3) establishing a
Joint Cyber Planning Office under CISA to facilitate comprehensive planning of
defensive cybersecurity campaigns across federal entities and the private sector;
(4) granting administrative subpoena authority to CISA to identify vulnerable
systems and notify public and private system owners; (5) mandating a comprehensive assessment of the threats and risks posed by quantum technologies to national
security systems; (6) requiring DOD to assess the need, models, and requirements
relative to a cyber reserve force and the resourcing required to establish threat
hunting and information sharing programs; (7) launching cyber exercises between
public and private sector entities; and (8) mandating the creation of a Continuity
of the Economy planning effort to ensure the rapid restart and recovery of the
U.S. economy after a major disruption. This finally created some momentum and
forward movement after twenty-five years. "NDAA Enacts 25 Recommendations
from the Bipartisan Cyberspace Solarium Commission," press release, January 2,
2021, https://www.solarium.gov/press-and-news/ndaa-override-press-release. See
also, "Executive Order on Enhancing the National Quantum Initiative Advisory
Committee," May 4, 2022, https://www.whitehouse.gov/briefing-room/pres

idential-actions/2022/05/04/executive-order-on-enhancing-the-national-quan
tum-initiative-advisory-committee; "National Security Memorandum on Promoting United States Leadership in Quantum Computing While Mitigating Risks to
Vulnerable Cryptographic Systems," May 4, 2022, https://www.whitehouse.gov
/briefing-room/statements-releases/2022/05/04/national-security-memorandum
-on-promoting-united-states-leadership-in-quantum-computing-while-mitigating
-risks-to-vulnerable-cryptographic-systems.

84. USCSC Report, i–iii.

85. Jan-Philipp Brauchle, Matthias Göbel, Jens Seiler, and Christoph von
Busekist, "Cyber Mapping the Financial System," Carnegie Endowment for International Peace, April 7, 2020, https://carnegieendowment.org/2020/04/07
/cyber-mapping-financial-system-pub-81414.

86. Brauchle et al., "Cyber Mapping the Financial System," 13–14.

87. Maurer and Nel, "International Strategy," 2, citing Financial Stability Board,
"Effective Practices for Cyber Incident Response and Recovery: Consultative
Document," April 20, 2020, https://www.fsb.org/2020/04/effectivepractices-for
-cyber-incident-response-and-recovery-consultative-document.

88. "Alert (AA20-106A), Guidance on the North Korean Cyber Threat," Cybersecurity and Infrastructure Security Agency, April 15, 2020, https://us-cert.cisa
.gov/ncas/alerts/aa20-106a.

89. Federal Bureau of Investigation, "Update on Sony Investigation," December 19, 2014, https://www.fbi.gov/news/pressrel/press-releases/update-on-sony
-investigation; U.S. Department of Justice's criminal complaint, "North Korean
Regime-Backed Programmer Charged with Conspiracy to Conduct Multiple Cyber Attacks and Intrusions," September 6, 2018, https://www.justice.gov/opa/pr
/north-korean-regime-backed-programmer-charged-conspiracy-conduct-multiple
-cyber-attacks-and.

90. U.S. Department of Justice, "North Korean Regime-Backed Programmer."

91. CISA's technical alert, "Indicators Associated with WannaCry Ransomware," May 12, 2017, https://www.us-cert.gov/ncas/alerts/TA17-132A; "White
House Press Briefing on the Attribution of WannaCry Ransomware," December
19, 2017, https://www.whitehouse.gov/briefings-statements/press-briefing-on
-the-attribution-of-the-wannacry-malware-attack-to-north-korea-121917; U.S.
Department of Justice's, "North Korean Regime-Backed Programmer"; and
U.S. Department of the Treasury, "Treasury Targets North Korea for Multiple Cyber-Attacks," September 6, 2018, https://home.treasury.gov/news/press
-releases/sm473.

92. CISA's alert on FASTCash campaign, October 2, 2018, https://www
.us-cert.gov/ncas/alerts/TA18-275A; CISA's malware analysis report, "FASTCash
-Related Malware," October 2, 2018, https://www.us-cert.gov/ncas/analysis
-reports/AR18-275A.

93. U.S. Department of the Treasury, "Sanctions against Individuals Laundering Cryptocurrency for Lazarus Group," March 2, 2020, https://home.treasury

.gov/news/press-releases/sm924; U.S. Department of Justice's indictment, "Two Chinese Nationals Charged with Laundering Cryptocurrency from Exchange Hack and Civil Forfeiture Complaint," March 2, 2020, https://www.justice.gov/opa /pr/two-chinese-nationals-charged-laundering-over-100-million-cryptocurrency -exchange-hack.

94. Aldasoro et al., "The Drivers of Cyber Risk."

95. Report of the DOE Quantum Internet Blueprint Workshop, "From Long-Distance Entanglement to Building a Nationwide Quantum Internet," SUNY Global Center, New York, NY, February 5–6, 2020, https://www.energy.gov /sites/prod/files/2020/07/f76/QuantumWkshpRpt20FINAL_Nav_0.pdf.

96. Davey Winder, "U.S. Government Says It's Building a 'Virtually Unhackable' Quantum Internet," *Forbes*, July 25, 2020, https://www.forbes.com/sites /daveywinder/2020/07/25/us-government-to-build-virtually-unhackable-quan tum-internet-within-10-years.

97. "Critical Infrastructure Protection: Treasury Needs to Improve Tracking of Financial Sector Cybersecurity Risk Mitigation Efforts," Government Account-ability Office, September 2020, https://www.gao.gov/assets/710/709455.pdf.

98. "Critical Infrastructure Protection," 16–20.

99. "Critical Infrastructure Protection," GAO Highlights.

100. European Commission, "Proposal for a Regulation of the European Parlia-ment and of the Council on Digital Operational Resilience for the Financial Sector and Amending Regulations (EC) No 1060/2009, (EU) No 648/2012, (EU) No 600/2014 and (EU) No 909/2014, Articles 28–29," September 24, 2020, https:// eur-lex.europa.eu/legal-content/EN/TXT/PDF/?uri=CELEX:52020PC0595.

101. Maurer and Nel, "International Strategy," 1 (footnotes omitted).

102. Adrian Nish, Saher Naumaan, and James Muir, "Enduring Cyber Threats and Emerging Challenges to the Financial Sector," Carnegie Endowment for International Peace, Cyber Policy Initiative Working Paper Series, Cybersecurity and the Financial System, No. 8, November 2020, https://carnegieendowment.org /files/NishNaumaan_FincyberThreatsChallenges_v3.pdf.

103. Nish, Naumaan, and Muir, "Enduring Cyber Threats," 18.

104. "G-7 Fundamental Elements of Cyber Exercises Programmes," November 2020, https://home.treasury.gov/system/files/216/G7-Fundamental-Elements-of -Cyber-Exercise-Programs-November-2020-Published.pdf.

105. Tim Maurer, Kathryn Taylor, and Taylor Grossman, "Capacity-Building Tool Box for Cybersecurity and Financial Organizations," Carnegie Endowment for International Peace, December 2020, https://swiftinstitute.org/wp-content /uploads/2021/01/CyberToolkit_EXECUTIVE-SUMMARY_11_25.pdf.

106. "Timeline of Cyber Incidents Involving Financial Institutions," Carnegie Endowment for International Peace, https://carnegieendowment.org/specialproj ects/protectingfinancialstability/timeline.

107. Dina Temple-Raston, "A 'Worst Nightmare' Cyberattack: The Untold Story of the SolarWinds Hack," NPR, April 16, 2021, https://www.npr.org

/2021/04/16/985439655/a-worst-nightmare-cyberattack-the-untold-story-of
-the-solarwinds-hack.

108. Temple-Raston, "A 'Worst Nightmare' Cyberattack"; "Timeline of Cyber Incidents."

109. Temple-Raston, "A 'Worst Nightmare' Cyberattack."

110. Michael Novinson, "SolarWinds Hack Could Cost Cyber Insurance Firms $90 Million," CRN, January 14, 2021, https://www.crn.com/news/security /solarwinds-hack-could-cost-cyber-insurance-firms-90-million.

111. Financial Stability Oversight Council, *2020 Annual Report*, 180, https:// home.treasury.gov/system/files/261/FSOC2020AnnualReport.pdf.

112. Maurer and Nel, "International Strategy," 3.

CHAPTER 14. WAKE UP AND SMELL
THE QUBITS: 2021 AND BEYOND

1. Arthur Herman, "The Quantum Computing Threat to American Society," *Wall Street Journal*, November 10, 2019, https://www.wsj.com/articles/thequan tum-computing-threat-to-american-security-11573411715.

2. Richard A. Clarke and Robert K. Knake, *Cyber War: The Next Threat to Cyber Security and What to Do about It* (New York: Ecco, 2010).

3. Jesús Fernández-Villaverde, Daniel Sanches, Linda Schilling, and Harald Uhlig, "Central Bank Digital Currency: Central Banking for All," Federal Reserve Bank of Philadelphia, August 2020, https://www.philadelphiafed.org/consumer -finance/payment-systems/central-bank-digital-currency-central-banking-for-all.

4. Nurjannah Ahmat and Sabrina Bashir, "Central Bank Digital Currency: A Monetary Policy Perspective," Bank Negara Malaysia, Central Bank of Malaysia, September 2017, https://www.bnm.gov.my/documents/20124/826874 /CB_Digital+Currency_Print.pdf/e8e3bee6-4763-9c92-cba0-f33d5b2065f5.

5. Tim Maurer and Arthur Nel, "International Strategy to Better Protect the Financial System against Cyber Threats," Carnegie Endowment for International Peace, in collaboration with the World Economic Forum, November 18, 2020, 1, https://carnegieendowment.org/files/Maurer_Nelson_FinCyber_final1.pdf, citing United Nations Security Council, "Letter Dated 31 July 2019 from the Panel of Experts Established Pursuant to Resolution 1874 (2009) Addressed to the Chair of the Security Council Committee Established Pursuant to Resolution 1718 (2006)." U.S. Government Joint Advisory, "Alert (AA20-239A) FASTCash 2.0: North Korea's BeagleBoyz Robbing Banks," August 26, 2020, https://us-cert.cisa.gov /ncas/alerts/aa20-239a.

6. Maurer and Nel, "International Strategy," 2–3.

7. Maurer and Nel, "International Strategy," 4–6.

8. "Transition Book for the Incoming Biden Administration," USSC, CSC White Paper No. 5, January 2021, https://www.solarium.gov/public-communi cations/transition-book.

9. David E. Sanger, Julian E. Barnes, and Nicole Perlroth, "White House Weighs New Cybersecurity Approach after Failure to Detect Hacks," *New York Times*, March 14, 2021, https://www.nytimes.com/2021/03/14/us/politics/us-hacks-china-russia.html.

10. Sanger, Barnes, and Perlroth, "White House Weighs New Cybersecurity Approach."

11. Sanger, Barnes, and Perlroth, "White House Weighs New Cybersecurity Approach."

12. Sanger, Barnes, and Perlroth, "White House Weighs New Cybersecurity Approach."

13. National Security Commission on Artificial Intelligence, "Final Report," March 2021, 1–2, https://www.nscai.gov/wp-content/uploads/2021/03/Full-Report-Digital-1.pdf.

14. National Security Commission on Artificial Intelligence, "Final Report," 8.

15. National Counterintelligence and Security Center, "Insider Threat Mitigation for U.S. Critical Infrastructure Entities: Guidelines from an Intelligence Perspective," March 2021, 2–3, https://www.dni.gov/files/NCSC/documents/news/20210319-Insider-Threat-Mitigation-for-US-Critical-Infrastru-March-2021.pdf.

16. National Counterintelligence and Security Center, "Insider Threat Mitigation," 18.

17. Shannon Vavra, "Fed Chair Deems Cyber Threat Top Risk to Financial Sector," CyberScoop, April 12, 2021, https://www.cyberscoop.com/federal-reserve-chairman-jerome-powell-cyberattack.

18. Kartikay Mehrotra and William Turton, "CNA Financial Paid $40 Million in Ransom after March Cyberattack," *Bloomberg*, May 20, 2021, https://www.bloomberg.com/news/articles/2021-05-20/cna-financial-paid-40-million-in-ransom-after-march-cyberattack.

19. William Turton and Kartikay Mehrotra, "Hackers Breached Colonial Pipeline Using Compromised Password," *Bloomberg*, June 4, 2021, https://www.bloomberg.com/news/articles/2021-06-04/hackers-breached-colonial-pipeline-using-compromised-password.

20. Collin Eaton and Dustin Volz, "Colonial Pipeline CEO Tells Why He Paid Hackers a $4.4 Million Ransom," *Wall Street Journal*, May 19, 2021, https://www.wsj.com/articles/colonial-pipeline-ceo-tells-why-he-paid-hackers-a-4-4-million-ransom-11621435636.

21. Kevin Collier, "Meat Supplier JBS Paid Ransomware Hackers $11 Million," June 9, 2021, https://www.cnbc.com/2021/06/09/jbs-paid-11-million-in-response-to-ransomware-attack-.html.

22. "Executive Order on Improving the Nation's Cybersecurity," White House, May 12, 2021, https://www.whitehouse.gov/briefing-room/presidential-actions/2021/05/12/executive-order-on-improving-the-nations-cybersecurity.

23. "Fact Sheet: Executive Order Addressing the Threat from Securities Investments that Finance Certain Companies of the People's Republic of China, White

House, June 3, 2021, https://www.whitehouse.gov/briefing-room/statements-re leases/2021/06/03/fact-sheet-executive-order-addressing-the-threat-from-securi ties-investments-that-finance-certain-companies-of-the-peoples-republic-of-china.

24. "Protecting Americans' Sensitive Data from Foreign Adversaries," Execu- tive Order 14034, White House, June 9, 2021, https://www.federalregister.gov /documents/2021/06/11/2021-12506/protecting-americans-sensitive-data-from -foreign-adversaries.

25. "Analysis of RegTech in the EU Financial Sector," European Banking Authority, June 2021, 5–6, https://www.eba.europa.eu/sites/default/documents /files/document_library/Publications/Reports/2021/1015484/EBA%20analy sis%20of%20RegTech%20in%20the%20EU%20financial%20sector.pdf.

26. "Chinese State-Sponsored Cyber Operations: Observed TTPs," Cyberse- curity Alert, National Security Agency, Cybersecurity and Infrastructure Security Agency, Federal Bureau of Investigation, July 19, 2021, https://media.defense.gov /2021/Jul/19/2002805003/-1/-1/1/CSA_CHINESE_STATE-SPONSORED _CYBER_TTPS.PDF.

27. "Chinese State-Sponsored Cyber Operations," 6–13.

28. "National Security Memorandum on Improving Cybersecurity for Critical Infrastructure Control Systems," White House, July 28, 2021, https:// www.whitehouse.gov/briefing-room/statements-releases/2021/07/28/national -security-memorandum-on-improving-cybersecurity-for-critical-infrastructure -control-systems.

29. Deborah Brown, "Cybercrime Is Dangerous, but a New UN Treaty Could Be Worse for Rights," Human Rights Watch, August 13, 2021, https:// www.hrw.org/news/2021/08/13/cybercrime-dangerous-new-un-treaty-could -be-worse-rights.

30. Drew Fitzgerald, "T-Mobile Hackers Stole Information on More than 40 Million People," *Wall Street Journal*, August 18, 2021, https://www.wsj .com/articles/t-mobile-says-hackers-stole-details-on-more-than-40-million-peo ple-11629285376.

31. Drew Fitzgerald and Robert McMillan, "T-Mobile Hacker Who Stole Data on 50 Million Customers: 'Their Security Is Awful,'" *Wall Street Journal*, August 26, 2021, https://www.wsj.com/articles/t-mobile-hacker-who-stole-data-on-50-mil lion-customers-their-security-is-awful-11629985105.

32. Lauren Feiner, "Biden to Host Tech, Finance and Energy CEOs for Secu- rity Summit at White House Following Wave of Cyberattacks," CNBC, August 25, 2021, https://www.cnbc.com/2021/08/25/biden-to-host-tech-finance-and -energy-ceos-for-white-house-cybersecurity-summit.html.

33. Authentication and Access to Financial Institution Services and Systems, Federal Financial Institutions Examination Council, August 11, 2021, https:// www.ffiec.gov/press/pdf/Authentication-and-Access-to-Financial-Institution-Ser vices-and-Systems.pdf.

34. "Federal Cybersecurity: America's Data Still at Risk," Committee on Homeland Security and Governmental Affairs, staff report, August 2021, https://

www.hsgac.senate.gov/imo/media/doc/2019-06-25%20PSI%20Staff%20Re
port%20-%20Federal%20Cybersecurity%20Updated.pdf.

35. Brian Dean, "iPhone Users and Sales Stats for 2021," Backlinko, May 28, 2021, https://backlinko.com/iphone-users#iphone-key-stats.

36. "Treasury Takes Robust Actions to Counter Ransomware," Department of Treasury, September 21, 2021, https://home.treasury.gov/news/press-releases /jy0364.

37. James Rundle and Kim S. Nash, "Justice Department to Fine Contractors for Not Reporting Cyber Incidents," *Wall Street Journal*, October 7, 2021, https:// www.wsj.com/articles/justice-department-to-fine-contractors-for-not-reporting -cyber-incidents-11633599001.

38. "Public Policy Principles for Retail Central Bank Digital Currencies," G7 UK 2021, October 14, 2021, https://assets.publishing.service.gov.uk/government /uploads/system/uploads/attachment_data/file/1025235/G7_Public_Policy_Prin ciples_for_Retail_CBDC_FINAL.pdf.

39. "Central Bank Digital Currencies: Foundational Principles and Core Features," Bank for International Settlements, September 2020, https://www.bis.org /publ/othp33.pdf.

40. See Thomas P. Vartanian, "No. 1 Rule for Central Bank Digital Currencies: Do No Harm," *American Banker*, April 21, 2021, https://www.americanbanker .com/opinion/no-1-rule-for-central-bank-digital-currencies-do-no-harm.

41. "Updated Guidance for a Risk-Based Approach: Virtual Assets and Virtual Asset Service Providers," Financial Action Task Force, October 2021, https:// www.fatf-gafi.org/media/fatf/documents/recommendations/Updated-Guidance -VA-VASP.pdf.

42. Chris Inglis, "National Cyber Policy Will Disrupt Crime and Instill Hope," *Wall Street Journal*, October 28, 2021, https://www.wsj.com/articles/national-cyber -policy-will-disrupt-crime-and-instill-hope-cyberattack-cyberspace-11635367439.

43. "Reducing the Significant Risk of Known Exploited Vulnerabilities," Cyberspace and Infrastructure Security Agency, Binding Operational Directive 22-01, November 3, 2021, https://cyber.dhs.gov/bod/22-01.

44. Peter Rudegeair, "Robinhood Hack Exposes Millions of Customer Names, Email Addresses," *Wall Street Journal*, November 8, 2021, https://www .wsj.com/articles/robinhood-hack-exposes-millions-of-customer-names-email-ad dresses-11636408263.

45. "Computer-Security Incident Notification Requirements for Banking Organizations and Their Bank Service Providers," OCC, 12 CFR Part 53 (Docket ID OCC-2020-0038), RIN 1557-AF02; Federal Reserve System, 12 CFR Part 225 (Docket No. R-1736), RIN 7100-AG06; FDIC, 12 CFR Part 304, RIN 3064-AF59, November 17, 2021, 86 FR 66424, 66429, https://www.govinfo .gov/content/pkg/FR-2021-11-23/pdf/2021-25510.pdf.

46. Joseph Marks and Aaron Schaffer, "Congress Can't Even Pass the Easy Cyber Stuff," *Washington Post*, December 8, 2021, https://www.msn.com/en-us /news/politics/congress-cant-even-pass-the-easy-cyber-stuff/ar-AARBm8U.

47. "Securing the Information and Communications Technology and Services Supply Chain; Connected Software Applications," Department of Commerce, RIN 0605-AA62 (Docket ID 2021-25329), 86 Fed. Reg. 67279–67385, November 26, 2021, https://www.federalregister.gov/documents/2021/11/26/2021-25329/securing-the-information-and-communications-technology-and-services-supply-chain-connected-software.

48. John D. McKinnon and Alex Leary, "U.S. Moving—Some Say Too Slowly—to Address TikTok Security Risk," *Wall Street Journal*, February 2, 2022, https://www.wsj.com/articles/tiktok-security-risk-china-biden-11643807751.

49. Robert McMillan and Dustin Volz, "Hackers Backed by China Seen Exploiting Security Flaw in Internet Software," *Wall Street Journal*, December 15, 2021, https://www.wsj.com/articles/hackers-backed-by-china-seen-exploiting-security-flaw-in-internet-software-11639574405.

50. Lily Hay Newman, "The Log4J Vulnerability Will Haunt the Internet for Years," *Wired*, December 13, 2021, https://www.wired.com/story/log4j-log4shell.

51. Patrick Howell O'Neill, "The Internet Runs on Free Open-Source Software. Who Pays to Fix It?," *MIT Technology Review*, December 17, 2021, https://www.technologyreview.com/2021/12/17/1042692/log4j-internet-open-source-hacking.

52. Financial Stability Oversight Council, *2021 Annual Report*, December 17, 2021, https://home.treasury.gov/system/files/261/FSOC2021AnnualReport.pdf.

53. Financial Stability Oversight Council, *2021 Annual Report*, 17.

54. Sumathi Reddy, "How Contagious Is Omicron? What Does That Mean for You?," *Wall Street Journal*, December 24, 2021, https://www.wsj.com/articles/how-contagious-is-omicron-what-does-that-mean-for-you-11640358001.

55. Jason Douglas, "New U.K. Study Reinforces Conclusion That Omicron Causes Less Severe Disease," *Wall Street Journal*, December 22, 2021, https://www.wsj.com/articles/u-k-study-finds-risk-of-hospitalization-with-omicron-50-to-70-lower-than-with-delta-11640281553.

56. "Coronavirus in the U.S.: Latest Map and Case Count," *New York Times*, December 25, 2021, https://www.nytimes.com/interactive/2021/us/covid-cases.html.

57. Scott Neuman, "Ukraine Is Hit by a Massive Cyberattack That Targeted Government Websites," NPR, January 14, 2022, https://www.npr.org/2022/01/14/1073001754/ukraine-cyber-attack-government-websites-russia.

58. "Memorandum for the Heads of Executive Departments and Agencies M-22-0," Executive Office of the President, Office of Management and Budget, January 26, 2022, https://www.whitehouse.gov/wp-content/uploads/2022/01/M-22-09.pdf.

59. Mike Isaac, "6 Reasons Meta Is in Trouble," *New York Times*, February 3, 2022, https://www.nytimes.com/2022/02/03/technology/facebook-meta-challenges.html.

60. Sarah N. Lynch and Chris Prentice, "FBI to Form Digital Currency Unit, Justice Dept Taps New Crypto Czar," Reuters, February 17, 2022, https://www.reuters.com/technology/fbi-form-new-digital-currency-unit-justice-dept-taps-new-crypto-czar-2022-02-17.

61. Chris Inglis and Harry Krejsa, "The Cyber Social Contract," *Foreign Affairs*, February 21, 2022, https://www.foreignaffairs.com/articles/united-states/2022-02-21/cyber-social-contract.

62. Patrick Howell O'Neill, "How a Russian Cyberwar in Ukraine Could Ripple Out Globally," *MIT Technology Review*, January 21, 2022, https://www.technologyreview.com/2022/01/21/1043980/how-a-russian-cyberwar-in-ukraine-could-ripple-out-globally.

63. Gillian Turner and Brooke Singman, "Hackers Targeted US Energy Companies ahead of Ukraine Invasion: Source," Yahoo! News, March 9, 2022, https://news.yahoo.com/hackers-targeted-us-energy-companies-182003869.html.

64. Jillian K. Melchior, "The Cyberspace Front in the Attacks on Ukraine," *Wall Street Journal*, February 18, 2022, https://www.wsj.com/articles/virtual-front-attacks-invasion-ukraine-cybersecurity-attack-threat-hackers-russia-ukraine-dos-denial-of-service-energy-notpetya-11645221274.

65. David Uberti, "Ukrainian Defense Ministry, Banks Hit by Suspected Cyberattacks, Officials Say," *Wall Street Journal*, February 15, 2022, https://www.wsj.com/articles/ukrainian-defense-ministry-banks-hit-by-suspected-cyberattacks-officials-say-11644960352.

66. David Uberti and Dustin Volz, "Destructive Malware Is Seen in Ukraine as the Risk of Cyber Spillover Looms," *Wall Street Journal*, February 24, 2022, https://www.wsj.com/livecoverage/russia-ukraine-latest-news/card/tdxMzFCXFSxCxBSRPIpX; Juan Andrés Guerrero-Saade, "HermeticWiper: New Destructive Malware Used in Cyber Attacks on Ukraine," SentinelLabs, February 23, 2022, https://www.sentinelone.com/labs/hermetic-wiper-ukraine-under-attack.

67. "Executive Order on Ensuring Responsible Development of Digital Assets," White House, March 9, 2022, https://www.whitehouse.gov/briefing-room/presidential-actions/2022/03/09/executive-order-on-ensuring-responsible-development-of-digital-assets.

68. Consolidated Appropriations Act of 2022, Rules Committee Print, 117–35, Text of the House Amendment to the Senate Amendment to H.R. 2471, March 8, 2022, https://rules.house.gov/sites/democrats.rules.house.gov/files/BILLS-117HR2471SA-RCP-117-35.pdf.

69. Robert McMillan, Kevin Poulsen, and Dustin Volz, "Secret World of Pro-Russia Hacking Group Exposed in Leak," *Wall Street Journal*, March 28, 2022, https://www.wsj.com/articles/trickbot-pro-russia-hacking-gang-documents-ukrainian-leaker-conti-11648480564.

70. Steven Zeitchik, "Hackers Hit Popular Video Game, Stealing More than $600 Million in Cryptocurrency," *Washington Post*, March 29, 2022, https://www.msn.com/en-us/news/technology/hackers-hit-popular-video-game-stealing-more-than-dollar600-million-in-cryptocurrency/ar-AAVDCQv.

71. Paul Vigna, "Crypto Thieves Get Bolder by the Heist, Stealing Record Amounts," *Wall Street Journal*, April 22, 2022, https://www.wsj.com/articles/crypto-thieves-get-bolder-by-the-heist-stealing-record-amounts-11650582598.

72. Amara Omeokwe and Andrew Duehren, "Biden's Budget Calls for Increase in Defense Spending, Including Funds for Ukraine," *Wall Street Journal*, March 28, 2022, https://www.wsj.com/articles/bidens-budget-calls-for-increase-in-defense-spending-aiding-ukraine-11648479687.

73. Maggie Miller, "The Hard Truth behind Biden's Cyber Warnings," Politico, March 27, 2022, https://www.politico.com/news/2022/03/27/bidens-cyber-warnings-00020638.

74. "Executive Order on Enhancing the National Quantum Initiative Advisory Committee," May 4, 2022, https://www.whitehouse.gov/briefing-room/presidential-actions/2022/05/04/executive-order-on-enhancing-the-national-quantum-initiative-advisory-committee; "National Security Memorandum on Promoting United States Leadership in Quantum Computing While Mitigating Risks to Vulnerable Cryptographic Systems," May 4, 2022, https://www.whitehouse.gov/briefing-room/statements-releases/2022/05/04/national-security-memorandum-on-promoting-united-states-leadership-in-quantum-computing-while-mitigating-risks-to-vulnerable-cryptographic-systems.

75. Executive Order on Enhancing the National Quantum Initiative Advisory Committee, May 4, 2022, https://www.whitehouse.gov/briefing-room/presidential-actions/2022/05/04/executive-order-on-enhancing-the-national-quantum-initiative-advisory-committee; National Security Memorandum on Promoting United States Leadership in Quantum Computing While Mitigating Risks to Vulnerable Cryptographic Systems, May 4, 2022, https://www.whitehouse.gov/briefing-room/statements-releases/2022/05/04/national-security-memorandum-on-promoting-united-states-leadership-in-quantum-computing-while-mitigating-risks-to-vulnerable-cryptographic-systems.

76. Board of Governors of the Federal Reserve System, *Financial Stability Report*, May 2022, https://www.federalreserve.gov/publications/files/financial-stability-report-20220509.pdf.

CHAPTER 15. VIRTUAL SHOOTOUTS

1. "Stakeholder Perspectives on the Cyber Incident Reporting for Critical Infrastructure Act of 2021," Testimony of Heather Hogsett, senior vice president, Technology and Risk Strategy for BITS, the Technology Policy Division of Bank Policy Institute, before the U.S. House Subcommittee on Cybersecurity, Infrastructure Protection, and Innovation, September 1, 2021, https://bpi.com/wp-content/uploads/2021/08/Testimony-Stakeholder-Perspectives-on-the-Cyber-Incident-Reporting-for-Critical-Infrastructure-Act-of-2021-BPI.pdf.

2. Gramm–Leach–Bliley Act, Pub. L. 106-102, 113 Stat. 1338 et seq., November 12, 1999; "Interagency Guidance on Response Programs for Unauthorized Access to Customer Information and Customer Notice," *Federal Register*, March 29, 2005, https://www.federalregister.gov/documents/2005/03/29/05-5980/inter

agency-guidance-on-response-programs-for-unauthorized-access-tocustomer-in
formation-and.

3. "Computer-Security Incident Notification Requirements for Banking Organizations and Their Bank Service Providers," Comptroller of the Currency, Federal
Reserve System, and Federal Deposit Insurance Corporation, Document No. 2020-
28498, https://www.federalregister.gov/documents/2021/01/12/2020-28498
/computer-security-incident-notification-requirements-for-banking-organizations
-and-their-bank.

4. New York Codes, Rules and Regulations (23 NYCRR 500), https://govt
.westlaw.com/nycrr/Browse/Home/NewYork/NewYorkCodesRulesandRegula
tions?guid=I5be30d2007f811e79d43a037eefd0011&o.

5. "Stakeholder Perspectives on the Cyber Incident Reporting," appendix A,
6–7.

6. Tim Maurer and Arthur Nelson, "International Strategy to Better Protect
the Financial System against Cyber Threats," Carnegie Endowment for International Peace, 2020, 49–52, https://carnegieendowment.org/files/Maurer_Nel
son_FinCyber_final1.pdf.

7. "FS-ISAC Exercises," https://www.fsisac.com/hubfs/Resources/FS
-ISAC_ExercisesOverview.pdf.

8. Financial Services Sector Coordinating Council for Critical Infrastructure
Protection and Homeland Security, Annual Report 2016–2017, 13, https://fsscc
.org/wp-content/uploads/2021/02/Annual_Report_2016-2017_final.pdf.

9. "Cybersecurity: Risks to the Financial Services Industry and Its Preparedness," Testimony of Bill Nelson, president and CEO, FS-ISAC, before the U.S.
Senate Committee on Banking, Housing and Urban Affairs, May 24, 2018,
4, https://www.fsisac.com/hubfs/Resources/FS-ISAC-Testimony_BillNelson
-2018-FIN.pdf; https://www.govinfo.gov/content/pkg/CHRG-115shrg31197
/html/CHRG-115shrg31197.htm.

10. *Quantum Dawn V After-Action Report*, Protiviti, February 26, 2020, https://
www.protiviti.com/US-en/insights/flash-report-quantum-dawn-v-after-action
-report. Organizations that have participated include the Securities Industry and
Financial Markets Association (SIFMA), the Association for Financial Markets in
Europe (AFME), the Asia Securities Industry and Financial Markets Association
(ASIFMA), the Securities and Exchange Commission (SEC), the U.S. Treasury's
Financial Services Information Sharing and Analysis Center (FS-ISAC), the Bank
of England, the Financial Conduct Authority, HM Treasury, the UK Finance
Sector Cyber Collaboration Centre (FSCCC), the Bank of Canada, and Canadian
provincial regulators.

11. SIFMA, *Quantum Dawn VI Fact Sheet*, December 2021, https://www.sifma
.org/wp-content/uploads/2019/11/QuantumDawn-VI-Fact-Sheet-2021.pdf.

12. https://www.fsisac.com/events/caps21-financial-institution1.

13. James Rundle, "NATO Wargame Examines Cyber Risk to Financial
System," *Wall Street Journal*, April 15, 2021, https://www.wsj.com/articles/nato
-wargame-examines-cyber-risk-to-financial-system-11618479000.

CHAPTER 16. CALL THE INTERNET REPAIRMAN

1. David D. Clark, *Designing an Internet* (Cambridge: Massachusetts Institute of Technology, 2018), 203.

2. Jon Danielsson, Robert Macrae, and Andreas Uthemann, "Artificial Intelligence, Financial Risk Management and Systemic Risk," London School of Economic and Political Science, SRC Special Paper No. 13, November 2017, 15, https://www.fmg.ac.uk/sites/default/files/publications/SP-13.pdf.

3. Joshua New, "Why the U.S. Could Fall Behind in the Global AI Race," *Fortune*, August 2, 2018, http://fortune.com/2018/08/01/artificial-intelligencerace -united-states.

4. Those experts include Bruce Schneier (*Click Here to Kill Everybody*), Richard A. Clarke and Robert K. Knake (*Cyber War*), Laura DeNardis (*The Global War for Internet Governance*), Nicole Perlroth (*This Is How They Tell Me the World Ends*), David Sanger (*The Perfect Weapon*), and Marc Goodman (*Future Crimes*).

5. "Europe Fit for the Digital Age: New Online Rules for Platforms," European Commission, https://ec.europa.eu/info/strategy/priorities-2019-2024/europe -fit-digital-age/digital-services-act-ensuring-safe-and-accountable-online-environ ment/europe-fit-digital-age-new-online-rules-platforms_en.

CHAPTER 17. THE INGREDIENTS OF A SAFER INTERNET

1. Walter Isaacson, "The Internet Is Broken. Starting from Scratch, Here's How I Would Fix It," LinkedIn, December 14, 2016, https://www.linkedin.com/pulse /internet-broken-starting-from-scratch-heres-how-id-fix-isaacson.

2. "What Is 'Authentication'?," *Economic Times*, https://economictimes.india times.com/definition/authentication.

3. Nikhilesh DeElon, "Musk Wants to Authenticate Every Twitter User. Crypto Twitter Should Take Notice," CoinDesk, April 26, 2022, https://www .coindesk.com/policy/2022/04/26/elon-musk-wants-to-authenticate-every-twit ter-user-crypto-twitter-should-take-notice.

4. "Memorandum for the Heads of Executive Departments and Agencies M-22-0," Executive Office of the President, Office of Management and Budget, January 26, 2022, 2, 5–9, https://www.whitehouse.gov/wp-content /uploads/2022/01/M-22-09.pdf.

5. "Memorandum for the Heads of Executive Departments," 7.

6. "Web Authentication, an API for Accessing Public Key Credentials Level 2, W3C Recommendation," April 8, 2021, https://www.w3.org/TR/webauthn-2.

7. Ben Cresitello-Dittmar, "Application of the Blockchain for Authentication and Verification of Identity," November 30, 2016, https://www.cs.tufts.edu /comp/116/archive/fall2016/bcresitellodittmar.pdf; "Top 5 Blockchain Security

Issues in 2019," LIFARS, December 6, 2019, https://lifars.com/2019/12/top
-5-blockchain-security-issues-in-2019.

8. Evan Sultanik, Alexander Remie, Felipe Manzano, Trent Brunson, Sam
Moelius, Eric Kilmer, Mike Myers, Talley Amir, and Sonya Schriner, "Are Blockchains Decentralized? Unintended Centralities in Distributed Ledgers," Trail of Bits,
June 2022, https://assets-global.website-files.com/5fd11235b3950c2c1a3b6df4/6
2af6c641ab72b3329b9a480_Unintended_Centralities_in_Distributed_Ledgers.pdf.

9. Diego Geroni, "Top 5 Blockchain Security Issues in 2021," 101 Blockchains, June 13, 2021, https://101blockchains.com/blockchain-security-issues.

10. Bruce Schneier, "Understanding Crypto 6: Security, Trust, and Blockchain," *The Rational Reminder Podcast*, July 8, 2022, https://rationalreminder
.libsyn.com/understanding-crypto-6-bruce-schneier-security-trust-and-blockchain.

11. Kai Strittmatter, *We Have Been Harmonized: Life in China's Surveillance State*
(New York: Custom House, September 1, 2020).

12. David D. Clark, *Designing an Internet* (Cambridge: Massachusetts Institute of
Technology, 2018), 219–20.

13. Oliver Wyman and the International Banking Federation, "Digital Trust:
How Banks Can Secure Our Digital Identity," 2022, https://www.oliverwyman
.com/content/dam/oliver-wyman/v2/publications/2021/nov/Digital-Trust
-Final.pdf.

14. Richard J. Danzig, "Surviving on a Diet of Poisoned Fruit: Reducing the
National Security Risks of America's Cyber Dependencies," Center for a New
American Society, July 2014, 13, https://www.jstor.org/stable/pdf/resrep06383.pdf.

15. Danzig, "Surviving on a Diet of Poisoned Fruit," 13–14.

16. See Cyber Reliant, press release, June 8, 2021, https://www.globenews
wire.com/news-release/2021/06/08/2243726/0/en/Cyber-Reliant-and-Cano
pius-Collaborate-to-Offer-Industry-Leading-Data-Protection-Warranty.html.

17. "What Is a Zero Trust Architecture?," Palo Alto Networks, https://www
.paloaltonetworks.com/cyberpedia/what-is-a-zero-trust-architecture.

18. Scott Rose, Oliver Borchert, Stu Mitchell, and Sean Connelly, "Zero Trust
Architecture," NIST Special Publication 800-207, August 2020, 2, https://nvl
pubs.nist.gov/nistpubs/SpecialPublications/NIST.SP.800-207.pdf.

19. "What Is a Zero Trust Architecture?"

20. "Memorandum for the Heads of Executive Departments."

21. Manny Barzilay, "Building a Brand New Internet," TechCrunch, March
13, 2016, https://techcrunch.com/2016/03/13/building-a-brand-new-internet.

22. Barzilay, "Building a Brand New Internet."

23. "What If All Workers Wrote Software, Not Just the Geek Elite?," *The
Economist*, January 29, 2022, https://www.economist.com/business/2022/01/29
/what-if-all-workers-wrote-software-not-just-the-geek-elite.

24. https://www.internetsociety.org.

25. https://www.iso.org/home.html.

26. https://www.enisa.europa.eu.

27. https://www.nist.gov.

28. https://www.isaca.org/resources/cobit.

29. https://www.iec.ch/homepage.

30. https://www.ieee.org.

31. Internet Society, "Code of Conduct," https://connect.internetsociety.org/codeofconduct.

32. "The Clipper Chip," Electronic Privacy Information Center, https://epic.org/crypto/clipper.

33. James Griffiths, *Firewall: How to Build and Control an Alternative Version of the Internet* (London: Zed Books, 2019), 156.

34. Laura DeNardis, *The Global War for Internet Governance* (New Haven, CT: Yale University Press, 2014), 199.

35. DeNardis, *The Global War for Internet Governance*, 203–5.

36. Michael Nieles, Kelley Dempsey, and Victoria Yan Pillitteri, "An Introduction to Information Security," NIST Special Publication 800-12, revision 1, National Institute of Standards and Technology, June 2017, https://nvlpubs.nist.gov/nistpubs/SpecialPublications/NIST.SP.800-12r1.pdf.

37. As examples, the Computer Security Act of 1987 required federal agencies to identify sensitive systems, conduct computer security training, and develop computer security plans, Pub. L. 100-235, 101 Stat. 1724, https://www.gpo.gov/fdsys/pkg/STATUTE-101/pdf/STATUTE-101-Pg1724.pdf; the Information Technology Management Reform Act of 1996, redesignated the Clinger-Cohen Act, is the primary regulation for the use, management, and acquisition of computer resources in the federal government, Pub. L. 107-217, 116 Stat. 1234, https://www.gsa.gov/graphics/staffoffices/Clinger.htm; the Gramm–Leach–Bliley Act of 1999 required that financial regulators require financial institutions to adopt safeguards for the security of "nonpublic personal information," Pub. L. 106-102, 113, Stat. 1338 § 501, 15 U.S.C. § 6801, https://www.govinfo.gov/content/pkg/PLAW-106publ102/pdf/PLAW-106publ102.pdf; the E-Government Act of 2002 is intended to enhance the management and promotion of electronic government services and processes by establishing a federal chief information officer (CIO) within OMB and by establishing the use of internet-based information technology to enhance citizens' access to government information, Pub. L. 107-347, 116 Stat. 2899, http://www.gpo.gov/fdsys/pkg/PLAW-107publ347/pdf/PLAW-107publ347.pdf; the Federal Information Security Management Act was enacted as part of the E-Government Act of 2002 to provide a comprehensive framework for ensuring the effectiveness of information security controls that support federal operations and assets, Pub. L. 107-347 (Title III), 116 Stat. 2946, https://www.gpo.gov/fdsys/pkg/CHRG-107hhrg86343/pdf/CHRG-107hhrg86343.pdf, and amended by the Federal Information Security Modernization Act of 2014 that

modernized federal security practices, Pub. L. 113-283, 128 Stat. 3073, http://www.gpo.gov/fdsys/pkg/PLAW-113publ283/pdf/PLAW-113publ283.pdf.

38. Bill O'Hern, "The Security Benefits of a Software Bill of Materials," *AT&T Technology Blog*, January 18, 2022, https://about.att.com/innovationblog/2022/security-benefits-software-bill-of-materials.html.

39. Secure and Trusted Communications Networks Act of 2019, Pub. L. 116-124, as amended through Pub. L. 116-260, enacted December 27, 2020, https://www.govinfo.gov/content/pkg/COMPS-15677/pdf/COMPS-15677.pdf.

40. David Shephardson, "Five Chinese Companies Pose Threat to U.S. National Security: FCC," Reuters, March 21, 2021, https://www.reuters.com/article/us-usa-china-tech-idUKKBN2B42DW.

41. See Danzig, "Surviving on a Diet of Poisoned Fruit," 17.

42. Joseph Marks and Aaron Schaffer, "Congress Can't Even Pass the Easy Cyber Stuff," *Washington Post*, December 8, 2021, https://www.msn.com/en-us/news/politics/congress-cant-even-pass-the-easy-cyber-stuff/ar-AARBm8U.

43. Department of Justice, Fair Lending Enforcement Program, August 6, 2015, https://www.justice.gov/crt/fair-lending-enforcement-program.

44. Clark, *Designing an Internet*, 201.

45. Danzig, "Surviving on a Diet of Poisoned Fruit," 7.

CHAPTER 18. IF YOU BUILD IT, THEY WILL CLICK

1. Patrick Howell O'Neill, "Inside the Plan to Fix America's Never-Ending Cybersecurity Failures," *MIT Technology Review*, March 18, 2022, https://www.technologyreview.com/2022/03/18/1047395/inside-the-plan-to-fix-americas-never-ending-cybersecurity-failures.

2. David Sanger, "Managing the Fifth Generation: America, China, and the Struggle for Technological Dominance," Aspen Institute, January 20, 2020, https://www.aspeninstitute.org/blog-posts/managing-the-fifth-generation-america-china-and-the-struggle-for-technological-dominance.

3. Aliya Sternstein, "Former CIA Director: Build a New Internet to Improve Cybersecurity," Nextgov, July 6, 2011, https://www.nextgov.com/cybersecurity/2011/07/former-cia-director-build-a-new-internet-to-improve-cybersecurity/49354.

4. Global Internet Forum to Counter Terrorism (GIFCT), https://gifct.org/about.

5. Oasis Consortium, https://www.oasisconsortium.com/usersafetystandards.

6. Tanya Basu, "This Group of Tech Firms Just Signed Up to a Safer Metaverse," *MIT Technology Review*, January 20, 2022, https://www.technologyreview.com/2022/01/20/1043843/safe-metaverse-oasis-consortium-roblox-meta.

7. Patrick Howell O'Neill, "The US Is Unmasking Russian Hackers Faster than Ever," *MIT Technology Review*, February 21, 2022, https://www.technolo gyreview.com/2022/02/21/1046087/russian-hackers-ukraine.

8. Richard J. Danzig, "Surviving on a Diet of Poisoned Fruit: Reducing the National Security Risks of America's Cyber Dependencies," Center for a New American Society, July 2014, 8, https://www.jstor.org/stable/pdf/resrep06383.pdf.

9. Danzig, "Surviving on a Diet of Poisoned Fruit," 21.

10. Anne Rawland Gabriel, "Network Latency for Financial Services: How Low Can You Go?," *BizTech*, October 3, 2012, https://biztechmagazine.com /article/2012/10/network-latency-financial-services-how-low-can-you-go.

11. Christopher S. Yoo, "When Are Two Networks Better than One? Toward a Theory of Optimal Fragmentation," in *A Universal Internet in a Bordered World: Research on Fragmentation, Openness and Interoperability*, Research Volume 1 (Waterloo, Ontario, Centre for International Governance Innovation, and London, Chatham House, Royal Institute of International Affairs, 2016), 128–29, https://www.cigion line.org/static/documents/documents/GCIG%20Volume%201%20WEB.pdf.

12. David D. Clark, *Designing an Internet* (Cambridge: Massachusetts Institute of Technology, 2018), 288–89, 304–6.

13. David Clark, Karen Sollins, John Wroclawski, Dina Katabi, Joanna Kulik, Xiaowei Yang, Robert Braden, Ted Faber, Aaron Falk, Venkata Pingali, Mark Handley, and Noel Chiappa, "New Arch: Future Generation Internet Architecture," June 30, 2000–December 31, 2003, 4, https://www.isi.edu/newarch /iDOCS/final.finalreport.pdf.

14. Clark et al., "New Arch," 6–7.

15. Steven M. Bellovin, David D. Clark, Adrian Perrig, and Dawn Song, "A Clean-Slate Design for the Next-Generation Secure Internet," July 2005, 2, https://people.eecs.berkeley.edu/~dawnsong/papers/bellovin_clark_perrig_song _nextGenInternet.pdf.

16. Rachel Ross, "A Fresh Start for the Internet," *MIT Technology Review*, March 19, 2007, https://www.technologyreview.com/2007/03/19/226188/a-fresh-start -for-the-internet.

17. Bellovin et al., "A Clean-Slate Design," 2.

18. Anja Feldmann, "Internet Clean-Slate Design: What and Why?," *ACM SIGCOMM Computer Communication Review* 37, no. 3 (July 2007): 62, https:// www.sigcomm.org/sites/default/files/ccr/papers/2007/July/1273445-1273453. pdf; see also, Jon Crowcroft and Peter Key, "Report from the Clean Slate Network Research Post-SIGCOMM 2006 Workshop," *ACM SIGCOMM Computer Communication Review* 37, no. 1 (January 2007): 75, http://ccr.sigcomm.org/online /files/ccr-report.pdf.

19. Constantine Dovrolis, "What Would Darwin Think about Clean-Slate Architectures?," *ACM SIGCOMM Computer Communication Review* 38, no. 1 (January 2008): 34, https://www.sigcomm.org/sites/default/files/ccr/papers/2008/January /1341431-1341436.pdf.

20. Gowtham Boddapati, John Day, Ibrahim Matta, and Lou Chitkushev, "Assessing the Security of a Clean-Slate Internet Architecture," Computer Science Department, Boston University, Technical Report BUCS-TR-2009-021, June 22, 2009, 1, https://hdl.handle.net/2144/1745.

21. Jonathan Ponniah, Yih-Chun Hu, and P. R. Kumar, "A Clean Slate Approach to Secure Wireless Networking," *Foundations and Trends in Networking* 9, no. 1 (2015): 1, https://experts.illinois.edu/en/publications/a-clean-slate-approach-to-secure-wireless-networking.

22. James Griffiths, *Firewall: How to Build and Control an Alternative Version of the Internet* (London: Zed Books, 2019), 219–22.

23. fTLD website, https://www.ftld.com/about.

24. Community Bank Connections, Federal Reserve System, https://communitybankingconnections.org/articles/2017/i1/considerations.

25. fTLD website, https://www.ftld.com/about.

26. "Deal on Digital Markets Act: EU Rules to Ensure Fair Competition and More Choice for Users," European Parliament News, March 24, 2022, https://www.europarl.europa.eu/news/en/press-room/20220315IPR25504/deal-on-digital-markets-act-ensuring-fair-competition-and-more-choice-for-users.

27. Bobby Allyn, "People Are Talking about Web3. Is It the Internet of the Future or Just a Buzzword?," NPR, November 21, 2021, https://www.npr.org/2021/11/21/1056988346/web3-internet-jargon-or-future-vision.

28. Charles Silver, "What Is Web 3.0?," *Forbes*, January 6, 2020, https://www.forbes.com/sites/forbestechcouncil/2020/01/06/what-is-web-3-0.

29. https://getdweb.net.

30. Noah Smith, "These Technologists Think the Internet Is Broken. So They're Building Another One," NBC News, July 21, 2019, https://www.nbcnews.com/tech/tech-news/these-technologists-think-internet-broken-so-they-re-building-another-n1030136.

31. Smith, "These Technologists Think the Internet Is Broken."

32. Frank McCourt, "The Internet Is Beyond Repair. It's Time to Stop Trying to Fix It," *Barron's*, November 22, 2021, https://www.barrons.com/articles/big-tech-has-broken-the-internet-dont-fix-it-rebuild-51637318700.

33. Emily Bobrow, "Frank McCourt Wants to Build a New Model for Social Media," *Wall Street Journal*, October 8, 2021, https://www.wsj.com/articles/frank-mccourt-wants-to-build-a-new-model-for-social-media-11633710479.

34. Bobrow, "Frank McCourt."

35. David Pierce, "One Man's Plan to Build a New Internet," Protocol, January 13, 2021, https://www.protocol.com/dfinity-interview.

36. https://solid.mit.edu; David Baker, "The Internet Is Broken," December 19, 2017, *Wired*, https://www.wired.co.uk/article/is-the-internet-broken-how-to-fix-it.

37. "The Web's Greatest Minds Explain How We Can Fix It," *Wired*, December 20, 2017, https://www.wired.co.uk/article/the-webs-greatest-minds-on-how-to-fix-it; see https://dfinity.org.

38. Allyn, "People Are Talking about Web3."

39. George Zarkadakis, "'Data Trusts' Could Be the Key to Better AI," *Harvard Business Review*, November 10, 2020, https://hbr.org/2020/11/data-trusts-could-be-the-key-to-better-ai.

40. Zarkadakis, "'Data Trusts' Could Be the Key."

41. Anouk Ruhaak, "How Data Trusts Can Protect Privacy," *MIT Technology Review*, February 24, 2021, https://www.technologyreview.com/2021/02/24/1017801/data-trust-cybersecurity-big-tech-privacy.

42. "Communication from the Commission to the European Parliament, the Council, the European Economic and Social Committee and the Committee of the Regions: A European Strategy for Data," European Commission, Brussels, February 19, 2020, https://ec.europa.eu/info/sites/default/files/communication-european-strategy-data-19feb2020_en.pdf.

43. Anna Artyushina, "The EU Is Launching a Market for Personal Data. Here's What That Means for Privacy," *MIT Technology Review*, August 11, 2020, https://www.technologyreview.com/2020/08/11/1006555/eu-data-trust-trusts-project-privacy-policy-opinion.

44. Ruhaak, "How Data Trusts Can Protect Privacy."

45. Sean McDonald, "Toronto, Civic Data, and Trust," October 17, 2018, https://medium.com/@digitalpublic/toronto-civic-data-and-trust-ee7ab928fb68.

46. https://www.ericsson.com/en/portfolio/digital-services/digital-bss/mediation.

47. Madhumita Murgia and Anna Gross, "Inside China's Controversial Mission to Reinvent the Internet," *Financial Times*, March 27, 2020, https://www.ft.com/content/ba94c2bc-6e27-11ea-9bca-bf503995cd6f.

48. Griffiths, *Firewall*, 26–28.

49. Griffiths, *Firewall*, 29.

50. Murgia and Gross, "Inside China's Controversial Mission."

51. Murgia and Gross, "Inside China's Controversial Mission."

52. James Ball, "Russia Is Risking the Creation of a "Splinternet"—and It Could Be Irreversible," *MIT Technology Review*, March 17, 2022, https://www.technologyreview.com/2022/03/17/1047352/russia-splinternet-risk.

53. "The Web's Greatest Minds."

54. Wikipedia, "TOR (network)," https://en.wikipedia.org/wiki/Tor_(network).

55. "The Dark Web: What Is It and Why Do People Use It?," CEOP Education, https://www.thinkuknow.co.uk/professionals/our-views/the-dark-web.

56. "Quantum Computing and Post-Quantum Cryptography," National Security Agency, Frequently Asked Questions, August 4, 2021, https://media.defense.gov/2021/Aug/04/2002821837/-1/-1/1/Quantum_FAQs_20210804.PDF.

57. NIST Computer Security Resource Center, "Post-Quantum Cryptography," https://csrc.nist.gov/projects/post-quantum-cryptography/round-3-submissions.

58. Report of the DOE Quantum Internet Blueprint Workshop, "From Long-Distance Entanglement to Building a Nationwide Quantum Internet," SUNY Global Center, New York, NY, February 5–6, 2020, https://www.energy.gov/sites/prod/files/2020/07/f76/QuantumWkshpRpt20FINAL_Nav_0.pdf.

59. "Department of Energy to Provide $25 Million toward Development of a Quantum Internet," Department of Energy, Office of Science, April 13, 2021, https://www.energy.gov/science/articles/department-energy-provide-25-million-toward-development-quantum-internet.

60. DOE Quantum Internet Blueprint Workshop, "From Long-distance Entanglement"; "U.S. Department of Energy Unveils Blueprint for the Quantum Internet at 'Launch to the Future: Quantum Internet' Event," July 23, 2020, https://www.energy.gov/articles/us-department-energy-unveils-blueprint-quantum-internet-launch-future-quantum-internet.

61. "The World's First Integrated Quantum Communication Network," Phys.org, January 6, 2021, https://phys.org/news/2021-01-world-quantum-network.html.

62. Daniel Garisto, "China Is Pulling Ahead in Global Quantum Race, New Studies Suggest," *Scientific American*, July 15, 2021, https://www.scientificamerican.com/article/china-is-pulling-ahead-in-global-quantum-race-new-studies-suggest.

63. Davey Winder, "U.S. Government Says It's Building A 'Virtually Unhackable' Quantum Internet," *Forbes*, July 25, 2020, https://www.forbes.com/sites/daveywinder/2020/07/25/us-government-to-build-virtually-unhackable-quantum-internet-within-10-years.

64. Clark, *Designing an Internet*, 107–9.

65. Clark, *Designing an Internet*, 120.

66. Clark, *Designing an Internet*, 113.

67. Clark, *Designing an Internet*, 110.

68. Menny Barzilay, "Building a Brand New Internet," *TechCrunch*, March 13, 2016, https://techcrunch.com/2016/03/13/building-a-brand-new-internet.

69. Barzilay, "Building a Brand New Internet."

70. https://www.intercomputer.com.

71. "Size of the Cybersecurity Market Worldwide, from 2021 to 2026," Statista, August 26, 2021, https://www.statista.com/statistics/595182/worldwide-security-as-a-service-market-size.

CHAPTER 19. THE NEW WATCHERS

1. Richard J. Danzig, "Surviving on a Diet of Poisoned Fruit: Reducing the National Security Risks of America's Cyber Dependencies," Center for a New American Society, July 2014, 21–22, https://www.jstor.org/stable/pdf/resrep06383.pdf.

2. Frank Pasquale, *New Laws of Robotics: Defending Human Expertise in the Age of AI* (Cambridge, MA: Belknap Press of Harvard University Press, 2020), 1.

3. Max Tegmark, *Life 3.0: Being Human in the Age of Artificial Intelligence* (New York: Knopf, 2017), 161–72.

4. Pasquale, *New Laws of Robotics*, 28.

5. Pasquale, *New Laws of Robotics*, 132, 136–37.

6. Sultan Meghji, "Why I Quit as FDIC Innovation Chief: Technophobia," *Bloomberg*, February 22, 2022, https://www.bloomberg.com/opinion/articles/2022 -02-22/i-quit-as-fdic-innovation-chief-because-of-regulators-technophobia.

7. Greg Baer and Rob Hunter, "A Tower of Babel: Cyber Regulation for Financial Services," Banking Perspectives, June 3, 2017, https://www.bankingper spectives.com/a-tower-of-babel-cyber-regulation-for-financial-services.

8. Baer and Hunter, "A Tower of Babel."

9. Baer and Hunter, "A Tower of Babel."

CHAPTER 20. THIS SHOULDN'T BE THE WAY THE WORLD ENDS

1. Nicole Perlroth, *This Is How They Tell Me the World Ends* (New York: Bloomsbury, 2021), 389–91.

APPENDIX A: TRADITIONAL MONEY AND PAYMENTS SYSTEMS

1. Board of Governors of the Federal Reserve System, FAQs, December 9, 2020, https://www.federalreserve.gov/faqs/currency_12773.htm.

2. "Coin and Currency: Making and Using Coin and Currency," Federal Reserve Bank of Richmond, https://www.richmondfed.org/faqs/coin_currency.

3. Andres Oppenheimer, "El Salvador Was the First Country to Adopt Bitcoin. It Might Be the First to Go Broke, Too," *Miami Herald*, January 26, 2022, https:// www.miamiherald.com/opinion/editorials/article257746438.html.

4. Carol Coye Benson, Scott Loftesness, and Russ Jones, *Payments Systems in the U.S.: A Guide for Payments Professionals* (Glenbrook Partners, 2012–2017), 29.

5. Benson, Loftesness, and Jones, *Payments Systems*, 30.

6. Benson, Loftesness, and Jones, *Payments Systems*, 31–32.

7. Benson, Loftesness, and Jones, *Payments Systems*, 44.

8. Federal Reserve 2019 Payments Study, January 6, 2020, https://www.federal reserve.gov/paymentsystems/2019-December-The-Federal-Reserve-Payments -Study.htm.

9. "Changes in U.S. Payments Fraud from 2012 to 2016: Evidence from the Federal Reserve Payments Study," Board of Governors of the Federal Reserve Sys-

tem, October 2018, 9, https://www.federalreserve.gov/publications/files/changes
-in-us-payments-fraud-from-2012-to-2016-20181016.pdf.

10. Federal Reserve 2019 Payments Study; Erica Sandberg, "The Average
Number of Credit Card Transactions per Day & Year," CardRates.com, November 9, 2020, https://www.cardrates.com/advice/number-of-credit-card-transactions-per-day-year.

11. Odysseas Papadimitriou, "How Credit Card Transaction Processing Works,"
WalletHub, December 20, 2020, https://wallethub.com/edu/cc/credit-card-transaction/25511.

12. John Boitnott, "Credit Card Fraud Statistics," Self, https://www.self.inc
/info/credit-card-fraud-statistics; see also, "Changes in U.S. Payments Fraud from
2012 to 2016."

13. Lyle Daly, "Identity Theft and Credit Card Fraud Statistics for 2021," The
Ascent, August 2021, https://www.fool.com/the-ascent/research/identity-theft
-credit-card-fraud-statistics.

14. "Payment, Clearing and Settlement Systems in the United States," Bank for
International Settlements, https://www.bis.org/cpmi/publ/d105_us.pdf.

15. Federal Reserve 2019 Payments Study.

16. "How Do ATMs Work," National ATM Systems, https://www.nasatm
.com/pages/how-do-atms-work.

17. "Changes in U.S. Payments Fraud from 2012 to 2016."

18. Tom Cronkright, "Why the Modern Transfer Wire System Is Flawed,"
CertifID, https://www.certifid.com/article/modern-wire-transfer-system-flawed.

19. "Scams Involving the Federal Reserve Name: Federal Open Market Committee (FOMC) Scam," Federal Reserve Bank of New York, June 2021, https://
www.newyorkfed.org/banking/frscams.html.

20. "Federal Reserve Announces FedNow Pilot Program Participants," Associated Press, January 25, 2021, https://apnews.com/press-release/business-wire
/technology-business-north-america-chicago-banking-and-credit-7f31feaa14eb
48a1b93dc37ab234a60f; Federal Reserve Board, "Frequently Asked Questions,"
https://www.federalreserve.gov/paymentsystems/files/fednow_faq.pdf.

21. Benson, Loftesness, and Jones, *Payments Systems*, 47–48.

22. "Changes in U.S. Payments Fraud from 2012 to 2016."

23. https://www.theclearinghouse.org.

24. https://www.theclearinghouse.org/payment-systems/chips.

25. CHIPS, Federal Reserve Bank of New York, https://www.newyorkfed
.org/aboutthefed/fedpoint/fed36.html.

26. SWIFT, https://www.swift.com/about-us/history.

27. Nicole Lindsey, "SWIFT Fraud on the Rise According to EastNets Survey
Report," *CPO Magazine*, December 11, 2019, https://www.cpomagazine.com
/cyber-security/swift-fraud-on-the-rise-according-to-eastnets-survey-report.

28. "The Value of Financial Market Transactions," DTCC, https://www.dtcc .com/dtcc-connection/articles/2021/february/02/the-value-of-financial-market -infrastructures.

29. Paolo Saguato, "The Unfinished Business of Regulating Clearinghouses," *Columbia Business Law* 449 (2020), review, https://journals.library.columbia.edu /index.php/CBLR/article/view/7219/3838.

30. Designated Financial Market Utilities, https://www.federalreserve.gov /paymentsystems/designated_fmu_about.htm.

INDEX